Contemporary Research

Contemporary Research

Models, Methodologies, and Measures in Distributed Team Cognition

Edited by Michael D. McNeese, Eduardo Salas, and Mica R. Endsley

CRC Press is an imprint of the
Taylor & Francis Group, an informa business

First edition published 2021
by CRC Press
6000 Broken Sound Parkway NW, Suite 300, Boca Raton, FL 33487–2742

and by CRC Press
2 Park Square, Milton Park, Abingdon, Oxon, OX14 4RN

CRC Press is an imprint of Taylor & Francis Group, LLC

Library of Congress Cataloging-in-Publication Data
Names: McNeese, Michael, 1954– editor. | Salas, Eduardo, editor. | Endsley, Mica R., editor.
Title: Contemporary research : models, methodologies, and measures in distributed team cognition / edited by Michael McNeese, Eduardo Salas, and Mica R. Endsley.
Description: First edition. | Boca Raton, FL : CRC Press, 2020. | Includes bibliographical references and index.
Identifiers: LCCN 2020013266 (print) | LCCN 2020013267 (ebook) | ISBN 9781138625693 (hardback) | ISBN 9780429459733 (ebook)
Subjects: LCSH: Group work in research. | Teams in the workplace. | Distributed cognition.
Classification: LCC Q180.55.G77 C65 2020 (print) | LCC Q180.55.G77 (ebook) | DDC 302.3/5—dc23
LC record available at https://lccn.loc.gov/2020013266
LC ebook record available at https://lccn.loc.gov/2020013267

ISBN: 978-1-138-62569-3 (hbk)
ISBN: 978-0-429-45973-3 (ebk)

Typeset in Times
by Apex CoVantage, LLC

Dedication

Just over four years ago I was sitting in an uncomfortable chair listening to the infusion pump load chemicals into Judy (my wife) to combat ovarian cancer. It was hard to concentrate on the ideas I was putting together and my mind wandered incessantly. I was on sabbatical for the fall semester from Penn State. Little did I know that I would be sitting in Mount Nittany Medical Center taking care of Judy as she went through chemotherapy. Fortunately, I was able to devote time as a caregiver simultaneous with focusing on academic goals set for myself while on sabbatical. One of those goals was to create an interdisciplinary handbook that would examine a spectrum of contemporary topics within the cross sections of distributed cognition and team cognition as they applied within given contexts of use. The result of that seed concept is this Handbook of Distributed Team Cognition. *I would like to dedicate this handbook to Judy for the strength, courage, character, and positive outlook she showed during her time of confronting cancer and fighting to get healthy. By the grace of God, she is now cancer-free and healthy again. We just had our 40th anniversary of marriage this May of 2020—we celebrate our mutual love and continued commitment to support each other in every way possible. We look at life as an adventure that may present unexpected yet beneficial experiences that make us better humans. I love you.*

Michael D. McNeese

Contents

Preface

As we examine contemporary research in the realm of distributed team cognition it is informative to look backward for perspective, as well as look forward to think about what future research might yield. Through these considerations one may add light and basic knowledge to interpret specific studies, experiments, models, analyses, and measures.

The focus of this book entails the intersection and melding of two distinct research areas: *team cognition* and *distributed cognition*. These areas, while typically being distinct in the past, have merged closer together in present research endeavors. The book's objective is to advance state-of-the-art knowledge in terms of real-world interactions among information, people, and technologies through explorations and discovery embedded within the research topics covered. It is our hope that each chapter will provide insight, comprehension, and differing, yet cogent, perspectives to topics relevant within distributed team cognition. In particular, given experts present their use of relevant models and frameworks, different and valuable approaches to study distributed team cognition, and new types of measures and indications of successful outcomes.

Because this book is part of the overall series of the *Handbook of Distributed Team Cognition*, it acts to develop and convey mutual understanding through the examination of intersections, layered integration, and diverse views of how people (and technologies) work together to achieve objectives and carry out actions according to their motivations. The research topics presented necessarily span the continuum of interdisciplinary philosophies, ideas, and concepts that underlie research investigation.

This book was sparked through much of my own discoveries in interdisciplinary cognitive/information sciences by examining cognition through the portals of (1) teamwork and cooperative action (2) computational innovation within the areas of human-computer interaction, computer-supported cooperative work, artificial intelligence, and learning technologies, and (3) in-depth exploration of real world contexts such as piloting and crisis management. Concomitantly, much of this research focuses on *cognition* underlying individual actions, teamwork, and organizational behavior and has been relevant through different points and stations in my research and academic life. In the early to mid-1980s the government and specific industry partners put together what was called the Decision Aiding Working Group (DAWG). Being a part of this group provided an initial inception into group decision making and the potential for technological aids that could enhance competent decisions in highly complex settings. Exposure to the DAWG stimulated my interests and desire to achieve interdisciplinary research into what would now be called distributed team cognition. My mentors during the 1980s that influenced this kind of interdisciplinary cognitive research were three social psychology professors (Dr. Clifford Brown of Wittenberg University, Dr. A. Rodney Wellens of the University of Miami, and Dr. Charles Kimble of the University of Dayton) who at different times worked with me as a research scientist at the Air Force Research Laboratory, Wright-Patterson

Air Force Base, Ohio, and my PhD advisor (Dr. John Bransford, then of Vanderbilt University).

While the book exemplifies the potential integrative power of studying cognition within contextual variation, how cognitive activities are stratified across levels of the team process, and how technology, information sciences, and computation have changed our definition of what distributed work means, the basic foundation of research remains the phenomena of cognition itself (see McNeese, 2017 for further elaboration). At times integration of research is not completely possible and therefore contemporary research may show intersections or coupling of two or more research strands in which they are informative for the greater good (multidisciplinary approaches). Other research may demonstrate unique ways that research strands mutually influence each other in dynamic formulation (i.e., interdisciplinary and transformative progress). Because the overall handbook increasingly encourages multiple perspectives on distributed team cognition it also may be possible that different chapters produce alternative interpretations of the world (i.e., diverse approaches). Because our overall intent is to provide a broad and comprehensive view of research all three of these characterizations may be present.

Michael D. McNeese
February 2020

SPECIAL NOTIFICATION

This book was produced during the outbreak of the COVID-19 crisis that caused many people around the world by necessity to participate in distributed team cognition in their everyday lives. Many of us were required to be separated by distance to avoid the spread of the virus, hence collaborative work/meetings, joint entertainment, church services, and other activities were conducted through the use of distributed tools, technologies, and apps. As such, the topics within this book are highly relevant for the times we live in and are experiencing. Technologies such as Zoom and Skype facilitated connectedness, teamwork, and social awareness that enabled life to continue in the best possible way. As we adapt to the circumstances of this virus, perhaps many elements of distribute team cognition will be inculcated as part of our permanent culture/society. As we face the summer of 2020, the trajectory of COVID-19 is uncertain and indeterminate. We wish all those affected and impacted by COVID-19 the best path forward.

Editors

Michael D. McNeese is a Professor (Emeritus) and was the Director of the MINDS Group (Multidisciplinary Initiatives in Naturalistic Decision Systems) at the College of Information Sciences and Technology (IST), The Pennsylvania State University, University Park, PA. Dr. McNeese has also been a Professor of Psychology (affiliated) in the Department of Psychology, and a Professor of Education (affiliated) in the Department of Learning Systems and Performance, at Penn State. Previously, he was the Senior Associate Dean for Research, Graduate Studies, and Academic Affairs at the College of IST. Dr. McNeese also served as Department Head and Associate Dean of Research and Graduate Programs in the College, and was part of the original ten founding professors in the College of IST. He has been the principal investigator and managed numerous research projects involving cognitive systems engineering, human factors, human-autonomous interaction, social-cognitive informatics, cognitive psychology, team cognition, user experience, situation awareness, and interactive modeling and simulations for more than 35 years. His research has been funded by diverse sources (NSF, ONR, ARL, ARO, AFRL, NGIA, Lockheed Martin) through a wide variety of program offices and initiatives. Prior to moving to Penn State in 2000, he was a Senior Scientist and Director of Collaborative Design Technology at the USAF Research Laboratory (Wright-Patterson Air Force Base, Ohio). He was one of the principal scientists in the USAF responsible for cognitive systems engineering and team cognition as related to command and control and emergency operations. Dr. McNeese received his Ph.D. in Cognitive Science from Vanderbilt University and an M.A. in Experimental-Cognitive Psychology from the University of Dayton, was a visiting professor at The Ohio State University, Department of Integrated Systems Engineering, and was a Research Associate at the Vanderbilt University Center for Learning Technology. He has over 250 publications in research/application domains including emergency crisis management; fighter pilot performance; pilot-vehicle interaction; battle management command, control, communication operations; cyber and information security; intelligence and image analyst work; geographical intelligence gathering, information fusion, police cognition, natural gas exploitation, emergency medicine; and aviation. His most recent work focuses on the cognitive science perspectives within cyber-security utilizing the interdisciplinary Living Laboratory Framework as articulated in this book.

Eduardo Salas is the Allyn R. and Gladys M. Cline Chair Professor and Chair of the Department of Psychological Sciences at Rice University. His expertise includes assisting organizations, including oil and gas, aviation, law enforcement, and healthcare industries, in how to foster teamwork, design and implement team training strategies, create a safety culture and minimize errors, facilitate learning and training effectiveness, optimize simulation-based training, manage decision making under stress, and develop performance measurement tools.

Dr. Salas has co-authored over 480 journal articles and book chapters and has co-edited 33 books and handbooks as well as authored one book on team training. He is a past president of the Society for Industrial and Organizational Psychology (SIOP) and the Human Factors and Ergonomics Society (HFES), and a fellow of the American Psychological Association (APA), Association for Psychological Science, and HFES. He is also the recipient of the 2012 Society for Human Resource Management Losey Lifetime Achievement Award, the 2012 Joseph E. McGrath Award for Lifetime Achievement for his work on teams and team training, and the 2016 APA Award for Outstanding Lifetime Contributions to Psychology. He received his PhD (1984) in industrial/organizational psychology from Old Dominion University.

Mica R. Endsley is the President of SA Technologies, a cognitive engineering firm specializing in the development of operator interfaces for advanced systems, including the next generation of systems for military, aviation, air traffic control, medicine, and power grid operations. Previously she served as Chief Scientist of the U.S. Air Force in where she was the chief scientific adviser to the Chief of Staff and Secretary of the Air Force, providing assessments on a wide range of scientific and technical issues affecting the Air Force mission. She has also been a Visiting Associate Professor at MIT in the Department of Aeronautics and Astronautics and Associate Professor of Industrial Engineering at Texas Tech University. Dr. Endsley is widely published on the topic of situation awareness and decision making in individuals and teams across a wide variety of domains. She received a PhD in industrial and systems engineering from the University of Southern California. She is a past president and fellow of the Human Factors and Ergonomics Society and a Fellow of the International Ergonomics Association.

Contributors

Peter Berggren
Department of Computer and Information Science
Linköping University
Linköping, Sweden

Jill L. Drury
Department of Collaboration and Social Computing
The MITRE Corp
Bedford, Massachusetts

Terri A. Dunbar
School of Psychology
Georgia Institute of Technology
Atlanta, Georgia

Daniel Duran
Florida Institute for Human and Machine Cognition
Pensacola, Florida

Mica Endsley
SA Technologies, Inc.
Gold Canyon, Arizona

Victor Finomore
Department of Neuroscience
West Virginia University
Morgantown, West Virginia

Gregory F. Funke
Applied Neuroscience Branch
Air Force Research Laboratory
Wright-Patterson Air Force Base, Ohio

Jamie C. Gorman
School of Psychology
Georgia Institute of Technology
Atlanta, Georgia

David A. Grimm
School of Psychology
Georgia Institute of Technology
Atlanta, Georgia

Rego Granlund
C2 Learning Labs
Sweden

Tobias Höllerer
Department of Computer Science
University of California Santa Barbara
Santa Barbara, California

James Humann
CCCD
Army Research Laboratory
Los Angeles, California

Björn J. E. Johansson
Department of Computer and Information Science
Linköping University
Linköping, Sweden

Matthew Johnson
Florida Institute for Human and Machine Cognition
University of West Florida
Pensacola, Florida

Sadaf Kazi
Armstrong Institute for Patient Safety and Quality,
Johns Hopkins University School of Medicine
Baltimore, Maryland

Salar Khaleghzadegan
Armstrong Institute for Patient Safety and Quality,
Johns Hopkins University School of Medicine
Baltimore, Maryland

Gary L. Klein
Department of Collaboration and Social Computing
The MITRE Corp
Bedford, Massachusetts

Vincent Mancuso
Cyber Operations and Analysis Technology Group
MIT Lincoln Laboratories
Lexington, Massachusetts

Michael D. McNeese
College of Information Sciences and Technology
The Pennsylvania State University
University Park, Pennsylvania

Neelam Naikar
Center for Cognitive Work and Safety Analysis
Defense Science and Technology Organisation
Melbourne, Australia

John O'Donovan
Department of Computer Science
University of California Santa Barbara
Santa Barbara, California

Karl Perusich
Polytechnic Institute
Purdue University
South Bend, Indiana

Mark S. Pfaff
Department of Collaboration and Social Computing
The MITRE Corp
Bedford, Massachusetts

Michael A. Riley
Department of Psychology
University of Cincinnati
Cincinnati, Ohio

Michael A. Rosen
Armstrong Institute for Patient Safety and Quality,
Johns Hopkins University School of Medicine
Baltimore, Maryland

James Schaffer
Sysco Labs
Sysco, Inc.
Houston, Texas

Michael T. Tolston
Ball Aerospace and Technologies Corporation
Dayton, Ohio

Micael Vignatti
Florida Institute for Human and Machine Cognition
Pensacola, Florida

Adam Werner
School of Psychology
Georgia Institute of Technology
Atlanta, Georgia

Primer (Introduction)

The second volume of this handbook, *Contemporary Research—Studies, Models Methods, and Measures*, builds on the girders presented in the first volume by examining distributed team cognition through different approaches that can be taken to simulate, model, measure, and analyze it. Like the diversity found in team composition, there are also many levels of diversity in the approaches taken to study distributed team cognition. A handbook should provide as many approaches deemed reasonable to communicate all the various dimensions of complexity that comprise interdisciplinary teamwork. Additionally, specific research studies have been undertaken to study the interconnections among teams, technologies, information, and context and are provided to demonstrate the value and worth of specific approaches and methodologies. While volume 2 does not capture every possible approach to contemporary research, the purpose is gather together relevant and representative information for fellow researchers to utilize and be cognizant of.

Volume 2 begins with Chapter 1 presented by my co-editor Mica R. Endsley and provides an insightful review of team situation awareness as related to and found evident in different domains. Her review portrays a lot of valuable information in conceptualizing and operationalizing the construct as applicable to distributed team cognition as processes, devices, and mechanisms contribute to the construct. In particular, the chapter elucidates research and findings relevant to measuring and modeling the construct for use in experiments and engineering design. Of great value is the idea of different devices that facilitate information sharing for team members and what requirements are salient.

The second chapter by Johansson, Granlund, and Berggren presents a very incisive review of C3Fire, the microsimulation that has been used for a wide variety of experiments and approaches to investigate distributed team cognition for the last 20 years. Within that time span the simulation has yielded a wide berth of discoveries in team cognition and distributed cognition that are highly relevant to understanding models, measures, and techniques. The chapter points out the versatility of a dynamic simulation and what it can mean for an integrated research program. The simulation itself is heavily coupled to the real-world domain of emergency response in addressing forest fires and becomes a bridge between experimental studies and field work. Many findings relative to coordination, dynamic decision making, and collaboration are highlighted throughout the chapter.

The third chapter by Dunbar, Gorman, Grimm, and Werner provides a nice review of framing team cognition through a dynamic systems lens. The chapter provides details as to how to apply methods of dynamic systems approach to specific domains, and what this might mean in terms of modeling and measuring team behavior. The chapter is valuable in the sense it provides yet another alternative worldview in framing and in turn researching distributed team cognition. The chapter also addresses the contemporary issues surrounding human-autonomy teaming and what may be done to assuage problem areas and address dynamic actions.

The fourth chapter by Naikar provides yet another alternative world view in approaching distributed team cognition: cognitive systems engineering. This view, like some of the others presented in volume 1, is highly aligned with the theories of ecological psychology but specifically makes use of models that capture interrelated elements of sociotechnical systems. The chapter's primary emphasis looks into the nature of self-organizing teams wherein work is both distributed and adaptive to the demands within the workplace. The chapter makes use of models of cognitive work analysis and design methods but specifically details the diagram of work possibilities which emphasizes computation work for both actors and artifacts.

The fifth chapter by Khaleghzadegan, Kazi, and Rosen explores the basis of team dynamics within the complexity of natural work environments through reviewing measurements directly valuable for distributed team cognition. The important area of measurement surrounding teamwork that involves physiological indicators within team cognition is of particular interest within the chapter. The authors review team cognition with special interest towards unobtrusive measurement strategies and look at shared physiology, linguistic and paralinguistic speech features, and activity tracking generating a very interdisciplinary approach to measurement. Their chapter weighs heavily towards the value of simulation, use of event-based methods, and technological advancement in distributed team cognition environments when considering unobtrusive markers.

The sixth chapter by Drury, Pfaff, and Klein addresses and reviews the strategic topic of causal mental models in distributed teams. The chapter explores techniques for assessing and comparing causal mental models to determine salient differences within their content. The authors utilize two rigorous modeling tools that derive from distributed cognition as part of the basis to compare models: (1) descriptive to simulation modeling and (2) distributed coordination space. The use and value of these two tools are highlighted in an example. DESIM is further elaborated in terms of limits and capabilities and the type of modeling situations for teamwork for which it is best suited.

The seventh chapter by Schaffer, Humann, O'Donovan, and Hollerer looks at some of the more advanced technological elements underlying complex teamwork by exploring areas of modeling and quantitative measurement in dynamic human-agent cognition. The focus is within recommender systems wherein human cognitive bandwidth may be required to interact with computational agents. The authors provide a statistical-based modeling technique to address this kind of interaction for two task paradigms. The results provide intriguing aspects related to cognitive traits and beliefs as relevant for agent interaction that are pertinent for considering global situation awareness, trust, and appropriate workload.

The eighth chapter by Karl Perusich delves into ways of considering complex and fuzzy relationships within distributed information that emerges across time. The fuzzy cognitive map is introduced as a qualitative modeling technique to capture changing situations that are present in most fields of practice that individuals and teams engage with. The modeling technique is conceptualized, explored and reviewed, and explained as relevant for a real-world modeling example. The chapter then shows how a cognitive map can be constructed to capture expert-situational knowledge. The chapter demonstrates how the fuzzy cognitive maps can be used for prediction of behavior.

The ninth chapter by Johnson, Vignatti, and Duran again draws focus to the important area of human-machine teaming and reviews the requirements necessary for effective performance. The chapter points out that traditional task analytical approaches are inadequate and instead concentrates on principles-based interdependence. Much of the chapter then lends credence to the Interdependence Analysis tool and how it can be used in addressing joint activity in understanding human factors considerations and technological constraints. The tool is explained and developed for use in designing human-machine teaming coupling for distributed team cognition applications.

The tenth chapter by Tolston, Funke, Riley, Mancuso, and Finomore explores the higher order relational properties inherent in team cognition by utilizing observables present in team activity. The chapter is hence predicated on measuring the emerging information processing that implicitly relies on knowledge structures utilized by interacting teammates as they uncover, share, and negotiate the meaning of their goal-directed behavior with a context-specific setting. The chapter reviews and presents a very critical measure for understanding team mental models and cooperative work: team communication analysis. The authors review the challenges present in using this type of analysis and address issues that are relevant for use of these kind of analyses. They then articulate the need and use of a specific kind of analysis—conceptual recurrence analysis—as a means for analyzing the structural semantic content of team communications. The work relies upon a bottom-up modeling technique that is based on natural language. This kind of analysis is applied to collaborative consensus building tasks for further understanding whereupon its value is derived.

In conclusion, volume 2 builds upon volume 1 but provides the reader examples of specific kinds of research studies, different methodologies, and measures used to comprehend constructs within distributed team cognition, and shows how unique modeling techniques can improve prediction, design, and understanding of salient variables that contribute to distributed team cognition. Volume 1 and volume 2 in turn provide a basis for thinking about and exploring solutions to challenges that arise in distributed team cognition involving technology innovation, design practice, and interface development within specific applied contexts (fields of practice) which will be addressed in volume 3.

Michael D. McNeese

1 Situation Awareness in Teams
Models and Measures

Mica R. Endsley

CONTENTS

INTRODUCTION

Situation awareness (SA) has been studied extensively in individuals and teams over the past three decades. SA, an understanding of what is happening in the current situation, has been shown to be critical for performance in a wide variety of domains, including aviation, air traffic control, military operations, emergency management, healthcare, and power grid operations (Endsley, 2015b; Parasuraman et al., 2008; Wickens, 2008). In each of these contexts people operating in various types of team settings must quickly understand the state of a complex and often rapidly changing environment in order to make good decisions, formulate effective plans, and carry out assigned duties.

At the level of the individual, SA has been defined as "the perception of the elements in the environment, within a volume of time and space, the comprehension of their meaning, and the projection of their status in the near future" (Endsley, 1988). Thus, it includes:

(1) *Level 1 SA—The perception of key information relevant to the decision maker's needs.* This may include the direct perception of information when the individual is embedded directly in the world (e.g. an infantry solider observing enemy movement or a pilot observing relevant terrain), but also often involves the receipt of information from other team members via verbal or nonverbal communications, written reports, and electronic information displays (Endsley, 1995a, 1995b). Thus, it includes information from natural, engineered, and human sources. For example, an air traffic controller who receives information from a controller in an adjacent sector or from an aircraft pilot is obtaining relevant information from other team members that is then compared to and combined with information from other sources.

(2) *Level 2 SA—The comprehension or understanding the significance of that information with regard to the decision makers' goals.* SA involves knowing more than just data; it also includes being able to put together disparate pieces of data to inform relevant decisions. For the air traffic controller, knowing that an aircraft is at a particular altitude, location and heading is level 1 SA; understanding that it is below its assigned altitude and therefore has a deviation is Level 2 SA. The formation of Level 2 SA is highly dependent on the goals and decision requirements of the individual, which may vary significantly, based on the person's role, within and across teams.

(3) *Level 3 SA—Projection of the current situation to inform likely or possible future situations.* Projection forms the third and highest level of SA, and is the hallmark of expertise in SA (Endsley, 1995b, 2018). Situation dynamics forms an important part of SA. By constantly projecting ahead, decision makers are able to act proactively instead of just reactively. For example, the air traffic controller is able to project that two aircraft will collide in the future, based on their current assigned trajectories. Similar to Level 2 SA, there can be considerable variance in Level 3 SA projections based on the differing goals and decision requirements of different team members.

It should be pointed out that these three levels of SA represent ascending levels of SA quality (i.e. a person who is able to make accurate projections about the situation has better SA than one who only knows lower-level information), but they are not necessarily linear in terms of process (Endsley, 1995b, 2004, 2015a). While Level 1 SA many generally lead to later integration and comprehension, in many cases Level 2 and 3 SA are also used to drive the search for low-level information or to compensate for data that is unknown (i.e., provide default values for missing data).

The cognitive processes and mechanisms involved in deriving SA have received considerable attention (Endsley, 1995b, 2015a). These include (1) the important role of goal-directed processing that drives the search for information, alternating with data-driven processing that helps drive the prioritization of goals, (2) limited attention and working memory that can act to constrain SA in complex environments, particularly for novices or those in novel situations, and (3) the formation of mental models and schema that provide mechanisms for rapidly comprehending and projecting information into the future, overcoming these limits to a large degree. This foundation sets the stage for understanding the factors that drive differences in SA across team members and the mechanisms available for supporting coordinated SA with and across teams.

SITUATION AWARENESS IN TEAMS

Team SA (TSA)

Teams are defined as "a distinguishable set of two or more people who interact dynamically, interdependently, and adaptively toward a common and valued goal/objective/mission, who have each been assigned specific roles or functions to perform, and who have a limited life span of membership" (Salas, Dickinson, Converse, & Tannenbaum, 1992). Critical features that define a team therefore include (1) a common goal, (2) interdependence, and (3) specific roles. The specific roles of the individual team member determine their individual goals, decisions, and SA needs.

In that the performance of team members is mutually interdependent for achieving the common goal, team SA (TSA) is defined as "the degree to which every team member possess the SA required for his or her responsibilities" (Endsley, 1995b). That is, every team member must have good SA for the information associated with his or her role, in order to support overall team performance. It is not sufficient for some members of the team to have a piece of information if the person on the team who needs it does not know it. For example, in the crash of Midland Flight 92 in Kegworth, UK, the flight attendants in the back knew that the pilots had shut off the wrong engine in response to an engine fire, but the pilots in front did not, contributing to the accident (United Kingdom Air Accidents Investigation Branch, 1990). In this way, the SA of individual team members are all relevant to the effective functioning of the team, as shown in Figure 1.1.

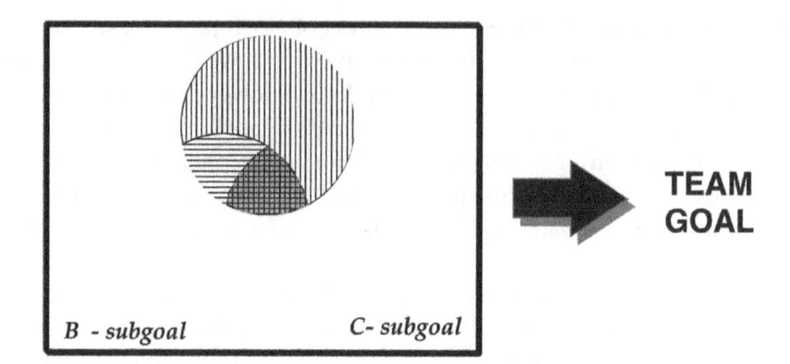

FIGURE 1.1 Team SA arises from the unique goals and SA requirements of all team members needed to achieve overall team performance.

Source: From Endsley & Jones, 1997, 2001. Reprinted with permission from a model of inter- and intrateam situation awareness: Implications for design, training and measurement, in *New trends in cooperative activities: Understanding system dynamics in complex environments*, 2001. Copyright 2001 by the Human Factors and Ergonomics Society. All rights reserved.

Shared SA (SSA)

Although the specific aspects of the situation that are relevant to each team member's SA may be different (in that they are determined by the unique goals of each role), because teams inherently involve some interdependence of their members, there also will exist a subset of information requirements that are common amongst the team members, as shown in Figure 1.2. It is this overlap in SA requirements that defines the need for shared SA (SSA). SSA is defined as "the degree to which team members possess the same SA on shared SA requirements" (Endsley & Jones, 1997, 2001). A consistent mental representation of the status of these overlapping requirements is essential for effective team coordination and performance, and drives much of the need for information sharing across teams. On these common SA requirements, two members may possess SA that is (1) shared and correct, (2) shared, but incorrect, (3) not shared, with one member correct and the other incorrect, or (4) not shared with both incorrect (Endsley & Jones, 1997, 2001).

As shown in Figure 1.2, not all the SA of the team members needs to be shared—just the SA associated with common SA requirements. So for example, the air traffic controller and the pilot do not need to share everything about their current situations, which will just create overload (Endsley & Jones, 1997, 2001). They do however, need to be on the same page with respect to the aircraft's current location, speed and altitude, clearance, the location of other aircraft that are traffic for it, its proximity to nearby terrain or restricted airspace, and the presence of turbulence or weather on its projected flight path that may create a problem for it (Endsley, Hansman, & Farley, 1998). In some cases the air traffic controller may have the more accurate information to share (such as in the case of other aircraft) and in some cases the pilot may have the more accurate information (such as is the case with weather information) (Farley, Hansman, Amonlirdviman, & Endsley, 2000).

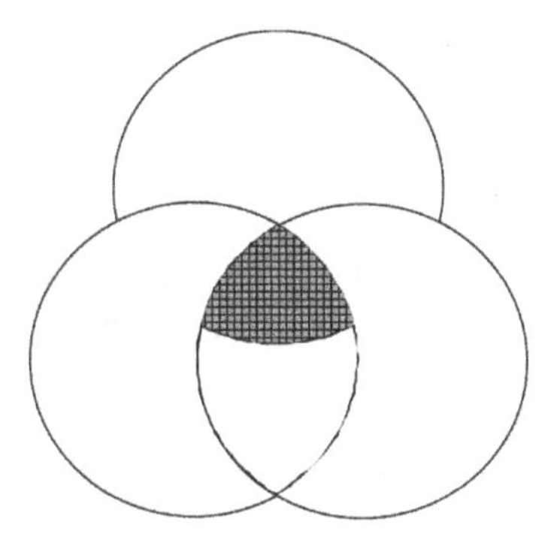

FIGURE 1.2 The need for shared SA is a function of the overlap in individual goals.

Source: From Endsley & Jones, 1997, 2001. Reprinted with permission from A model of inter- and intrateam situation awareness: Implications for design, training and measurement, in *New trends in cooperative activities: Understanding system dynamics in complex environments*, 2001.

Relevance of TSA and SSA

SSA has been shown to be predictive of team performance in a number of studies (Bonney, Davis-Sramek, & Cadotte, 2016; Cooke, Kiekel, & Helm, 2001; Coolen, Draaisma, & Loeffen, 2019; Rosenman et al., 2018). In Bonney et al.'s (2016) study of business markets, team performance was predicted by both SSA on the team (all three levels of SA contributing over 34% of the variance) and by having a shared team strategy. Similarly, in the medical domain, Rosenman et al. (2018) demonstrated that SSA was predictive of performance, and Coolen et al. (2019) showed that SSA on both the problem and the diagnosis were highly predictive of good team performance. Cooke et al. (2001) found that both TSA and SSA of level 1 and level 3 queries was highly predictive of team performance in a study involving the operation of unmanned air vehicles (UAVs).

Overall TSA (based on combined or average SA across the team) has also been found to be predictive of overall team performance (Cooke et al., 2001; Crozier et al., 2015; Gardner, Kosemund, & Martinez, 2017; Parush et al., 2017; Prince, Ellis, Brannick, & Salas, 2007), however, some studies have not found this to be the case (Brooks, Switzer, & Gugerty, 2003; Morgan et al., 2015; Sorensen, Stanton, & Banks, 2010). For example, Prince et al. (2007) demonstrated that combined TSA scores collected on pilots in low-fidelity simulations were predictive of performance in high-fidelity simulations. Gardner et al. (2017) showed that combined TSA scores were correlated with teamwork ratings in a medical trauma simulation. Similarly, Crozier et al. (2015) showed that a combined TSA score correlated with the experience level of medical trauma teams and was predictive of checklist performance measures in their simulation. Price and LaFiandra (2017) found that stress doubled the level of overconfidence in teams and negatively affected TSA scores, but that efforts to increase team engagement helped to moderate these stress effects. Sorensen et al. (2010), however, were not successful in comparing median TSA across the team to performance, and Morgan et al. (2015) found only a weak correlation between a combined TSA score and performance.

MODEL OF TEAM SA

Endsley and Jones (1997, 2001) developed a model of TSA that includes four major contributors to achieving TSA and SSA. This includes (1) *TSA requirements* as needed to support the SA and SSA of the team; (2) *TSA devices* that provide methods for sharing information across the team; (3) *TSA mechanisms* that support the ability of the team to achieve accurate SSA; and (4) *TSA processes* that have been found to be important for effective communication and coordination in teams for supporting high levels of TSA and SSA. Each of these will be discussed in more detail.

TEAM SA REQUIREMENTS

At the basic level, the SSA requirements of teams can be defined in terms of the overlap in SA requirements between any two roles, as shown in Figure 1.2. As an example of this, Bolstad, Riley, Jones, and Endsley (2002) analyzed the SA requirements associated with army brigade command and control officers. Using a goal-directed task analysis (GDTA) process, we showed how each role had different goals creating the need for very unique sets of information requirements associated with terrain, and how they needed to make very different assessments and projections (Level 2 and 3 SA) based on that information (TSA). A comparison across different roles, however, provided a clear indication of which terrain information needed to be shared across officers to support a common understanding of the battlefield (SSA).

In another example we compared the goals and SA requirements of pilots and air traffic controllers based on GDTAs of each role (Endsley, Hansman et al., 1998; Farley et al., 2000; Farley, Hansman, Endsley, Amonlirdviman, & Vigeant-Langlois, 1998). While we discovered much commonality between these roles in terms of their high-level goals (such as assuring flight safety, avoiding conflicts, and handling perturbations such as weather and emergencies), at the lower levels there are a number of sources of divergence that create conflict for Level 2 and 3 SA. With respect to re-routing decisions, for example, pilots assess potential changes in terms of time or fuel efficiency, while controllers primarily consider their effect on separation and traffic flows. In general, pilots' aircraft-centered goals were often in conflict with controllers' system-centered goals, creating the potential for less collaborative negotiations. The analysis was able to identify not only where the SA needs of these two members of the aviation team overlapped, but also the ways in which their higher-level SA assessments were significantly different from each other.

In addition to each team member needing the SA required for his or her individual job, a number of other aspects related to the team form a part of their shared SA needs. For instance, information regarding the actions other team members have taken and their capabilities (e.g., as affected by training, injuries, workload, or fatigue for example) may be important to another team member's SA (Endsley, Farley, Jones, Midkiff, & Hansman, 1998). Team members need to know the status of all team member activities in terms of how they impact on their own goals and requirements. For instance, one maintenance technician may need to alert other team members that he is opening a valve that would affect the operations or safety of other team members (Endsley & Robertson, 1996; Endsley & Rodgers, 1994). A shared understanding of the impact of the other team members' task status on one's own functions, and thus the team overall goal, is important. Similarly, team members need to know how their own task status and actions impact on other team members so that they can coordinate appropriately (Endsley & Robertson, 1996). Taken together, this set of shared SA requirements forms the basis for a significant amount of information sharing that must occur within teams, based on one or more TSA devices.

Team SA Devices

Several classes of devices are available for supporting information sharing associated with SSA requirements across team members. This includes:

(1) *Communications*—Direct communication, both verbal and nonverbal, is a central method for sharing information across teams and for facilitating a common understanding and projection of events. Team communication has been emphasized as having an important role in supporting the development of SSA across the team in many domains (Endsley, 1995b; Salas, Prince, Baker, & Shrestha, 1995). Communications can occur face-to-face or via phone, radio, or other electronic means. Nonverbal communication has also been shown to be important for SSA in both aviation (Segal, 1994) and medical teams (Xiao, Mackenzie, & Patey, 1998). See Tiferes and Bisantz (2018) for a review on situational and team factors affecting team communications.

(2) *Shared environment*—Being a part of a shared environment that allows team members to directly observe the same information can also provide SSA. For example, the pilot and co-pilot are in a shared in environment in a cockpit, so they can both directly observe the aircraft takeoff and may not need to verbally communicate that event to each other. They also have the advantage of sharing a view of the same displays. Additional information could be communicated verbally, or through nonverbal cues, such as a look of confusion or fatigue.

(3) *Shared displays*—The development of effective SSA displays has been a theme of interest in many domains. Not all devices are equivalent in terms of their ability to support SSA, however. Bolstad and Endsley (2005) compared a number of tools for support collaboration in teams including face-to-face, video conferencing, audio conferencing, telephones/radios, chat/instant messaging, white boards, file transfers, program sharing, email, groupware, bulletin boards, geographic information systems, and domain-specific tools. These devices were significantly different in terms of their ability to support synchronous vs. asynchronous collaboration, scheduled vs. unscheduled interactions, collocated vs. distributed, and low intensity vs. high intensity collaboration. They also differed in

their level of traceability, identifiability of participants, and ability to support unstructured as well as structured communication. Overall, highly flexible tools such as phones or video/audio conferencing were found to be highly beneficial to developing SSA, particularly for Level 2 and 3 SA in that they are good for handling unstructured, distributed interactions. While computer-based tools such as chat, email, and file transfers have become very popular because they support distributed, unscheduled, and asynchronous communications, they often create substantial data overload to process large volumes of unstructured data. Domain-specific tools consist of customized displays that structure the needed information to support SA and SSA and which are updated in real time. This type of shared display device is the most efficient in that needed information can be integrated and displayed in ways that are easier to comprehend, substantially reducing the workload compared to other TSA devices. These displays require more up-front work to design and build, however, and often may need to be augmented with additional tools for unstructured communications if all the information needed for SSA is not included.

Strater, Cuevas, Connors, Ungvarsky, and Endsley (2008) examined the use of collaborative tools in an army command and control exercise, and found that participants rated face-to-face communication the highest for both routine and non-routine conditions. They rated chat and shared maps/domain tools the next most effective, although not for critical communications; however, they felt these tools contributed the most to SSA, even over face-to-face communications.

In the command and control environment, there has been significant focus on the development of common operating pictures (COP) to support the shared SA of a widely distributed team (i.e., domain-specific shared displays). It is widely recognized that information on the COP must be tailored to each team member's SA needs, even when working in teams. Bolstad and Endsley (1999, 2000) showed that when the COP was carefully tailored to each team member's SA needs performance was enhanced, however, when everyone's SA needs were provided to team members, SA quickly became degraded due to information overload. Javed, Norris, and Johnston (2012) demonstrated improvements in both TSA and SSA with a new display designed to improve SA in an emergency management scenario.

Endsley and Jones (2012) provide a number of guidelines for the design of displays for supporting TSA and SSA.

(1) *Build a common picture to support team SA*—Shared displays should be fed by a common database of information, or if different team members' displays come from different sensors and sources, information should be provided so that team members are aware of what each other knows.
(2) *Avoid overload in shared displays*—Common operating pictures should be carefully tailored to the SA needs of each team member, avoiding the overload associated with providing all of the team's information.
(3) *Provide flexibility to support shared SA across functions*—The perspectives and information provided to each team member need to be flexible to

support not only their current SA needs, but also their need to understand each other's tasks and perspectives, including shifts in physical vantage points, goal orientations, and semantics.

(4) *Support transmission of different comprehensions and projections across teams*—Most shared displays provide only low-level data. However, in that different team members will create very different Level 2 and 3 SA, even from the same basic data, shared displays should assist individuals in making cross-team assessments of information to support their higher SSA requirements (e.g., understanding task status and usage of common resources, as well as projections of task timing across the group).

(5) *Limit non-standardization of display coding techniques*—While it is very common to want to allow individuals to customize their displays, in team settings this can act to lower SA in that different team members who share displays may interpret displays very differently (e.g., if red means bad to one team member, but good to another, miscommunications are more likely). Thus, standardization of iconology, color coding, and other user interface features can help support SSA.

(6) *Support transmission of SA within positions by making status of elements and states overt*—In many systems considerable local knowledge may be needed, for example, who has what expertise, or who has been contacted about a problem. When this knowledge only exists in a person's head, the ability for other team members to assist in problem solving or take over when needed is significantly lowered. SSA can be improved when hidden, implicit information is made explicit to support team communication and coordination.

Supporting the SA required for team operations first requires that systems are designed to support the SA of each individual team member (see Endsley & Jones, 2012). In addition, display features should be provided to ensure that data, comprehension, and projections to support SSA requirements are quickly and accurately communicated across team members without creating overload. Endsley, Bolstad, Jones, and Riley (2003) show an example of applying these design principles to create effective shared displays for army command and control. Based on a GDTA of each role on the brigade staff, tools for supporting a common understanding of unit status, mission schedule, mission planning, and the geo-spatial location of friendly, enemy, and civilian units were developed. These tools were highly flexible, providing each officer with the ability to customize views and information content to address their various SA needs, as well as to support SSA during mission planning and execution.

Tradeoffs across TSA Devices

Individuals may draw from any one of these TSA devices for forming SA, resulting in dynamic tradeoffs in the degree to which different people rely on different TSA devices at different times. Further, as new TSA devices become available, significant shifts in their reliance on other devices can occur. For example, Bolstad and Endsley (1999, 2000) show that when shared SA displays were provided to a team, their verbal communications decrease significantly. Artman (1999) discovered information overload problems that hampered SSA when information flowed independently to team members, possibly discouraging team communication and coordination. Parush and Ma (2012) demonstrated an advantage for team displays, particularly in overcoming communication breakdowns.

In that new, shared displays are being provided in many domains, it is important to consider that some subtle information embedded in verbal communications may be inadvertently lost in the process. For example, controllers routinely assess the experience level of pilots based on how they sound (Midkiff & Hansman, 1992), and military officers may gage the level of fatigue or stress of their troops based on voice communications (Endsley et al., 2000). These important cues may be lost when shared displays become the norm and voice communications subsequently decline, requiring new forms of support.

Team SA Mechanisms

In addition to team SA devices, many teams can be aided in developing SSA based on the presence of shared mental models (SMM) (Salas et al., 1995; Stout, Cannon-Bowers, & Salas, 2017). SMM are generally considered a consistent understanding and representation of how systems work (components, content, structural relationships, and cause and effect). Since people's comprehension and projection are largely dependent on mental models, the degree to which two team members have an SSM will help drive similar interpretations of information that is perceived, as well as agreement in the drive for relevant information from the environment.

When people have an SMM they are more likely to arrive at a common understanding of the current situation without needing as much verbal communication. For example, Mosier and Chidester (1991) showed that better-performing aircrews communicated less than poorer-performing ones, most likely due to SMM. In contrast, teams without SMM will most likely require a great deal of real-time coordination and communication to ensure that their activities are carried out properly and will be far more susceptible to lapses in this process. The National Transportation Safety Board (1994) found that 73% of commercial aircraft accidents happen on the first day of a new crew pairing, most likely because strong SMM has not yet been developed to support SSA and team coordination. Bolstad and Endsley (1999) demonstrated that teams with SMM (developed through cross training) performed significantly better on a team task than those without. Bolstad, Cuevas, Gonzalez, and Schneider (2005) and Saner, Bolstad, Gonzalez, and Cuevas (2009) calculated SSA for members of a military team performing exercises and found that both shared knowledge (SMM) and organizational hub distance were predictive of SSA.

SMM are thought to be developed through common training (joint training or cross training), shared experiences in working together as a team, and direct communications between team members as needed to develop SMM in advance of operations (Endsley & Jones, 2001). They provide for standardized communication patterns and vernacular for interactions (Foushee, Lauber, Baetge, & Acomb, 1986; Kanki, Lozito, & Foushee, 1989), create an understanding of who has what information, and put everyone on the same page with respect to the problem being addressed (Orasanu & Salas, 1993).

Bolstad and Endsley (1999) and Gorman, Cooke, and Amazeen (2010) both found that cross training resulted in higher levels of SSA and team performance. Cooke et al. (2003), however, found that while cross training improved shared knowledge, it failed to improve team performance in their study. Espevik, Johnsen, and Eid (2011) showed that naval teams who had trained together over a long period of time developed much better SSM, which provided better team performance than teams who had not trained together, even though they had experienced cross-training sessions. The teams with SMM were better at communicating and providing back-up behaviors. The researchers believe that the superior knowledge of other team member's characteristics, abilities, and tendencies, developed by training together were important to this finding.

In that mental models are important mechanisms for directing people's attention to critical data and for forming the higher levels of SA, often almost automatically or subconsciously, they form both a powerful tool for SSA and simultaneously an opportunity for the divergence of SA across teams of individuals working together whose mental models may be different. While it can be argued that teams may benefit from divergent mental models (DMM), in order to bring in new and different ideas or to support the different types of tasks that people need to do on teams, this depends significantly on the types of tasks being performed. Just as SA does not need to be identical among teammates, with SSA needed on only the aspects of the situation that are common to both teammates due to their interdependencies, this is also true of SMM. They only need to be consistent with regard to the degree to which they will drive consistent comprehension and projections of the situation to support shared tasks, as long as they are accurate. As SMM are often not present in many teams (due to the divergent roles, responsibilities, and training of heterogeneous team members, for example), these teams will need considerable communication and coordination to resolve inconsistencies in interpretations and understanding so as to achieve common team goals.

This discussion has primarily focused on the issue of SMM focused around the systems and environmental information associated with a domain. In addition, Mohammed, Ferzandi, and Hamilton (2010) discuss team mental models (TMM) as a "shared, organized understanding of the key elements in a team's environment." They also include in TMM common ideas of "taskwork (what the team must do in order to complete goals), teamwork (who team members interact with and how they work together collectively), and . . ., timework (when members interact with each other)" (Marhefka et al., this book). These aspects of TMM are additionally important for team coordination and performance, in addition to the formation of SMM to support SSA. SMM of teamwork has been found to predict team performance (Cooke et al., 2001).

These concepts are depicted in Figure 1.3, which shows the different aspects of TMM and how they contribute to both effective team processes and to TSA and SSA, each of which contribute to effective team performance. This model is largely consistent with Cooke, Salas, Kiekel, and Bell (2004), but stresses the independent role that team processes can have on team performance in addition to its role on TSA and SSA, as well as the additional factors that affect TSA and SSA per Endsley and Jones (1997, 2001).

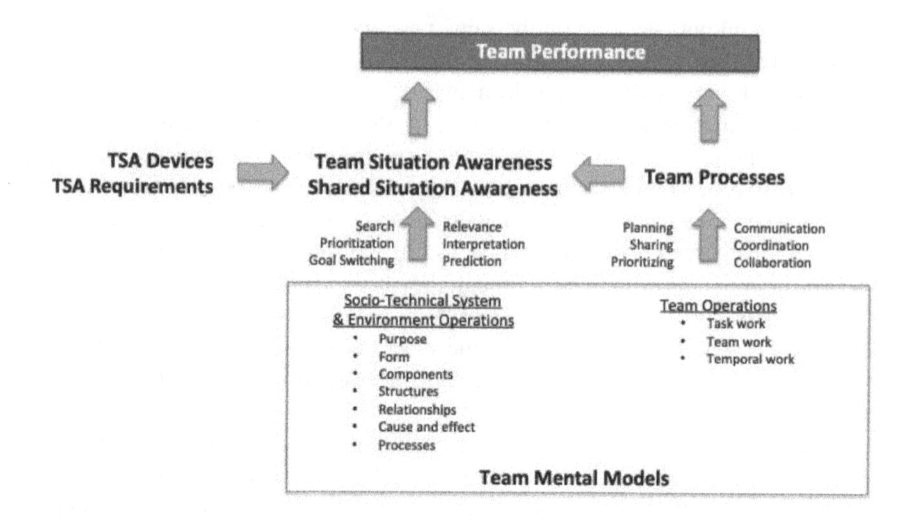

FIGURE 1.3 Team mental models, including the degree to which team members share a common understanding of the sociotechnical systems, environment, and team operations, provide TSA mechanisms for supporting TSA, SSA, and team processes.

Team SA Processes

Considerable research has been conducted on the types of processes that teams use and their impact on team coordination and TSA. Orasanu and Salas (1993) summarize a number of studies involving aircrew and military teams to show that effective teams (1) engage in contingency planning that helps setup shared mental models for emergencies, (2) have leaders who establish a democratic environment that supports better sharing of information, and who explicitly state more plans, strategies, and intentions, consider more options, provide more explanations, and give more warnings or predictions (Chidester, Kanki, Foushee, Dickinson, & Bowles, 1990; Orasanu, 1990), and (3) develop a shared understanding of the problem prior to looking for solutions, thus avoiding getting bogged down (Hirokawa, 1983), which is particularly important the more diverse the team members' backgrounds (Citera et al., 1995). Klein, Zsambok, and Thordsen (1993) additionally point to the importance of (1) clear delineation and understanding of tasks, goals, and roles, (2) avoidance of micro-managing but willingness to help other teammates, (3) avoidance of fixation and willingness to examine various factors of the situation, (4) encouragement of different opinions and a process to come to convergence, and (5) the ability to manage time and make changes as needed.

Taylor, Endsley, and Henderson (1996) found that effective teams established a group norm of information sharing and self-checking to make sure everyone was on the same page at each step. They coordinated as a group, delegated tasks, and gathered information from each other. They imagined possible events in the future and came up with contingency plans for addressing them. They also actively prioritized goals, so that overall performance was not sacrificed due to distractions or unexpected problems. In contrast, poorly performing teams had a group norm in which pertinent information was not shared, so that they went along with the group without contributing important, but conflicting information. They were easily distracted by unexpected problems and were unable to prioritize tasks effectively. They tended to rely more on expectations, which may have been incorrect, and had no team processes in place for detecting this. In some cases, a strong personality acted to lead the others astray based on a strong, but erroneous, picture of the situation, creating what one aviator described as an "SA black hole."

Other researchers have studied the ways in which TSA varies between high- and low-performing teams. Hallbert (1997) found that high-performing teams had better SSA that he attributed to the importance of team cohesion and coordination. Working with trauma resuscitation teams, Crozier et al. (2015) showed that SSA, complementary SA (knowledge needed by individuals that is not shared), and overall TSA all increased with team experience. Sulistyawati, Chui, and Wickens (2008) also found that good teamwork behaviors (such as coordinating when necessary and providing feedback and support) were significantly correlated with good SA of both a pilot's own tasks and the teammate's tasks.

In studying the processes used by control room teams to solve challenging scenarios, Patrick, James, Ahmed, and Halliday (2006) found differences in planning, problem-solving techniques, team coordination, attention, communication, and knowledge that they felt contributed to differences in TSA. Gross and Kluge (2014) demonstrated that effective team processes for knowledge sharing directly improved SMM in teams in the steel industry. Stout, Cannon-Bowers, Salas, and Milanovich (1999) also showed that team planning could improve SMM among team members, creating more efficient communications patterns and more coordinated team performance. Berggren, Prytz, Johansson, and Nählinder (2011) showed a strong relationship between teamwork (coordination, support, and communication) and subjective SA in a simulated firefighting experiment, with workload negatively affecting both teamwork and SA. Altogether, a strong body of research supports the importance of team processes for achieving effective SSA and TSA in teams and for achieving overall team performance.

CHALLENGES FOR TEAM SA

A number of challenges have been found that inhibit good TSA and SSA. These are often particularly problematic as the teams become distributed (separated by space, time, or obstacles), and when teams of teams are involved. For example, Endsley and Robertson (1996; Robertson & Endsley, 1997a) investigated TSA in aircraft maintenance teams conducting both an SA requirements analysis of the multiple teams involved and a contextual inquiry to examine information flow across the organization and found:

(1) *Poor shared mental models*—The teams from different parts of the organization had poor mental models of the activities and information needs of the other teams leading to not fully understanding the implications of transmitted information, which contributed to errors and inefficiencies.

(2) *Poor verbalization of decisions*—Teams were poor at passing on to other teams why they chose a particular course of action, resulting in misunderstandings and sub-optimizing performance as the knowledge of other teams could not effectively come into play.
(3) *Inefficient shift meetings and teamwork*—Team leaders failed to adequately convey common goals for the team, provide a clear understanding of task assignments, point out the inter-relationships between different people's tasks, and provide clear expectations regarding teamwork.
(4) *Poor feedback on the effect of actions across the distributed organization*—As the aircraft would move on to other geographic locations, people rarely got good feedback on the result of repair actions they had taken, which significantly inhibited the learning needed for SA and diagnostics.
(5) *Problems with poor individual SA*—Many common SA challenges were found, such as forgetting steps due to interruptions, missing critical information due to other task distractions, misinterpreting information due to inaccurate expectations, and failing to pass on important information or communicating poorly. These SA failures at the individual level would then propagate across the team to result in TSA failures.

A number of other challenges for TSA and SSA have been found.

Poor Support for Distributed Teams

Distributed teams need to achieve the same SSA as co-located teams, but are separated by time, space, or obstacles (Endsley & Jones, 2001). They frequently do not have many TSA devices available to them. For example, a pilot and an air traffic controller generally do not share the same environment, cannot communicate via nonverbal cues, and historically have a limited amount of information about each other on shared displays. Thus, communications channels can be highly overloaded as the main pathway for building SSA.

As technology has improved, new sensors have been added that allow controllers to see basic information about each aircraft and allow pilots to gather some information about each other (Jones, 1997), however, much of their SSA must still be achieved through verbal communications via the radio. Farley and colleagues showed that the addition of datalink significantly improved the SSA of pilots and controllers with respect to weather and air traffic (Farley et al., 2000; Farley et al., 1998). This was accomplished with less radio communication, even though the rate of route negotiations increased, demonstrating the tradeoffs between TSA devices that can occur.

A Lack of Shared Displays and Information Overload

In many environments, such as command and control or emergency management, for example, information is not shared well across team members via shared displays. Instead, different team members each see a subset of information, often generated via different sensors or information sources. This situation places a very high demand on the use of voice communications or chat rooms to try to compensate for the inadequacies of the information systems, resulting in a heavy manual load and often poor SA as information falls through the cracks or is interpreted differently across the distributed team. When distributed teams also need to deal with poor voice communications lines, such as may be common in many of these same settings, the problems are compounded.

Lack of Temporal Overlap

In many environments work occurs across a 24/7 time schedule or moves across different geographic units, with different teammates assuming responsibility over time. For example, different air traffic controllers must pass aircraft across sectors of responsibility, and must pass responsibility for their sector to other controllers across shifts. The same situation occurs in other settings, including healthcare in hospital settings, aircraft maintenance, and command and control of military, space, and power systems.

The need for people to coordinate across multiple shifts and locations adds to the difficulty of fully communicating an understanding of what is happening as work shifts across teammates. In particular, the challenge of communicating pertinent status information, watch items, concerns, and task statuses can be significant, particularly as shift turnovers are often hurried and poorly supported by information displays. The use of well-structured shift turnover practices and tools have been extended from air traffic control to many other areas, including aviation maintenance (Parke & Kanki, 2008) and healthcare (Jeffs et al., 2013).

Problems with Social and Cultural Differences in Teams of Teams

Robertson and Endsley (1997b) found a lack of trust across distributed teams, as well as the development of different cultures across distributed teams that contributed to challenges with teamwork and developing SSA. Given that many distributed teams are often not functionally or organizationally integrated, resolving these differences can be quite difficult.

Teams of Teams and Organizational Structures

Wellens (1993) investigated distributed teams involved in responding to emergencies, including police and firefighter units. He found that as the workload increased, the degree to which people communicated across team boundaries decreased. This challenge is often exacerbated if shared SA devices are poor, such as was the case in the response on 9/11.

In a study of SSA in military command and control teams, Saner, Bolstad, Gonzalez, and Cuevas (2010) found that the most important predictor of SSA in a team was the organizational proximity of the team members. While communication frequency was related to better Level 1 SA, for more complex information higher SSA was associated with a higher cognitive workload that was required to achieve it. They also showed that greater similarity in experience was related to better SSA.

Buchler et al. (2016) discovered that a few individuals dominated information sharing across the staff in a command and control setting, but that those who engaged in high levels of email output had lower SA. Higher SA on the other hand was correlated with more information inputs and fewer information outputs on the email system. While this may be due to the workload demands of email generation, these findings may also reflect the hierarchical nature of military command and control where less experienced personnel are assigned more mundane tasks (i.e. a division of labor decision strategy). They also found that team members with high SA were more likely to communicate with others with high SA and low SA people were more likely to communicate more with other low SA people.

Team Composition

A number of researchers have demonstrated that team composition significantly affects TSA. Some teams are much better at sharing and promoting SA across the team, and being in such a team can improve the SA of its members. Sætrevik (2012) showed that team membership significantly influenced individual SA in a study of emergency handling in teams in the energy industry. Several researchers have also demonstrated that SA is sensitive to which teams people belong to in military command control exercises (Bolstad & Endsley, 2003; Leggatt, 2004; Seet, Teh, Soo, & Teo, 2004). So, an individual's SA is impacted by not only his or her own cognitive skills, but also the skills and knowledge of the team they are a part of.

Other research, however, found that individuals working in teams had lower SA than when working alone on a simulated firefighting task, presumably due to the overhead costs of communicating to share information within the team (Parush, Hazan, & Shtekelmacher, 2017). Thus, some tasks may be better performed by individuals. But for those that inherently require the contributions of multiple team members, TSA and SSA are critical to effective team performance.

Leadership

Team leaders also have a marked effect on TSA. Cuevas and Bolstad (2010) showed over three separate studies that team leader SA was positively correlated with the SA of the team members, accounting for between 12% and 49% of the variance in SA between teams.

Ad-hoc Teams

While many teams remain relatively fixed, in many cases people may flow onto and off of teams in a more dynamic fashion, for example, design teams or teams formed to solve special problems in an organization (often called *tiger teams* or *kaizen*). Strater et al. (2010, 2008) studied ad-hoc teams set up by the military. We found that ad-hoc teams were challenged by being largely distributed and lacking common training which made it difficult to formulate trust and team cohesion. They did not possess the common knowledge and common background that was needed to develop effective SMM. Their goals tended to be more abstract and ill defined, and they often worked in unfamiliar environments and with poorly defined roles and responsibilities. Further, as they worked across different temporal timelines, they experienced the challenges of regaining SA across shift changes on a frequent basis. Often the members of these ad-hoc teams were required to balance other competing duties associated with their regular jobs in addition to the ad-hoc team, creating SA problems due to multi-tasking. As knowledgeable team members leave the team, TSA often suffered as well, since there may be poor documentation or information to cover that person's knowledge.

MEASUREMENT OF TEAM AND SHARED SA

TSA and SSA have been assessed in a number of ways including inferring it from team processes, communications, group transactions, and information sharing, or by directly measuring the SA state of team members and comparing it within and/or across teams.

Team SA Process Measures

Some researchers have focused on individual and TSA by examining the processes used to achieve it. See Cooke and Gorman (2009) for a review. This research records team communications as a window into what teams are thinking and how they are interacting. These communications are then analyzed using techniques including network models (Buchler et al., 2016; Gorman, Weil, Cooke, & Duran, 2007), time series analysis (Kitchin & Baber, 2017), and latent semantic analysis (Bolstad et al., 2007; Cooke et al., 2004). Others have focused on team behaviors and interactions to investigate the ways in which teams coordinate to achieve SA. Cooke, Stout, and Salas (2001) review a number of possible techniques, including verbal protocols, structured interviews, and process tracing to gain insights into TMM and TSA.

As examples of this work, Gorman, Cooke, and Winner (2006) examined team behaviors in responding to unexpected situational changes, with an emphasis on team coordination. Similarly, Gorman, Cooke, Pederson, Connor, and DeJoode (2005) classified the effectiveness of team communications following a problem with a scenario, confirming that it is important that relevant team members have SA of a problem, but not necessarily all team members.

Patrick et al. (2006) subjectively observed team behaviors in a process control room and evaluated how effective the team processes were in handling programmed disturbances. Hauland (2008) used eye-tracking to assess the attentional strategies of air traffic controllers as a means of inferring TSA. Salmon et al. (2008) focused on the transactions between team members and team artifacts to examine information flow using network-based analyses. This approach, however, combines a consideration of SA sources (which can vary significantly between individuals and between situations) and team members, making its interpretation limited.

Overall, while there are valuable insights to be gained from studying team communications and the processes that teams use to develop SA, it is ultimately difficult to say how successful such processes are in contributing to TSA or SSA without an independent objective measure of their effectiveness. While factors such as communication and teamwork are undoubtedly important for good SA, accurate SSA can also be achieved outside of these processes, such as by team members who gather accurate shared understanding of the world by virtue of both getting the same information directly from displays or the environment that they share, or via SMM. Therefore, team process measures may only provide a partial understanding of SSA. Further, even teams attempting to communicate and coordinate may end up doing so ineffectively. And much of SA (particularly comprehension and projection) may be largely cognitive and not revealed in typical team communications or interactions. An objective measure of TSA and SSA provides the relevant outcome metric for comparing to team process measures in order to fully appreciate their effectiveness for achieving TSA.

Team SA Objective State Measures

Another approach has been to directly assess the SA of team members and to use that to draw comparisons across team members, or across different teams. The Situation Awareness Global Assessment Technique (SAGAT) (Endsley, 1995a) provides for a simultaneous assessment of each team member's SA during periodic freezes in team operations. It is then possible to identify problems due to inadequate tools or team processes by comparing the SA of different team members or sub-teams at the same point in time.

Endsley (2019) conducted a review of 24 studies that used SAGAT to evaluate team SA. Researchers were found to use a number of approaches in examining team SA, including: (1) forming an overall team SA score based on the total or average SA across the team (11 studies), (2) allowing for a collaborative team response to SA queries (two studies), (3) assessing SSA (or SA similarity) based on the degree of concurrence between teammates on information elements relevant to both roles (14 studies), (4) team meta-SA, examining the degree to which team members are aware of the SA of each other (three studies), and (5) determining the degree of correlation between the SA of different team members, or sub-teams (six studies). Each of these scoring approaches will be considered separately.

Combined TSA

Bolstad and Endsley (2003) created a combined TSA score for different teams involved in a large command and control exercise, showing significant differences between teams associated with poor team collaboration tools. Gardner et al. (2017) calculated composite TSA scores by calculating the accuracy of each team member's SA on each SAGAT query and summing the scores across all team members. They showed that TSA for medical teams involved in trauma care was significantly related to team work scores ($r^2 = .50$ and $.55$), as well as team performance scores ($r^2 = .30$ and $.38$).

Cooke, Stout, Rivera, and Salas (1998) created a combined TSA score in a helicopter simulation and showed it was significant correlated with team knowledge accuracy. Shared knowledge decreased over time in their study, however. Cooke et al. (2001) also showed that TSA was predictive of team performance, and Crozier et al. (2015) similarly demonstrated that TSA was correlated with checklist performance.

Collaborative TSA

In a different approach, some researchers have allowed team members to collaborate when answering SAGAT queries, rather than providing independent scores that are later compared (Hallbert, 1997; Price & LaFiandra, 2017). While this may make sense in some operational settings, it does not allow for a comparison of SA within teams, only across teams.

SSA

Endsley, Bolte, and Jones (2003) recommend directly comparing the SA of different team members on SA queries that they share in common in order to get a measure of their shared SA. Using this approach in an air operations center, they show how the SSA of two team members falls into one of the four categories: (1) both are correct, (2) one team member correct and the other incorrect, (3) both incorrect in the same way, and (4) both incorrect in different ways. This provides for a scoring of SSA for each relevant SAGAT query between any set of teammates.

Using this approach, Bonney et al.'s (2016) study of business markets showed that team performance was predicted by both SSA of the team (all three levels of SA contributing over 34% of the variance) and by having a shared team strategy. Similarly, in the medical domain, Rosenman et al. (2018) demonstrated that SSA was predictive of performance, and Coolen et al. (2019) showed that SSA on both the problem and the diagnosis were highly predictive of good team performance. Artman (1999) demonstrated differences in SSA between teams in a simulated firefighting task that was significantly related to the types of communications that were provided (serial or parallel). Javed et al. (2012) demonstrated improvements in SSA with new displays for emergency management.

Other researchers have created a combined SSA score. Bolstad et al. (2005) investigated the SSA of team members as measured by SAGAT scores in military teams by assigning a 1 for all queries where team members answered the same and a 0 for all queries where they were different to create a SSA similarity score across SAGAT queries. They then compared the teams' SSA scores to their physical distance, social network distance as calculated from frequency of communications, rank similarity, and branch of service similarity. Only physical distance was a significant predictor of SSA between team members, accounting for about half of the variance, as well as the vast majority of the difference in social network distance.

Sætrevik and Eid (2014) proposed examining SSA by comparing each team member's answer to the average of all team member answers. Thus, the measure did not reflect SA correctness, but only how cohesive the teams were in their perceptions. They also compared how similar each team member's SA was to the team leader. These assessments did not show any sensitivity to team membership, numbers of team meetings, or the time since the last meeting, leaving the value of this approach questionable.

In contrast, Saner et al. (2009) argue that there is no shared SA unless it is also accurate. Therefore, their SSA similarity index was calculated on the basis of SSA accuracy for any two team members. They showed that both organizational hub distance and the levels of shared knowledge of the team members predicted SSA. It should be noted that these approaches create a combined SSA score across different SA queries. In a meta-analysis of SA metrics, Endsley (2019) found that combined SAGAT scores are generally less sensitive that those that make comparisons by query or by SA level.

Team Meta-SA

Some researchers have examined the degree to which team members are aware of the accuracy of their own SA and the level of SA their teammates. Sulistyawati et al. (2008) distinguish between the need to (1) have SA related to one's own goals, (2) have SA required to back up team member SA, (3) have an accurate understanding of how good one's own SA is (meta-SA), and (4) awareness of a teammate's workload and SA levels. They found that pilot teams with good SA were significantly less likely to have over-confidence bias ($r = .85$, $p < .01$), demonstrating that those with good SA also had good meta-SA of their own knowledge. However, they did not find any relationship between accurate SA of one's own SA requirements and that of the teammate's SA or workload. Pilots in their study were generally poor at estimating the SA and workload of their teammate.

Sulistyawati et al. (2008; Sulistyawati, Wickens, & Chui,2009) showed that individual's SAGAT scores were highly predictive of survivability ($r = .69$), but awareness of the teammates' SA was not predictive. Yuan, She, Li, Zhang, and Wu (2016) found that awareness of teammate SA was negatively correlated with own SA in their study, likely due to competing task demands. Thus, the value of team meta-SA appears not to be supported.

SA Correlation

A number of researchers examined the correlations between the SA of different team members, or sub-teams. For example, Sætrevik (2012) and Cuevas and Bolstad (2010) examined correlations between team members' and team leaders' SA. In analyzing the correlation between SA of one's own requirements and of the teammates' situation as measured by SAGAT, Sulistyawati et al. (2008) found a marginal correlation ($r = .60$, $p = .06$).

These studies demonstrate that TSA and SSA can be derived from SAGAT data which is both objective and highly validated as a measure of individual SA (Endsley, 2000, 2019). These measures provide an independent assessment of the quality of TSA and SSA to support research on the various factors that effect it, and to provide a clear outcome measure for comparison to TSA process measures.

CONCLUSIONS

TSA has become widely recognized as critical for effective team performance in a wide variety of domains. Considerable research has been conducted over the past 30 years demonstrating the importance of team SA processes, team SA devices, and team SA mechanisms for supporting team SA requirements. This body of research provides a solid foundation for the development of training programs and display design guidelines for improving SSA and TSA. Further, existing TSA measurement approaches have demonstrated utility for supporting research on TSA and SSA, and can be used to evaluate the effectiveness of design and training interventions. Overall, the study of TSA has achieved a considerable degree of maturity, however, extensive work is still needed to extend these findings into many real-world training and design applications.

REFERENCES

Artman, H. (1999). Situation awareness and co-operation within and between hierarchical units in dynamic decision making. *Ergonomics*, *42*(11), 1404–1417.

Berggren, P., Prytz, E., Johansson, B., & Nählinder, S. (2011). The relationship between workload, teamwork, situation awareness, and performance in teams: A microworld study. In *Proceedings of the proceedings of the human factors and ergonomics society annual meeting* (pp. 851–855). Los Angeles, CA: Sage Publications.

Bolstad, C. A., Cuevas, H., Gonzalez, C., & Schneider, M. (2005). Modeling shared situation awareness. *Paper presented at the 14th conference on Behavior Representation in Modeling and Simulation (BRIMS).* Los Angles, CA.

Bolstad, C. A., & Endsley, M. R. (1999). Shared mental models and shared displays: An empirical evaluation of team performance. In *Proceedings of the 43rd annual meeting of the human factors and ergonomics society* (pp. 213–217). Santa Monica, CA: Human Factors and Ergonomics Society.

Bolstad, C. A., & Endsley, M. R. (2000). The effect of task load and shared displays on team situation awareness. In *Proceedings of the 14th triennial congress of the international ergonomics association and the 44th annual meeting of the human factors and ergonomics society* (pp. 189–192). Santa Monica, CA: Human Factors and Ergonomics Society.

Bolstad, C. A., & Endsley, M. R. (2003). Measuring shared and team situation awareness in the army's future objective force. In *Proceedings of the human factors and ergonomics society 47th annual meeting* (pp. 364–373). Santa Monica, CA: Human Factors and Ergonomics Society.

Bolstad, C. A., & Endsley, M. R. (2005, Fall). Choosing team collaboration tools: Lessons learned fron disaster recovery efforts. *Ergonomics in Design*, 7–13.

Bolstad, C. A., Foltz, P., Franzke, M., Cuevas, H. M., Rosenstein, M., & Costello, A. M. (2007). Predicting situation awareness from team communications. In *Proceedings of the human factors and ergonomics society 51st annual meeting* (pp. 789–793). Santa Monica, CA: Human Factors and Ergonomics Society.

Bolstad, C. A., Riley, J. M., Jones, D. G., & Endsley, M. R. (2002). Using goal directed task analysis with Army brigade officer teams. In *Proceedings of the 46th annual meeting of the human factors and ergonomics society* (pp. 472–476). Santa Monica, CA: Human Factors and Ergonomics Society.

Bonney, L., Davis-Sramek, B., & Cadotte, E. R. (2016). Thinking about business markets: A cognitive assessment of market awareness. *Journal of Business Research*, *69*(8), 2641–2648.

Brooks, J. O., Switzer, F. S., & Gugerty, L. (2003). Effects of situation awareness training on novice process control plant operators. In *Proceedings of the human factors and ergonomics society annual meeting* (pp. 606–609). Los Angeles, CA: Sage.

Buchler, N., Fitzhugh, S. M., Marusich, L. R., Ungvarsky, D. M., Lebiere, C., & Gonzalez, C. (2016). Mission command in the age of network-enabled operations: Social network analysis of information sharing and situation awareness. *Frontiers in Psychology*, *7*, 937.

Chidester, T. R., Kanki, B. G., Foushee, H. C., Dickinson, C. L., & Bowles, S. V. (1990). *Personality factors in flight operations: Volume I. Leadership characteristics and crew performance in a full-mission air transport simulation* (NASA Tech Memorandum No. 102259). Moffett Field, CA: NASA Ames Research Center.

Citera, M., McNeese, M. D., Brown, C. E., Selvaraj, J. A., Zaff, B. S., & Whitaker, R. D. (1995). Fitting information systems to collaborating design teams. *Journal of the American Society for Information Science*, *46*(7), 551–559.

Cooke, N. J., & Gorman, J. C. (2009). Interaction-based measures of cognitive systems. *Journal of Cognitive Engineering and Decision Making*, *3*(1), 27–46.

Cooke, N. J., Kiekel, P. A., & Helm, E. E. (2001). Measuring team knowledge during skill acquisition of a complex task. *International Journal of Cognitive Ergonomics*, *5*(3), 297–315.

Cooke, N. J., Kiekel, P. A., Salas, E., Stout, R., Bowers, C., & Cannon-Bowers, J. A. (2003). Measuring team knowledge: A window to the cognitive underpinnings of team performance. *Group Dynamics, Theory, Research and Practice*, *7*(3), 179–199.

Cooke, N. J., Salas, E., Kiekel, P. A., & Bell, B. (2004). Advances in measuring team cognition. In *Team cognition: Understanding the factors that drive process and performance* (pp. 83–106). Washington, DC: American Psychological Association.

Cooke, N. J., Stout, R. E., Rivera, K., & Salas, E. (1998). Exploring measures of team knowledge. In *Proceedings of the human factors and ergonomics society annual meeting* (pp. 215–219). Los Angeles, CA: Sage.

Cooke, N. J., Stout, R. J., & Salas, E. (2001). A knowledge elicitation approach to the measurement of team situation awareness. In M. McNeese, E. Salas, & M. R. Endsley (Eds.), *New trends in cooperative activities: Understanding system dynamics in complex environments* (pp. 114–139). Santa Monica, CA: Human Factors and Ergonomics Society.

Coolen, E., Draaisma, J., & Loeffen, J. (2019). Measuring situation awareness and team effectiveness in pediatric acute care by using the situation global assessment technique. *European Journal of Pediatrics*, 1–14.

Crozier, M. S., Ting, H. Y., Boone, D. C., O'Regan, N. B., Bandrauk, N., Furey, A., . . . Hogan, M. P. (2015). Use of human patient simulation and validation of the Team Situation Awareness Global Assessment Technique (TSAGAT): A multidisciplinary team assessment tool in trauma education. *Journal of Surgical Education*, *72*(1), 156–163.

Cuevas, H. M., & Bolstad, C. A. (2010). Influence of team leaders' situation awareness on their team's situation awareness and performance. In *Proceedings of the human factors and ergonomics society annual meeting* (pp. 309–313). Los Angeles, CA: Sage.

Endsley, M. R. (1988). Design and evaluation for situation awareness enhancement. In *Proceedings of the human factors society 32nd annual meeting* (pp. 97–101). Santa Monica, CA: Human Factors Society.

Endsley, M. R. (1995a). Theoretical underpinnings of situation awareness: A critical review. In D. J. Garland & M. R. Endsley (Eds.), *Experimental analysis and measurement of situation awareness* (pp. 17–23). Daytona Beach, FL: Embry-Riddle University.

Endsley, M. R. (1995b). Toward a theory of situation awareness in dynamic systems. *Human Factors*, *37*(1), 32–64.

Endsley, M. R. (2000). Direct measurement of situation awareness: Validity and use of SAGAT. In M. R. Endsley & D. J. Garland (Eds.), *Situation awareness analysis and measurement* (pp. 147–174). Mahwah, NJ: LEA.

Endsley, M. R. (2004). Situation awareness: Progress and directions. In S. Banbury & S. Tremblay (Eds.), *A cognitive approach to situation awareness: Theory, measurement and application* (pp. 317–341). Aldershot: Ashgate Publishing.

Endsley, M. R. (2015a). Situation awareness misconceptions and misunderstandings. *Journal of Cognitive Engineering and Decision Making*, *9*(1), 4–32.

Endsley, M. R. (2015b). Situation awareness: Operationally necessary and scientifically grounded. *Cognition, Technology and Work*, *17*(2), 163–167.

Endsley, M. R. (2018). Expertise and situation awareness. In K. A. Ericsson, R. R. Hoffman, A. Kozbelt & A. M. Williams (Eds.), *Cambridge handbook of expertise and expert performance* (2nd ed., pp. 714–744). Cambridge: Cambridge University Press.

Endsley, M. R. (2019). A systematic review and meta-analysis of direct, objective measures of SA: A comparison of SAGAT and SPAM. *Human Factors*. https://doi.org/10.1177%2F0018720819875376

Endsley, M. R., Bolstad, C. A., Jones, D. G., & Riley, J. M. (2003). Situation awareness oriented design: From user's cognitive requirements to creating effective supporting technologies. In *Proceedings of the 47th annual meeting of the human factors and ergonomics society* (pp. 268–272). Santa Monica, CA: Human Factors and Ergonomics Society.

Endsley, M. R., Bolte, B., & Jones, D. G. (2003). *Designing for situation awareness: An approach to human-centered design*. London: Taylor & Francis.

Endsley, M. R., Farley, T. C., Jones, W. M., Midkiff, A. H., & Hansman, R. J. (1998). *Situation awareness information requirements for commercial airline pilots* (ICAT-98–1). Cambridge, MA: Massachusetts Institute of Technology International Center for Air Transportation.

Endsley, M. R., Hansman, R. J., & Farley, T. C. (1998). Shared situation awareness in the flight deck—ATC system. In *Proceedings of the AIAA/IEEE/SAE 17th digital avionics systems conference*. Bellevue, WA: Isogen International Corporation.

Endsley, M. R., Holder, L. D., Leibrecht, B. C., Garland, D. C., Wampler, R. L., & Matthews, M. D. (2000). *Modeling and measuring situation awareness in the infantry operational environment (1753)*. Alexandria, VA: Army Research Institute.

Endsley, M. R., & Jones, D. G. (2012). *Designing for situation awareness: An approach to human-centered design* (2nd ed.). London: Taylor & Francis.

Endsley, M. R., & Jones, W. M. (1997). *Situation awareness, information dominance, and information warfare* (AL/CF-TR-1997-0156). Wright-Patterson AFB, OH: United States Air Force Armstrong Laboratory.

Endsley, M. R., & Jones, W. M. (2001). A model of inter- and intrateam situation awareness: Implications for design, training and measurement. In M. McNeese, E. Salas, & M. Endsley (Eds.), *New trends in cooperative activities: Understanding system dynamics in complex environments* (pp. 46–67). Santa Monica, CA: Human Factors and Ergonomics Society.

Endsley, M. R., & Robertson, M. M. (1996). Team situation awareness in aviation maintenance. In *Proceedings of the 40th annual meeting of the human factors and ergonomics society* (pp. 1077–1081). Santa Monica, CA: Human Factors and Ergonomics Society.

Endsley, M. R., & Rodgers, M. D. (1994). Situation awareness information requirements for en route air traffic control. In *Proceedings of the human factors and ergonomics society 38th annual meeting* (pp. 71–75). Santa Monica, CA: Human Factors and Ergonomics Society.

Espevik, R., Johnsen, B. H., & Eid, J. (2011). Outcomes of shared mental models of team members in cross training and high-intensity simulations. *Journal of Cognitive Engineering and Decision Making*, *5*(4), 352–377.

Farley, T. C., Hansman, R. J., Amonlirdviman, K., & Endsley, M. R. (2000). Shared information between pilots and controllers in tactical air traffic control. *Journal of Guidance, Control and Dynamics*, *23*(5), 826–836.

Farley, T. C., Hansman, R. J., Endsley, M. R., Amonlirdviman, K., & Vigeant-Langlois, L. (1998). The effect of shared information on pilot/controller situation awareness and re-route negotiation. In *Proceedings of the 2nd FAA/Eurocontrol ATM R&D seminar*. Washington, DC: FAA.

Foushee, H. C., Lauber, J. K., Baetge, M. M., & Acomb, D. B. (1986). *Crew factors in flight operations: III. The operational significance of exposure to short-haul air transport operations* (NASA Tech Memorandum 88322). Moffett Field, CA: NASA Ames Research Center.

Gardner, A. K., Kosemund, M., & Martinez, J. (2017). Examining the feasibility and predictive validity of the SAGAT tool to assess situation awareness among medical trainees. *Simulation in Healthcare*, *12*(1), 17–21.

Gorman, J. C., Cooke, N. J., & Amazeen, P. G. (2010). Training adaptive teams. *Human Factors*, *52*(2), 295–307.

Gorman, J. C., Cooke, N. J., Pederson, H. K., Connor, O. O., & DeJoode, J. A. (2005). Coordinated Awareness of Situation by Teams (CAST): Measuring team situation awareness of a communications glitch. In *Proceedings of the human factors and ergonomics society 49th annual meeting* (pp. 274–277). Santa Monica, CA: Human Factors and Ergonomics Society.

Gorman, J. C., Cooke, N. J., & Winner, J. L. (2006). Measuring team situation awareness in decentralized command and control environments. *Ergonomics*, *49*(12–13), 1312–1325.

Gorman, J. C., Weil, S. A., Cooke, N., & Duran, J. (2007). Automatic assessment of situation awareness from electronic mail communication: Analysis of the Enron dataset. In *Proceedings of the proceedings of the human factors and ergonomics society annual meeting* (pp. 405–409). Los Angeles, CA: Sage Publications.

Gross, N., & Kluge, A. (2014). Predictors of knowledge-sharing behavior for teams in extreme environments: An example from the steel industry. *Journal of Cognitive Engineering and Decision Making*, *8*(4), 352–373.

Hallbert, B. P. (1997). Situation awareness and operator performance: Results from simulator-based studies. In *Proceedings of the IEEE sixth conference on human factors and power plants* (pp. 18/11–18/16). New York: IEEE.

Hauland, G. (2008). Measuring individual and team situation awareness during planning tasks in training of en route air traffic control. *The International Journal of Aviation Psychology*, *18*(3), 290–304.

Hirokawa, R. Y. (1983). Group communication and problem solving effectiveness: An investigation of group phases. *Human Communication Research*, *9*, 291–305.

Javed, Y., Norris, T., & Johnston, D. (2012). Evaluating SAVER: Measuring shared and team situation awareness of emergency decision makers. *Paper presented at the 9th international ISCRAM conference*. Vancouver, Canada.

Jeffs, L., Acott, A., Simpson, E., Campbell, H., Irwin, T., Lo, J., . . . Cardoso, R. (2013). The value of bedside shift reporting enhancing nurse surveillance, accountability, and patient safety. *Journal of Nursing Care Quality*, *28*(3), 226–232.

Jones, W. M. (1997). Enhancing team situation awareness: Aiding pilots in forming initial mental models of team members. In *Proceedings of the ninth international symposium on aviation psychology* (pp. 1436–1441). Columbus, OH: The Ohio State University.

Kanki, B. G., Lozito, S., & Foushee, H. C. (1989). Communication indices of crew coordination. *Aviation, Space and Environmental Medicine*, *60*, 56–60.

Kitchin, J., & Baber, C. (2017). The dynamics of distributed situation awareness. In *Proceedings of the human factors and ergonomics society annual meeting* (pp. 277–281). Los Angeles, CA: Sage Publications.

Klein, G. A., Zsambok, C. E., & Thordsen, M. L. (1993, April). Team decision training: Five myths and a model. *Military Review*, pp. 36–42.

Leggatt, A. (2004). Objectively measuring the promulgation of commander's intent in a coalition effects based planning experiment (MNE3). In *Proceedings of the 9th international command and control research and technology symposium* (pp. 143–177). Copenhagen: Defense Command Denmark.

Midkiff, A. H., & Hansman, R. J. (1992). Identification of important party-line information elements and implications for situational awareness in the datalink environment. In *Proceedings of the SAE Aerotech conference and exposition*. Anaheim, CA: Sage.

Mohammed, S., Ferzandi, L., & Hamilton, K. (2010). Metaphor no more: A 15-year review of the team mental model construct. *Journal of Management*, *36*(4), 876–910.

Morgan, P., Tregunno, D., Brydges, R., Pittini, R., Tarshis, J., Kurrek, M., . . . Ryzynski, A. (2015). Using a situational awareness global assessment technique for interprofessional obstetrical team training with high fidelity simulation. *Journal of Interprofessional Care*, *29*(1), 13–19.

Mosier, K. L., & Chidester, T. R. (1991). Situation assessment and situation awareness in a team setting. In Y. Queinnec & F. Daniellou (Eds.), *Designing for everyone* (pp. 798–800). London: Taylor and Francis.

National Transportation Safety Board. (1994). *A review of flight crews involved in major accidents of U.S. air carriers 1978–1990*. Washington, DC: Author.

Orasanu, J. (1990, July). Shared mental models and crew decision making. In *Proceedings of the 12th annual conference of the cognitive science society*. Cambridge, MA: Cognitive Science Society.

Orasanu, J., & Salas, E. (1993). Team decision making in complex environments. In G. A. Klein, J. Orasanu, R. Calderwood, & C. E. Zsambok (Eds.), *Decision making in action: Models and methods* (pp. 327–345). Norwood, NJ: Ablex.

Parasuraman, R., Sheridan, T. B., & Wickens, C. D. (2008). Situation awareness, mental workload and trust in automation: viable empirically supported cognitive engineering constructs. *Journal of Cognitive Engineering and Decision Making*, *2*(2), 140–160.

Parke, B., & Kanki, B. G. (2008). Best practices in shift turnovers: Implications for reducing aviation maintenance turnover errors as revealed in ASRS reports. *The International Journal of Aviation Psychology*, *18*(1), 72–85.

Parush, A., Hazan, M., & Shtekelmacher, D. (2017). Individuals perform better in teams but are not more aware-performance and situational awareness in teams and individuals. In *Proceedings of the human factors and ergonomics society annual meeting* (pp. 1173–1177). Los Angeles, CA: Sage.

Parush, A., & Ma, C. (2012). Team displays work, particularly with communication breakdown: Performance and situation awareness in a simulated forest fire. In *Proceedings of the proceedings of the human factors and ergonomics society annual meeting* (pp. 383–387). Los Angeles, CA: Sage Publications.

Parush, A., Mastoras, G., Bhandari, A., Momtahan, K., Day, K., Weitzman, B., . . . Calder, L. (2017). Can teamwork and situational awareness (SA) in ED resuscitations be improved with a technological cognitive aid? Design and a pilot study of a team situation display. *Journal of Biomedical Informatics*, *76*, 154–161.

Patrick, J., James, N., Ahmed, A., & Halliday, P. (2006). Observational assessment of situation awareness, team differences and training implications. *Ergonomics*, *49*(4), 393–417.

Price, T. F., & LaFiandra, M. (2017). The perception of team engagement reduces stress induced situation awareness overconfidence and risk-taking. *Cognitive Systems Research*, *46*, 52–60.

Prince, C., Ellis, E., Brannick, M. T., & Salas, E. (2007). Measurement of team situation awareness in low experience level aviators. *The International Journal of Aviation Psychology*, *17*(1), 41–57.

Robertson, M. M., & Endsley, M. R. (1997a). Development of a situation awareness training program for aviation maintenance. In *Proceedings of the human factors and ergonomics society 41st annual meeting* (pp. 1163–1167). Santa Monica, CA: Human Factors and Ergonomics Society.

Robertson, M. M., & Endsley, M. R. (1997b). Impact of distributed teams on organizational effectiveness. In *Proceedings of the 13th triennial congress of the international ergonomics association* (pp. 519–521). Helsinki: Finnish Institute of Occupational Health.

Rosenman, E. D., Dixon, A. J., Webb, J. M., Brolliar, S., Golden, S. J., Jones, K. A., . . . Chao, G. T. (2018). A simulation-based approach to measuring team situational awareness in emergency medicine: A multicenter, observational study. *Academic Emergency Medicine*, *25*(2), 196–204.

Sætrevik, B. (2012). A controlled field study of situation awareness measures and heart rate variability in emergency handling teams. In *Proceedings of the human factors and ergonomics society annual meeting* (pp. 2006–2010). Los Angeles, CA: Sage.

Sætrevik, B., & Eid, J. (2014). The "similarity index" as an indicator of shared mental models and situation awareness in field studies. *Journal of Cognitive Engineering and Decision Making*, *8*(2), 119–136.

Salas, E., Dickinson, T. L., Converse, S., & Tannenbaum, S. I. (1992). Toward an understanding of team performance and training. In R. W. Swezey & E. Salas (Eds.), *Teams: Their training and performance* (pp. 3–29). Norwood, NJ: Ablex.

Salas, E., Prince, C., Baker, D. P., & Shrestha, L. (1995). Situation awareness in team performance: Implications for measurement and training. *Human Factors*, *37*(1), 123–136.

Salmon, P., Stanton, N., Walker, G., Baber, C., Jenkins, D., McMaster, R., & Young, M. (2008). What is really going on? Review of situation awareness models for individuals and teams. *Theoretical Issues in Ergonomics Science*, *9*(4), 297–323.

Saner, L. D., Bolstad, C. A., Gonzalez, C., & Cuevas, H. M. (2009). Measuring and predicting shared situation awareness in teams. *Journal of Cognitive Engineering and Decision Making*, *3*(3), 280–308.

Saner, L. D., Bolstad, C. A., Gonzalez, C., & Cuevas, H. M. (2010). Predicting shared situation awareness in teams: A case of differential SA requirements. In *Proceedings of the human factors and ergonomics society annual meeting* (pp. 314–318). Los Angeles, CA: Sage Publications.

Seet, A. W., Teh, C. A., Soo, J. K., & Teo, L. (2004). *Constructible assessment for situation awareness in a distributed C2 environment*. Singapore: DSO National Laboratories.

Segal, L. D. (1994). Actions speak louder than words. In *Proceedings of the human factors and ergonomics society 38th annual meeting* (pp. 21–25). Santa Monica, CA: Human Factors and Ergonomics Society.

Sorensen, L. J., Stanton, N., & Banks, A. P. (2010). Back to SA school: Contrasting three approaches to situation awareness in the cockpit. *Theoretical Issues in Ergonomics Science*, *12*(6), 451–471.

Stout, R. J., Cannon-Bowers, J. A., & Salas, E. (2017). *The role of shared mental models in developing team situational awareness: Implications for training situational awareness* (pp. 287–318). Abingdon: Routledge.

Stout, R. J., Cannon-Bowers, J. A., Salas, E., & Milanovich, D. M. (1999). Planning, shared mental models, and coordinated performance: An empirical link is established. *Human Factors*, *41*(1), 61–71.

Strater, L. D., Cuevas, H. M., Connors, E. S., Gonzalez, C., Ungvarsky, D. M., & Endsley, M. R. (2010). An investigation of technology-mediated ad hoc team operations: Consideration of components of team situation awareness. In K. Mosier & U. Fischer (Eds.), *Informed by knowledge: Expert performance in complex situations* (pp. 153–168). Boca Raton, FL: Taylor and Francis.

Strater, L. D., Cuevas, H. M., Connors, E. S., Ungvarsky, D. M., & Endsley, M. R. (2008). Situation awareness and collaborative tool usage in ad hoc command and control teams. In *Proceedings of the proceedings of the human factors and ergonomics society annual meeting* (pp. 468–472). Los Angeles, CA: Sage Publications.

Sulistyawati, K., Chui, Y. P., & Wickens, C. D. (2008). Multi-method approach to team situation awareness. In *Proceedings of the human factors and ergonomics society annual meeting* (pp. 463–467). Los Angeles, CA: Sage.

Sulistyawati, K., Wickens, C. D., & Chui, Y. P. (2009). Exploring the concept of team situation awareness in a simulated air combat environment. *Journal of Cognitive Engineering and Decision Making*, *3*(4), 309–330.

Taylor, R. M., Endsley, M. R., & Henderson, S. (1996). Situational awareness workshop report. In B. J. Hayward & A. R. Lowe (Eds.), *Applied aviation psychology: Achievement, change and challenge* (pp. 447–454). Aldershot: Ashgate Publishing Ltd.

Tiferes, J., & Bisantz, A. M. (2018). The impact of team characteristics and context on team communication: An integrative literature review. *Applied Ergonomics*, *68*, 146–159.

United Kingdom Air Accidents Investigation Branch. (1990). *Report on the accident to Boeing 737–400 G-OBME near Kegworth, Leicestershire on 8 January 1989* (Report No. 4/90 (EW/C1095)). Farnborough, UK: Department for Transport.

Wellens, A. R. (1993). Group situation awareness and distributed decision making: From military to civilian applications. In N. J. Castellan Jr. (Ed.), *Individual and group decision making* (pp. 267–291). Hillsdale, NJ: Lawrence Erlbaum.

Wickens, C. D. (2008). Situation awareness: Review of Mica Endsley's 1995 articles on situation awareness theory and measurement. *Human Factors*, *50*(3), 397–403.

Xiao, Y., Mackenzie, C. F., & Patey, R. (1998). Team coordination and breakdowns in a real-life stressful environment. In *Proceedings of the human factors and ergonomics society 42nd annual meeting* (pp. 186–190). Santa Monica, CA: Human Factors and Ergonomics Society.

Yuan, X., She, M., Li, Z., Zhang, Y., & Wu, X. (2016). Mutual awareness: Enhanced by interface design and improving team performance in incident diagnosis under computerized working environment. *International Journal of Industrial Ergonomics*, *54*, 65–72.

2 Studying Team Cognition in the C3Fire Microworld

Björn J. E. Johansson, Rego Granlund, and Peter Berggren

CONTENTS

INTRODUCTION

The aim of this chapter is to provide a description of how the C3Fire microworld has contributed to knowledge development concerning various aspects of team cognition and how the platform has been and can be utilized for research on team cognition. C3Fire has been used for more than two decades by several research institutes and universities for a wide range of studies, resulting in more than a hundred different publications. Some of the main research tracks that have used C3Fire as a research tool include distributed decision making, command and control, the effects of new technologies, how cultural differences manifest in team decision making, and the development of a new measure of shared understanding in teams. In this chapter, we focus on C3Fire studies concerning aspects of team cognition, such as how organizational arrangements affect communication and performance, how new technologies affect the outcome of teamwork, and how the microworld has been used to support the development of a new measure of sharedness in teams. This endeavor will be undertaken in a chronological fashion with some selected work performed by some of the main Swedish C3Fire user groups, beginning with early work at Linköping University concerning different organizational structures and their impact on team decision making, followed by a presentation of some of the investigations of how geographical information systems can support decision making, then proceeding towards the most recent publications on method development conducted at the Swedish Defence Research Agency.

The chapter is organized as follows: Firstly, we will provide a background to the microworld research tradition related to team cognition. Then, we will provide a summary of the previously mentioned research tracks that have been investigated with the C3Fire microworld. After that, we continue by discussing pros and cons of the microworld approach in relation to team cognition research. Lastly, we present our conclusions and recommendations based on the from the conducted studies.

MICROWORLDS

Microworlds have been used in research on decision making since the late 1970s (Brehmer, 2004; Dörner, 1980; Dörner, Kreuzig, Reiter, & Stäudel, 1983; Funke, 1993, 2001), although the etymology of the term remains hidden in history. Other terms that have been used are scaled worlds (Gray, 2002; Schiflett et al., 2004), synthetic environments (Cooke & Shope, 2004), and simulations (McNeese et al., 2005). In this chapter, we refer to computer-based simulations that share some properties that are common to all these notions when using the term *microworld*. Gonzalez, Vanyukov, and Martin (2005) provide a detailed list of the most commonly used microworlds, which is informative for anyone seeking a broad perspective on the types of problems and simulations that have been developed. Microworlds were originally developed to allow researchers to investigate complex aspects of decision making (Funke, 2010). The German tradition of *komplexes problemlösung* (complex problem solving; Dörner & Schaub, 1992), initiated by Dietrich Dörner of the Bamberg University, used various simulations to investigate how individuals cope with highly complex decision tasks in an explorative fashion. In Sweden, Berndt Brehmer and colleagues pursued the task of understanding how people cope with the control of dynamic systems (Brehmer, 1987; Brehmer, 2004, 2005), using simulated environments in traditional experiments. Brehmer & Dörner (1993) jointly suggested that microworlds bridge the gap between (psychological) laboratory studies and the "deep blue sea" of field research. A microworld is implemented in a computer and can be seen as a computer-based simulation. This is partly true if we by simulation mean any computer program that has some similarity with a real-world task. That would however be a misuse of the term "simulation," since a simulation often is interpreted to be a model-based implementation of a more or less exact representation of a real-world task (Grüne-Yanoff & Weirich, 2010). For example, a flight simulator for professional training may be very advanced, providing an almost entirely realistic interaction (Liu, Macchiarella, & Vincenzi, 2009). This is not the purpose of a microworld.

> In experiments with microworlds, subjects are required to interact with and control computer simulations of systems such as forest fires, companies, or developing countries for some period of time. Microworlds are not designed to be high fidelity simulations. Instead, they are related to the systems that they represent in the same manner as wood cuts are related to what they represent. That is, it is possible to recognise what is being represented, but there is little detail. However, microworlds always have the

> fundamental characteristics of decision problems of interest, here, viz., complexity and in-transparency.
>
> **(Brehmer, 2000, pp. 7–8)**

The purpose of a microworld is thus to present a recognizable problem to the subjects taking part in a study. However, the microworld must still be complex enough so that the subjects experience a dynamic situation presenting a certain degree of uncertainty. A recognizable task often used is forest fire fighting (Svenmarck & Brehmer, 1991; Granlund, 2002; Gray, 2002). When implementing something like a forest fire in a microworld, the important point is not to preserve detailed real-world characteristics, but rather to have a level of fidelity that is high enough to be acceptable by the participants in the study but yet low enough to be easily manageable and analyzable by the researcher(s). It should thus trigger essential cognitive processes that are representative of the processes taking place in a real-world task. According to Brehmer and Dörner (1993) microworlds are characterized by the fact that they are *complex*, *dynamic*, and *opaque*, suggesting that they will trigger cognitive processes taking place when coping with such challenges. They are *complex* as the subjects have to consider a number of aspects, like different courses of actions or contradicting goals. Secondly, they are *dynamic* in the sense that subjects, at least in the forest firefighting example, have to consider different time scales and sudden, most likely, unforeseen, effects since the relationships between different variables are difficult to understand. The *opaqueness* comes from the fact that parts of the simulation are invisible to the subject, who has a "black box" view of it. Participants in the microworld thus have to make hypotheses and test them in order to handle the situation (Brehmer & Dörner, 1993). These three characteristics are representative to many real-world situations, and definitely to many situations in which teams work, such as crisis management or military operations.

In this chapter we define computer-based simulations intended to present research participants with a recognizable problem, triggering cognitive processes, that reflect core aspects of a real-world phenomena while allowing for experimental control at a level of resolution possible to interpret and analyze from a scientific point of view. Thus, microworlds are research tools aligning themselves between "analogue" experimentations and high-fidelity simulation or field studies.

Therefore, it could be assumed that microworlds are suitable not only for investigating how individuals behave when they are confronted with a dynamic problem, but also for how teams handle such tasks. Brehmer and Svenmarck (1995) developed such a microworld, based on the forest firefighting task, called D3Fire. It was designed to allow for studies on "distributed decision making," which presents problems that cannot be managed by individual decision makers, but rather must be handled by a team. D3Fire was mainly used to study how the problem of extinguishing the forest fire was affected by different configurations of the communication paths between participants, such as hierarchical structures versus networked (Brehmer & Svenmarck, 1995; Svenmarck & Brehmer, 1994). The D3Fire simulation was later used as inspiration for the C3Fire microworld developed by Rego Granlund during the late 1990s (Granlund, 2002), although the latter was designed to be platform independent and highly configurable, properties that the D3Fire microworld lacked. During the same time period, Australian researchers developed a platform called the Networked Fire Chief (Omodei, Elliott, Walshe, & Wearing, 2005) at La Trobe University in Melbourne, mainly to "investigate the underlying causes of unsafe decisions in context of wildland firefighting" (Omodei et al., 2005, p 1). The Networked Fire Chief platform held similar characteristics as C3Fire, although with a narrower research focus, and will not be described further in this chapter. Instead, we will carry on by providing a general introduction to team research in microworlds.

Team Research in Microworlds

For the purpose of this chapter, the relation between team research, mainly team cognition, and microworlds will be discussed in this section. Even though teams have been studied in social psychology in terms of groups, the area of team research has since the 1990s called for empirical studies of teams (Swezey & Salas, 1992; Baker & Salas, 1992; Salas & Fiore, 2004; Salas, Fiore, & Letsky, 2012; Tannenbaum, Mathieu, Salas, & Cohen, 2012; Wildman, Salas, & Scott, 2014). Often these calls have asked for validated metrics, a larger empirical foundation, or to move outside of the laboratory. As is pointed out by these authors, there are several difficulties to overcome. The primary problem with team research in the wild is the possibility to collect data, especially when it comes to less frequently occurring events. Therefore, much knowledge is based on case-like descriptions of, for example, crisis response, emergency management, and similar events (Comfort, 2007; Boin & McConnell, 2007; Bodeau, Fedorowicz, Markus, & Brooks, 2009; Ouyang & Wang, 2015; Johansson, Trnka, & Berggren, 2016). In summary, some of the problems associated with team research are number of cases, level of analysis, metrics issues, experimental control, learning effects, etc. As described previously, many of these problems where identified by Brehmer and Dörner (1993), whereas they proposed the microworld approach as a sound way of overcoming these concerns.

A broad definition of *team cognition* is the cognitive activity that occurs within a team (Cooke, Gorman, & Rowe, 2009). How team members manage information, communicate, coordinate actions, and collaborate are central aspects of team cognition. In addition, how this can be assessed, measured, and modeled are other concerns. However, many of these aspects can be studied in microworld settings. The challenge is to find a suitable microworld platform that allows for such studies. Next, we describe the C3Fire platform, which is a microworld based on the forest firefighting task that was developed specifically to study team decision making.

C3FIRE

The C3Fire microworld was developed over several years, and it is therefore difficult to provide a just description of all functionality of the system. This section will hence only give an overview of the general properties and functions of the system. The purpose of C3Fire is, as in most microworlds (Brehmer & Dörner 1993), that the participants should experience a task environment that has some of the important complex dynamic behavior of a real system, but without the nitty-gritties of the actual task. Decisions are made in an environment where the researcher can manipulate dynamics, complexity, and the degree of opaqueness. Both the forest fire and the firefighting organization can be configured to exhibit varying degrees of complexity and dynamics. The forest fire will change both autonomously and as a consequence of actions taken by the research participants (Granlund, 1997, 2001, 2002, 2003). The decision making can be configured to be distributed over a number of persons and can be viewed as team decision making where the members have different roles, tasks, and items of information available for their decision process. The organization can be designed to mimic a hierarchy where the decision makers work on different time scales. The firefighting unit chiefs (often referred to as "ground chiefs") are responsible for the low-level operation, such as the fire fighting, which is done in a short time frame. The staff works at a higher time level and are responsible for the co-ordination of the firefighting units and the strategic thinking (Granlund, 2002, 2003).

Computer-based monitoring is used to collect data about participants actions and communication in the C3Fire system. The monitoring is integrated in the simulation as well as all the information tools used by the participants, such as text messages and shared geographical information systems (Granlund, 2003; Johansson, Persson, Granlund, & Mattsson, 2003). During a session, the C3Fire system creates a data log. The log process receives data from the simulation about the current activities in the forest-fire simulation. All events added to the data log file are time-stamped and can be easily imported into Microsoft Excel and most of the common software for statistical analysis. The outline of the simulation and its components can be found in Figure 2.1.

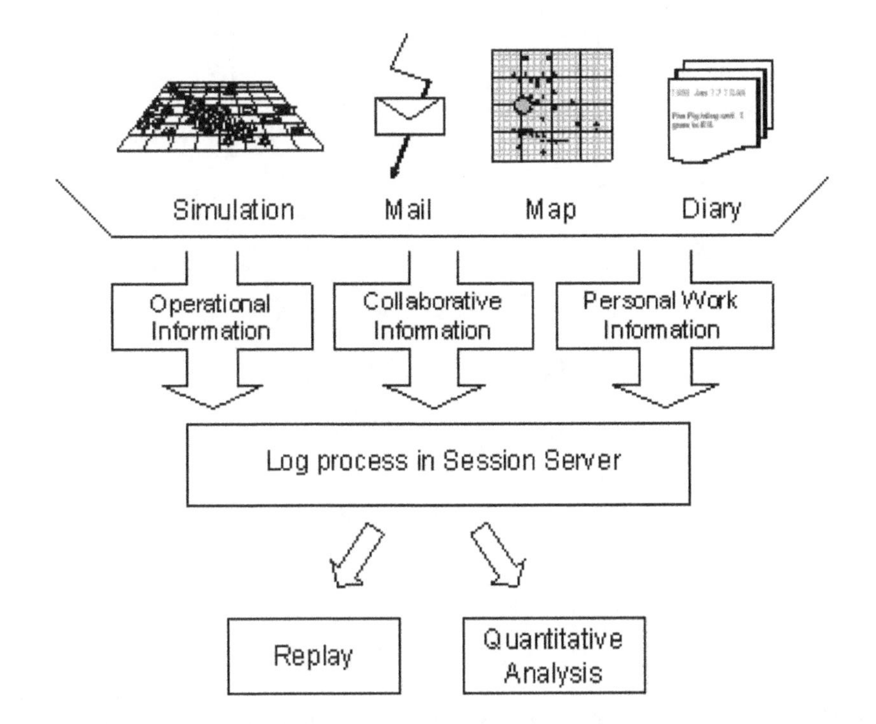

FIGURE 2.1 Basic components of the C3Fire simulation system.

Source: Johansson, Trnka, & Granlund, 2007.

The participants interact with the simulation through an interface that can be configured in different ways. Typically, it presents a map view of the environment, in which the participants can control the units they are responsible for. Mostly, it also includes a messaging system that allows for communication with other participants. The communication patterns that are possible are defined in a configuration file. The type of information that is available to the participants, for example how large parts of the simulated area are visible from each unit, or at all, is also defined in advance. An example of the C3Fire interface can be found in Figure 2.2.

Organizational aspects of the participating team can be configured in different ways, partly by simply assigning participants to certain roles, but also by dictating communication paths and providing exclusive access to specific resources to certain roles. For example, a simple hierarchy can be constructed by assigning some participants as commanders, while other participants are assigned as ground chiefs, responsible for controlling the firefighting units. If the communication paths mimic these roles, this is enforced, as ground chiefs only can communicate "vertically" with their commanders, and not with their fellow ground chiefs, assuring that their activities are coordinated by the commanders in the simulation (see Figure 2.3).

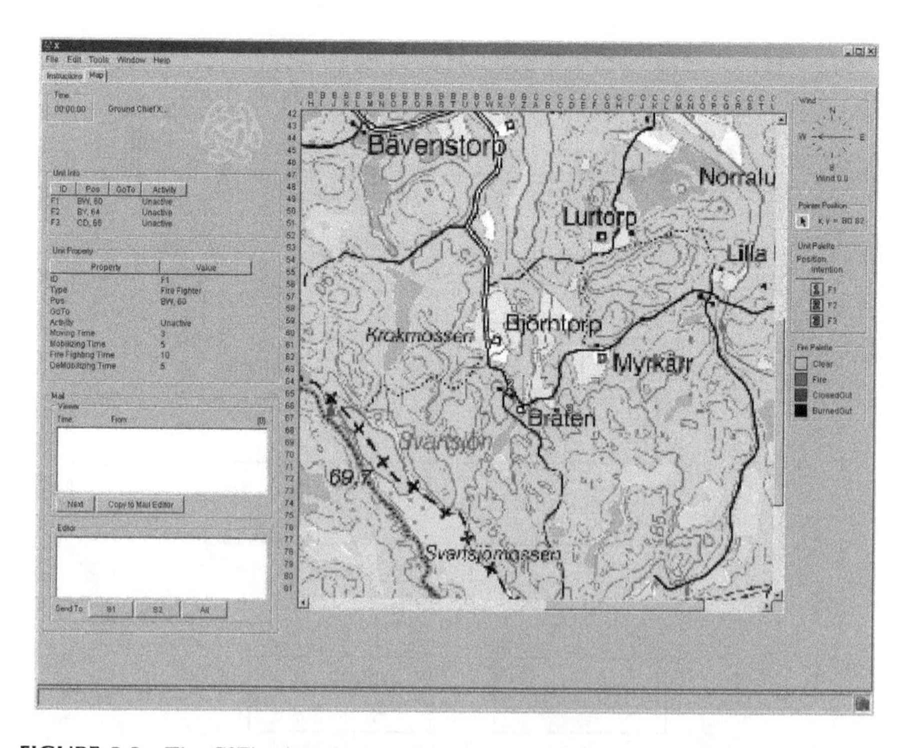

FIGURE 2.2 The C3Fire interface as shown to a participant responsible for controlling three firefighting units. The messaging system can be seen in the lower left part of the figure.

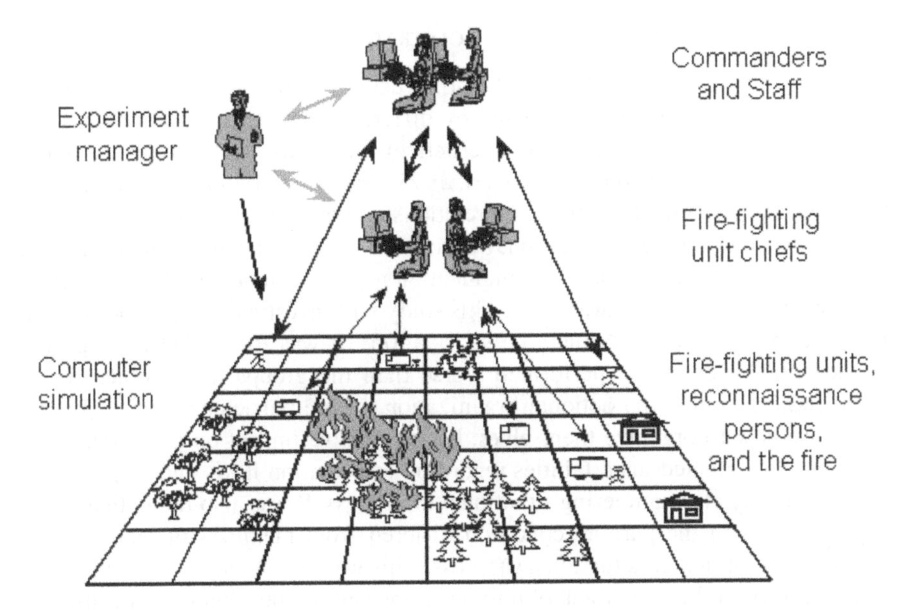

FIGURE 2.3 An example of a hierarchical arrangement of a C3Fire experiment.

As C3Fire can be configured to represent a number of different roles and organizational structures, it allows for the investigation of a wide range of team-related research. Participant performance in the C3Fire microworld can easily be extracted from the log files (Granlund & Johansson, 2003). For example, performance can be calculated in terms of the number of cells affected by fire, number of structures affected, or whether a cell has been fully burnt out or if the participants have stopped the fire in time. As these measures are recorded dynamically (not truly dynamically, but with a high enough sampling rate to capture any changes) the participants' progress can easily be tracked and analyzed. This performance can be connected to individual participants, teams, or the whole system and related to, for example, communication messages sent. The generated log files also allows for playback of the scenario development, to use for training and feedback to the participants. The temporal aspects of simulation can be used as a performance measure, for example, how long did it take for team member X to respond to a message or how long did it take until all fires were controlled? Time can also be set to frame the experimental design, for example, how much fire was controlled before the scenario was stopped after 20 minutes? Our experience with scenario length is that most scenarios are run for no more than 45 minutes (even though a scenario could continue for many hours). Obviously, this is dependent on the research questions asked. The speed by which different types of terrain is consumed by the fires can also be manipulated, as well as the speed of fire fighting and the movement speed of vehicles.

Next, we will describe the most important research efforts that have been conducted using the C3Fire simulation.

Studying Situation Awareness and Cooperation in Teams

Artman (1999) performed the initial studies using C3Fire with the aim of investigating how teamwork was impacted by different organizational configurations (serial, parallel, and an "optional" alternative). This was especially evaluated in terms of situational awareness (Endsley, 1995) and co-operation (by communication analysis). This work postulated a relationship between communication within a team and the degree of situational awareness that the team could reach. The specific formulation was that "communication must of course be relevant in order to support team situation awareness. This study will investigate what kind of communication is relevant" (Artman, 1999, p. 1406). It was found that most groups in the optional condition performed worse than the groups in the other conditions, indicating that the optional organization, where participants were allowed to choose how to organize themselves, has a negative impact on performance as that approach created uncertainties regarding information flows and responsibilities, eventually also hampering situational awareness. It was also found that members of teams in the parallel condition differed from members of teams in the serial condition in their situation awareness. This was attributed to the fact that the responsibility for keeping track of unit positions and fire development was distributed between different persons in the parallel organization, while it was handled by a single individual in the serial configuration. This distribution may have led to a situation were no members of the team really were able to create a picture of the situation as a whole. Further, more successful commander teams produced more planning in relation to hypothesis, as well as sending fewer messages in total between the units, than the less successful team (Artman, 1999). The Artman studies showed how C3Fire can be used to investigate several aspects of team decision making as well as how technology can affect team decision making. This inspired a set of further studies at both Linköping University and the Swedish National Defence College (SNDC).

C3Fire as a Tool for Studying Command and Control Teams

The aftermath of the United States victory in the 1991 Gulf War against Iraq created a case for the importance of information and communication technology (ICT) on the battlefield, as it seemed to provide a way to create unsurpassed situational awareness as well as superior coordination of resources (Carlerby & Johansson, 2017). This "information superiority" promised a "revolution in military affairs" (RMA), which created a huge interest from military actors, industrial suppliers, as well as researchers and command and control concept developers. This development eventually led to the introduction of concepts such as network-centric warfare (NCW; Cebrowski & Garstka, 1998; Osinga, 2010). Most military research institutes initiated studies concerning how such concepts could be utilized, including the SNDC. Alberts, Garstka, and Stein (1999) defined NCW in the following way:

> We define NCW as an information superiority-enabled concept of operations that generates increased combat power by networking sensors, decision makers, and shooters

> to achieve shared awareness, increased speed of command, higher tempo of operations, greater lethality, increased survivability, and a degree of self-synchronisation.
>
> **(Alberts, Garstka, & Stein, 1999, p. 2)**

A series of experiments were launched at the SNDC with the aim of testing some of these suggestions, in particular the promises of increased shared awareness and increased performance. This was done in the context of the development of a new concept for command and control environments, the Joint Mobile Command and Control Post of the Future (Sundin & Friman, 1998, 2000). The purpose of the Joint Mobile Command and Control Post project was to create a command environment (a staff room) where teams of decision makers could work jointly around a large table-like screen, promoting a shared view as well as an environment that enabled discussion and creativity to a larger extent than a traditional staff environment, thus increasing the quality of decisions made (Sundin & Friman, 1998). A prototypical environment was built at the SNDC, where such a table-like screen, and additional computers and screens, were available for experimentation. It was soon concluded that such conceptual propositions would be difficult to study through full-scale exercises or high-fidelity simulation as it would require extensive implementation of technologies that were not yet available. Instead, this was approached by utilizing the C3Fire microworld as the basic research platform. Several studies were conducted, where the effects of rapid information flows through geographical information systems could be utilized to increase shared awareness. For example, Parush and Ma (2012) showed how a team display supported team performance during communication breakdowns in a command and control task using the C3Fire environment. Another example is how the effect of updating maps directly from sensors in the field contrasted with the traditional approach of having a staff member process and deliver information to decision makers (Johansson, Granlund, & Waern, 2000; Granlund et al., 2001; Persson & Johansson, 2001; Johansson, Persson, Granlund, & Mattsson, 2003). These studies showed that although the errors in information clearly were reduced in when maps were updated directly from the field rather than by a staff member, overall team performance did not improve. This was partly attributed, as in the case of the Artman (1999) study, that the work division in the staff team was less clear in the condition where information was presented directly in the shared map than when a specific staff member added data to the shared map. Video analysis suggested that the military commanders in the manual updating condition recognized the similarities in working with traditional paper-based maps. In the direct update, the command team rather gathered around the shared map, waiting for something to happen. A hypothesis is that this hampered performance in the direct-updating condition since the tool eliminated the need to constantly monitor the emails coming from the ground chiefs, but it also forced them to look at the shared map (Johansson, 2005). In manual update the commander and the assistant completed the picture based on verbal reports from the two communications officers. These verbal reports were spoken out in the room (by the communication officers), meaning that anyone in the room also could hear what was going on. This can possibly have minimized the need to observe the shared screen for assessing the situation. There is also a risk that the direct updating condition, with its higher rate of data input on the screen, creates a situation where the commanders are "chasing" the situation rather than handling it.

These studies initiated further research concerning the effects of new technologies on team decision making, as will be described next.

Studies of the Impact of GIS on Emergency Response Management

These are studies conducted at Linköping University funded by KBM (Swedish acronym for *Krisberedskapsmyndigheten*, the crisis preparedness authority which later merged into the Swedish Civil Contingencies Agency; MSB) during the period of 2005 to 2011.

During the time from 2005 to 2011, three research projects were performed at Linköping University with the goal to investigate the impact of a decision support system that presents real-time information through a global positioning system (GPS) to decision makers in crisis management organizations. The goal was to compare the performance between teams that had access to unit positions and sensor information in the command post with teams that received information through text messages and had to keep track of situation development using paper maps. The method used was controlled experiments with the C3Fire microworld. All three projects had the same experimental approach, although three different types of participants took part in the experiments: university students, municipal crisis management organizations, and rescue service personnel.

A total of 304 participants took part in the three studies. In the first study, conducted in 2006, the research was tested on non-professional participants, 132 students formed into 22 groups (Johansson et al., 2007, 2010). In the second study, conducted from 2008 to 2009, the participants were professionals belonging to different municipal crisis management organizations, including both rescue service personnel and other municipal employees. In this study, 108 professionals formed 18 teams (Granlund et al., 2010; Granlund & Granlund 2011a, 2011b). In the third study, conducted from 2010 to 2011, a total of 64 professionals formed 8 rescue service teams.

The organizations of interest were Swedish municipal crisis management organizations and their crisis management teams. The goal was to understand differences in the collaboration and work processes of teams that had access to unit and sensor information in their digital map systems at command center and command post level, compared to teams that had only paper maps in the command posts.

The participating professionals belonged to different municipalities in Sweden. Currently, many of these have made, or are about to make, investments in information and communication technologies that give the decision makers in the command centers and command posts access to GPS information. All this is done in order to enhance performance and control of work.

Digital maps are seen as support tools for crisis management. Management is understood to be more efficient with the introduction of these new technical supports. The reality is that the tools distribute large amounts of information automatically to the decision makers. All users at all levels of management have in many situations access to the same information simultaneously. What was originally seen as an aid in the management work can have unsuspected consequences. The tools change the requirements for leading and organizing emergency efforts.

The experiment designs of Study 1 and Study 2 were identical. For the third study the original design was amended with to the aim of increasing complexity and respond to rescue service professionals training needs. Study 1 and Study 2 had the same between-group design with one factor: (a) crisis management teams receiving unit positions and sensor information through a digital map, and (b) crisis management teams receiving information in the form of text messages, using paper maps (Figures 2.4 and 2.5).

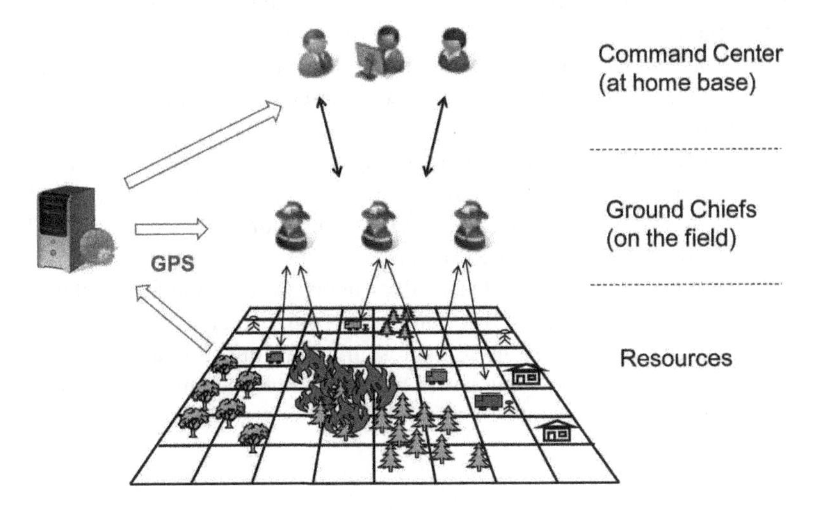

FIGURE 2.4 Design Study 1 and 2, GPS condition.

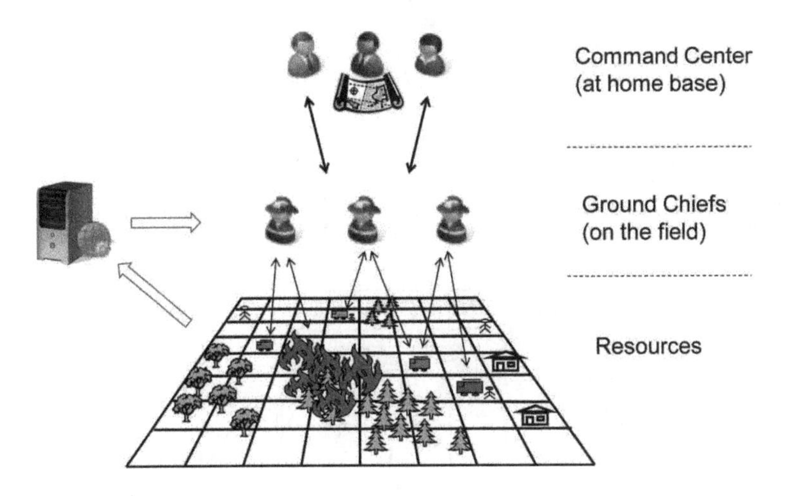

FIGURE 2.5 Design Study 1 and 2, paper map condition.

In each group, three participants worked as a command center (CC) with one commanding officer and two liaison officers. Three participants worked as ground chiefs (GC). The CC had no direct contact with the simulation and controlled the simulated world indirectly via the GC. The GC directly controlled three units (fire brigades) each in the simulation, a total of nine units. Each unit could "see" a limited part of the world in the immediate surroundings. In the first condition, unit positions and the sensor information received by that unit were transmitted directly to the command post. The difference between the two conditions was the *complexity* of the type of support the commanding officer obtained, in terms of GPS with access to exact positioning of the resources in the simulated world, or a paper map without any automatic information.

Study 3 had the same between-group design with one factor as Study 1 and Study 2: (a) crisis management teams receiving information about unit positions and sensor information through a digital map, and (b) crisis management teams using paper maps. The differences between the designs were based on six points requested by the participating rescue service personnel.

(1) The organization of the participant group had four levels of command instead of three: command center (CC) at home base, command post (CP) on the field, and four ground chiefs (GC) directing units in the simulation (Figures 2.6 and 2.7). (2) Each group had eight participants. (3) The resources in the simulated world were ten firefighters, five water logistic trucks that supply water to the fire fighters, three excavators for digging firebreaks, three units for evacuating citizens, and one unit for reconnaissance purposes. (4) None of the resources are linked to any of the ground chiefs. A structure for who is using which resources and when must be set up by the

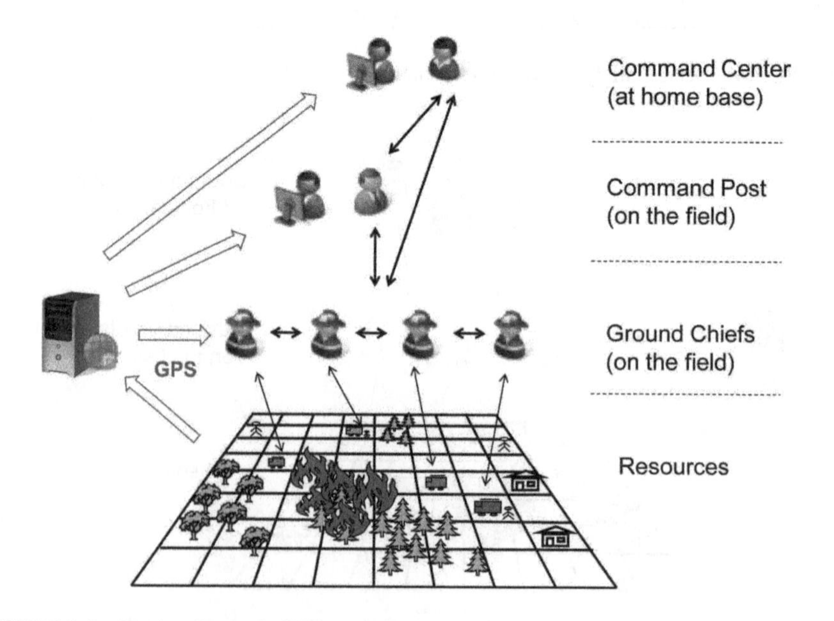

FIGURE 2.6 Design Study 3, GPS condition.

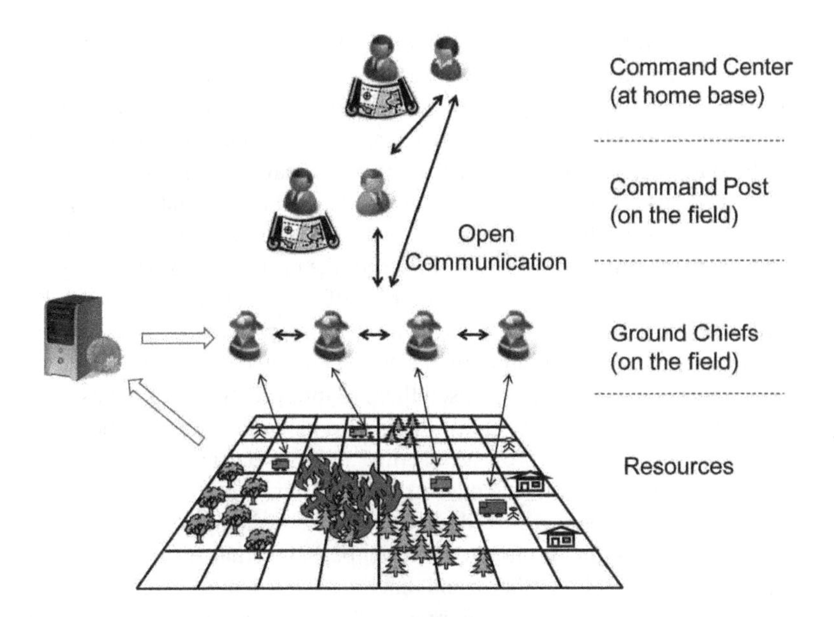

FIGURE 2.7 Design Study 3, paper map condition.

team. The complexity of the task of extinguishing the forest fire was increased. Also, the ability to communicate was not restricted: everyone could send messages to anyone. This forced the team to form a structure describing how interactions should take place and in what manner. (5) The simulation time was increased from 20 minutes to 40 minutes, but the whole training day has been reduced to one full training session and three full simulation cycles. (6) The fire spread at a slower pace and the firefighting units moved slower.

For the first study, where students participated, results showed no significant difference between teams receiving information directly in the digital map and teams working with paper maps (Johansson et al., 2010). The results showed that groups with an access to real-time support had a better performance than those who had to keep track of the situation using paper maps, in terms of the area lost to the forest fire. This suggested that it was easier to understand the situation and coordinate resources when receiving rapid feedback about unit positions and sensor information from the immediate surroundings of the units. This was also reflected in the communication between the participants, which revealed that a larger number of messages were sent in the condition without direct access to unit positions and sensor information. Specifically, the amount of questions differed between the two conditions, where the teams (paper map condition, see Figure 2.5) having to keep track of unit positions and fire development on paper maps sent a significantly larger amount of questions about unit positions and the development of the forest fire(s) (Johansson et al., 2010).

For the second study, where municipality crisis managers participated, there was no overall performance difference between automated information transfer and manual information management using paper maps (Granlund, Granlund, & Dahlbäck,

2011). The municipality crisis managers had an inconsistent result compared to participants in the first study. Observations suggested that the crisis managers recognized the simulations as training in crisis management or in communication, not as a computer game, which the student participants in the first study may have done. Further, the observations indicated that the professional participants perceived the simulation as more complex than the students, although the simulation setting itself was not changed. The professionals, as well as the students, knew they were observed and analyzed in a research project. They behaved like professionals when they solved their problems. They used strategies from real life, for example to discuss before they acted, not using all the resources directly, etc. They saw the simulated event's resemblance to a real situation rather than a computer game. Hence, the professional background of the crisis managers caused them to interpret the situation as training, and they tried to do what they normally would have done in a real situation.

In the third study, where professional fire fighters participated (Granlund & Granlund, 2011), the results indicated a surprising difference in the sense that teams where the commanders only received messages and worked with paper maps performed significantly better. It should however be noted that the amount of trials conducted was lower in this condition and that the experimental design was not identical to Studies 1 and 2 (as pointed out previously). A reason for this could be complexity. The automated information flow was directed to both CC and CP without prior classification as important or unimportant. This classification is in actual work conducted by ground chiefs in the field. The teams in the condition where information flows were automatized had to allocate time to handle the information. The teams thus did not have time to adjust their ordinary management methods, basically adapted to paper maps, to the situation with direct information flows. Their amount of burned-out area was in the last simulation trial still increasing. That is a sign of the three simulation trials not being enough for learning to handle the complexity of receiving automated information flows.

The CC and CP in teams in the condition where information had to be handled manually had a situation that much resembled what rescue service commanders experience in reality. The complexity was experienced as lower and better adjusted to their ordinary methods. The commanders did not have to refine or alter their known methods too much, thus they performed better (Granlund & Granlund, 2011).

Using C3Fire to Support Role-Playing Exercises in Emergency Response Research

These studies were conducted at Linköping University and differed from the previously mentioned studies in the sense that they did not follow the experimental tradition with contrasting conditions and controlled variables. Instead, a role-playing exercise approach was applied, focusing on utilizing components from C3Fire to create an immersive situation where professional participants could interact with each other (Trnka & Johansson, 2009). The C3Fire components were used to support data collection in terms of messaging, information dissemination, etc. (Trnka & Jenvald, 2006). An example of how this approach can be applied is described in Trnka & Jenvald (2006) where dispatchers from the county emergency call center, incident commanders, fire and rescue dispatch officers, police on-site incident commanders, and police dispatch officers participated in a scenario where a major forest fire threatens a local zoo and its visitors. In this simulation, the participants did not interact directly with the C3Fire interface, instead information about the progress of the fire and firefighting units was presented to the participants in the study by exercise facilitators (Trnka & Jenvald, 2006). The exercise facilitators injected information into the exercise either verbally or by text messages to the participants, mimicking the kind of information the emergency responders normally would get from units in the field and from the public. The focus was to understand how participants acted under time pressure and uncertainty with limited resources. Using C3Fire as input for the role-playing exercise included both planned and unplanned variations, thus requiring the participants to handle situations that changed over time in a nonpredictable way. Trnka and Jenvald (2006) showed that high realism of the simulation content, low-fidelity ecological settings, co-localization of the participants, and the use of domain, modeling, and simulation experts in planning and execution were necessary for a successful role-playing exercise. Trnka and Johansson (2009) used the same material and applied communication analyses such as episodic analysis (Korolja & Linell, 1996), socio-metric status (Houghton et al., 2006), and communication roles to assess coordination among participants. The results suggested that the participants used informal procedures within the command and control organization to perform various informal functions and roles across organizational and domain boundaries. Utilizing C3Fire in this way created a hybrid simulation where specific processes were represented within the microworld while other aspects of the simulation were conducted as role playing (Trnka & Jenvald, 2006).

Developing the Shared Priorities Instrument of Shared Strategic Understanding

Berggren and colleagues (Berggren, 2016; Berggren, Johansson, & Baroutsi, 2017) used C3Fire to develop a new measure of shared understanding in teams handling various dynamic tasks. Mainly using students as research participants, the instrument was developed through several iterations, where the main platform used was C3Fire. A brief summary of this development is discussed next.

The demand for a comprehensible, easy to use, useful instrument for assessing a team's shared mental models has been identified over and over again (Cannon-Bowers, Salas, & Converse, 1993; Klimoski & Mohammed, 1994; Smith-Jentsch, Campbell, Milanovich, & Reynolds, 2001; Langan-Fox, Wirth, Code, Langfield-Smith, & Wirth, 2001; Mohammed, Ferzandi, & Hamilton, 2010; Mohammed, Klimoski, & Rentsch, 2000). One example of this is the authors' own experience of a military brigade HQ meeting where the chief of staff (COS) provided a briefing and gave orders to the staff officers who acknowledged the orders and then left the staff meeting. Outside of the staff tent they were then asked individually what the intent of the COS was and what the shared goals of the brigade were. Responding to this question, the officers provided almost as many different answers as there were individuals. This feedback was then presented to the COS who recognized the need for an instrument that could help him evaluate if the command team had a shared understanding of the shared goals. Most existing instruments were either very difficult to utilize, took a long time to prepare, were only useful for a context-dependent setup, were more or less intrusive, and/or took a long time to analyze (Langan-Fox, Code, & Langfield-Smith, 2000; Mohammed, Klimoski, & Rentsch, 2000; Resick, Murase, Bedwell, Sanz, Jiménez, & DeChurch, 2010; DeChurch and Mesmer-Magnus, 2010; Wildman, Salas, & Scott, 2014).

Hence, the mission was to develop an instrument that was useful, understandable, easy to apply, and cost effective. The first iteration was an experiment in C3Fire (reported in Berggren, Alfredson, Andersson, & Modéer, 2007 and in Berggren, Alfredson, Andersson, & Granlund, 2008). Here, the instrument used lists of pre-defined items relating to different aspects of teamwork in the C3Fire scenario. The pre-defined lists were tested using three-member teams (with different roles) who were required to collaborate in order to perform well in the microworld setting. Four different conditions were tested (ranging from easy to difficult by manipulating the amount of information that was available for the participants). The participants were asked to rank order the lists, which were then compared among team members (within teams) to analyze degree of correspondence. The results revealed no differences. There were several factors identified explaining this: untrained participants, a too-weak relationship between items and behavior, and too difficult tasks.

The following iteration involved military teams working as tank crews (reported in Berggren, Svensson, Hörberg, Jonsson, & Höglund, 2009 and in Berggren & Johansson, 2010). The experimental organization included three tank crews that scouted an area in a peace-enforcing scenario. In this study, two different conditions were used. The first condition was that the teams communicated in the same command and control (C2) system, and the other condition was that one team could only use radio and paper maps. For this study the participants were asked to generate the items in the lists individually, that is, individually defined items. Within the teams they were then asked to rank order these lists. The outcome from these ranked lists was analyzed using Kendall's measure of concordance (Kendall & Babington Smith, 1939). Here, a difference between the two conditions was present in the outcome of the instrument assessing shared understanding. That is, a difference between the two conditions could be seen regarding degree of shared understanding.

In the next iteration these findings were again tested in the C3Fire microworld (reported in Berggren, Johansson, Svensson, Baroutsi, & Dahlbäck, 2014 and Berggren et al., 2017). This time six teams were trained over ten occasions in collaborating in the C3Fire microworld. These teams were then tested against six teams that were co-trained once prior to the experimental data collection. The participants were asked to generate the items in the lists individually, that is, individually defined items. Within the teams they were then asked to rank order these lists. The outcome from these ranked lists was analyzed using Kendall's measure of concordance (Kendall & Babington Smith, 1939). Here, a difference between the trained and novice teams conditions was present in the outcome of the instrument. In addition, the generated items were analyzed regarding content, and the trained teams' items were focused on team-level concerns, whereas the novice teams' items were focused on the individual's needs.

The shared priorities instrument assesses shared strategic understanding, that is "the ability of multiple agents to exploit common bodies of causal knowledge over time for the purpose of accomplishing common (or shared) goals" (Berggren, 2016, p. 125). The instrument was found to be usable, understandable, objective, flexible, and self-explanatory (Berggren, 2016), thus meeting several criteria that were asked for by the practitioners, i.e., military commanders. After the development phase of the instrument, it has been tested in several other domains: with nuclear power plant control room crews (Berggren, Johansson, & Ekström, 2016), emergency rooms (Berggren, Johansson, Allard, & Torensjö, 2016), at an emergency exercise (Prytz, Rybing, Jonson, Pettersson, Berggren, & Johansson, 2015), and for training of pre-hospital ambulance personnel (Berggren, Herrera Velasquez, Pettersson, Henning, Lidberg, and Johansson, 2018).

CONCLUSION

The aim of this chapter was to provide a description of how the C3Fire microworld has contributed to knowledge development concerning various aspects of team cognition and how the platform has been and can be utilized for research on team cognition. We have presented how the C3Fire microworld has been used to conduct different types of studies and how this has supported the development of team cognition research. Using microworlds to conduct research has, as has been argued (cf. Brehmer & Dörner, 1993), many advantages. Microworlds provide a platform where teams can be studied regarding structure, size, organization, roles, and different types of interaction. In this chapter we have shown how several aspects of team cognition have been explored in the C3Fire microworld, for example, team situation awareness, communication, shared mental models, coordination, and collaboration (in terms of command and control).

There are several results relating to team research that the content of this chapter has touched upon. Artman (1999) showed how microworlds can be used to investigate how various organizational configurations of teams affect performance and decision making. Microworlds were also utilized for the study of future command and control environments, as this required an approach that allowed for research on concepts that were too immature to be fully implemented from a technical point of view (Johansson, Persson, Granlund, & Mattsson, 2003). The three studies of how directed information flows concerning unit positions and sensor information presented in digital maps influences emergency management teamwork showed a number of interesting results. In novice teams, direct information improved performance, while this was not shown in studies involving professional participants, although communication was affected for all types of participants (Granlund & Granlund, 2011). The studies performed by Trnka and colleagues showed that a microworld can be used as a part of a hybrid simulation (Trnka & Jenvald, 2006; Trnka & Johansson, 2009). Further, the studies performed by Berggren and colleagues show how new measures of sharedness in teams can be developed and validated using microworlds as a research platform (Berggren, 2016).

Thus, microworlds can be used for aspects of team research such as method development, training, and testing of different psychological concepts (mental workload, situational awareness, motivation, expertise, trust, leadership, etc.), or as part of a larger experimental setup.

An important experience from working with microworld research is the lack of useful scenario descriptions in order to reproduce research findings. Hence, we call for better descriptions of scenario and design configurations to be able to replicate studies. This would be easy in these kinds of reproducible technical systems, where little is left to chance, except the behaviors of the participants. For research purposes, a standardized way of describing scenarios would make it easier to replicate studies and provide confirmations or refutations of earlier results. Such standardized descriptions could be implemented into any microworld platform, regardless of developer, for future researchers to use for replication so that findings generalize across time and across situations.

We also agree with McNeese and Pfaff (2012) on the importance of training participants prior to the actual data collection. This cannot be stressed enough, since reaching the sufficient capacity level of the participants will affect the outcome of the study. The situations and complexities require that the participants understand their part and have the skills necessary for controlling and interacting with the simulation.

A general observation is that participant background, i.e. whether the participants have an occupation or experience of situations resembling the microworld, seems to affect how they view the problems presented even when they are greatly simplified. For such participants, microworlds might be frustrating as they are overly simplistic and limit the possibility to utilize professional knowledge and tactics that could have been appropriate in a real-world situation or a more advanced simulation. This was observed for example in the studies conducted by Granlund, Granlund, and Dahlbäck (2011).

Using microworlds is an appreciative platform for team research. There is a learning effect of using a microworld platform for the participants, however, as the learning curve is quite steep the participants are easily up and running for the actual experiments in a short time. This compared to training teams of helicopter pilots or command teams, something that requires years of training and experience. Alas, as a researcher, you don't want to waste valuable professional teams on testing new methods and instruments. When the methods and instruments have been validated and tested, then it is time to go into "the wild" with the research. This way you have feasible and useful hypothesises, concept models, and methods to use with the hard to access professional teams. Of course there is a difference between the real world of professionals and naïve subjects and teams in microworld studies, but the benefits of testing material and methods to adapt them provides a cost-effective way of developing our team theories in a controlled and replicable environment.

REFERENCES

Alberts, D. S., Garstka, J. J., & Stein, F. P. (1999). *Network centric warfare. Developing and leveraging information superiority* (2nd ed.). Washington DC: CCRP Publication Series.

Artman, H. (1999). Situation awareness and co-operation within and between hierarchical units in dynamic decision making. *Ergonomics, 42*(11), 1404–1417.

Baker, D. P., & Salas, E. (1992). Principles for measuring teamwork skills. *Human Factors, 34*(4), 469–475.

Berggren, P. (2016). *Assessing Shared Strategic Understanding.* Linköping Studies in Arts and Science, Dissertation No. 677. Linköping, Sweden: Linköping University Electronic Press.

Berggren, P., & Johansson, B. J. E. (2010). Developing an instrument for measuring shared understanding. In S. French, B. Tomaszewski, & C. Zobel (Eds.), *Proceedings of ISCRAM 2010* (pp. 1–8). Seattle, WA: ISCRAM.

Berggren, P., Alfredson, J., Andersson, J., & Granlund, R. (2008). Assessing shared situational awareness in dynamic situations. *NATO RTO HFM-142 Symposium*, 1–7. Copenhagen, DK: NATO RTO.

Berggren, P., Alfredson, J., Andersson, J., & Modéer, B. (2007). Comparing measurements of shared situational awareness in a microworld with a developmental environment. *IFAC Proceedings Volumes, 10* (PART 1).

Berggren, P., Herrera Velasquez, M., Pettersson, J., Henning, O., Lidberg, H., & Johansson, B. J. E. (2018). Reflection in Teams for Training of Prehospital Command and Control Teams. *ISCRAM 2018.* Rochester, NY.

Berggren, P., Johansson, B. J. E., & Baroutsi, N. (2017). Assessing the quality of Shared Priorities in teams using content analysis in a microworld experiment. *Theoretical Issues in Ergonomics Science, 18*(2), 128–146.

Berggren, P., Johansson, B. J. E., & Ekström, E. (2016). Resilience through training - assessing cognition in teams. *Proceedings of ISCRAM 2016.*

Berggren, P., Johansson, B. J. E., Allard, O., & Torensjö, E. (2016). Training resilient medical teams. *ECCE 2016.* Nottingham.

Berggren, P., Johansson, B. J. E., Svensson, E., Baroutsi, N., & Dahlbäck, N. (2014). Statistical modelling of team training in a microworld study. *Proceedings of the Human Factors and Ergonomics Society 58th Annual Meeting*, (2011), 894–898. Chicago, Il.

Berggren, P., Svensson, J., Hörberg, U., Jonsson, S., & Höglund, F. (2009). Shared priorities as a measure of shared understanding. *Proceedings of the Europe Chapter Human Factors and Ergonomics Society (HFES) Conference 2009.* Linköping.

Bodeau, D., Fedorowicz, J., Markus, L., & Brooks, J. (2009). Characterizing and Improving Collaboration and Information-Sharing Across Emergency Preparedness and Response Communities. *International Conference on E-Government*, 192–200.

Boin, A., & McConnell, A. (2007). Preparing for Critical Infrastructure Breakdowns: The Limits of Crisis Management and the Need for Resilience. *Journal of Contingencies and Crisis Management, 15*(1), 50–59.

Brehmer, B. (1987). System Design and the Psychology of Complex Systems. In J. Rasmussen & P. Zunde (Eds.), *Empirical foundations of Information and Software Science* (pp. 21–32). New York: Plenum Publishing Cooperation.

Brehmer, B. (2000). Dynamic Decision Making in Command and Control. In C. McCann & R. Pigeau (Eds.), *The Human in Command: Exploring the Modern Military Experience* (pp. 233–248). New York: Academic/Plenum Publishers.

Brehmer, B. (2004). Some Reflections on Microworld Research. In L. R. Elliott & M. D. Coovert (Eds.), *Scaled Worlds: Development, Validation and Applications* (pp. 22–36). Aldershot: Ashgate.

Brehmer, B. (2005). Micro-worlds and the circular relation between people and their environment. *Theoretical Issues in Ergonomics Science, 6*(1), 73–93.

Brehmer, B., & Allard, R. (1991). Dynamic decision making: The effects of task complexity and feedback delay. In J. Rasmussen, B. Brehmer, & J. Leplat (Eds.), *Distributed Decision Making: Cognitive models for cooperative work* (pp. 319–334). Chichester: John Wiley & Sons.

Brehmer, B., & Dörner, D. (1993). Experiments with Computer-Simulated Microworlds: Escaping both the Narrow Straits of the Laboratory and the Deep Blue Sea of the Field Study. *Computers in Human Behavior, 9*, 171–184.

Brehmer, B., & Svenmarck, P. (1995). Distributed Decision Making in Dynamic Environments: Time Scales and Architectures of Decision Making. In J. P. Caverni, M. Bar-Hillel, F. H. Barron, & H. Jungermann (Eds.), *Contributions to decision making*. Elsevier Science Ltd.

Cannon-Bowers, J. A., Salas, E., & Converse, S. A. (1993). Shared mental models in expert team decision making. In N. J. Castellan (Ed.), *Individual and group decision making* (pp. 221–246). Hillsdale, NJ: Lawrence Erlbaum Associates.

Carlerby, M., & Johansson, B. J. E. (2017). The lack of convergence between C2 theory and practice. *The 22nd International Command and Control Research and Technology Symposium*, 1–20.

Cebrowski, A. K., & Garstka, J. J. (1998). Network-Centric Warfare: Its Origin and Future. *US Naval Institute Proceedings, 124*, 28-35.

Comfort, L. K. (2007). Crisis management in hindsight: Cognition, communication, coordination, and control. *Public Administration Review, 67*, 189–197.

Cooke, N. J., & Shope, S. M. (2004). Designing Synthetic Task Environments. In S. G. Schiflett, L. R. Elliott, E. Salas, & M. D. Coovert (Eds.), *Scaled worlds: Development, validation, and application* (pp. 263–278). Aldershot: Ashgate.

Cooke, N. J., Gorman, J. C., & Rowe, L. J. (2009). An ecological perspective on team cognition. In E. Salas, G. F. Goodwin, & C. S. Burke (Eds.), *Team effectiveness in complex organizations: Cross-disciplinary perspectives and approaches* (pp. 157–182). New York: Routledge/Taylor & Francis Group.

DeChurch, L. A., & Mesmer-Magnus, J. R. (2010). Measuring shared team mental models: A meta-analysis. *Group Dynamics: Theory, Research, and Practice, 14*(1), 1–14.

Dörner, D. (1980). On the difficulties people have in dealing with complexity. *Simulation & Gaming, 11*(1), 87–106.

Dörner, D., & Schaub, H. (1992). Spiel und Wirklichkeit: Über die Verwendung und den Nutzen computersimulierter Planspiele. *Kölner Zeitschrift Für "Wirtschaft Und Pädagogik, 7*(12), 55–78.

Dörner, D., Kreuzig, H., Reiter, F., & Stäudel, T. (1983). *Lohhausen. Vom Umgang mit Unbestimmtheit und Komplexität*. Bern: Huber.

Endsley, M. R. (1995). Toward a theory of situation awareness in dynamic systems. *Human Factors, 37*(1), 32–64.

Funke, J. (1993). Microworlds based on linear equation systems: A new approach to complex problem. In G. Strube & K. F. Wender (Eds.), *The cognitive psychology of knowledge* (pp. 313–330). Amsterdam: Elsevier Science Publishers B.V.

Funke, J. (2001). Dynamic systems as tools for analysing human judgement. *Thinking & Reasoning, 7*(1), 69–89.

Funke, J. (2010). Complex problem solving: A case for complex cognition? *Cognitive Processing, 11*(2), 133–142.

Gonzalez, C., Vanyukov, P., & Martin, M. K. (2005). The use of microworlds to study dynamic decision making. *Computers in Human Behavior, 21*(2), 273–286.

Granlund, R. (1997). *C3Fire A microworld supporting emergency management training*. Licentiate Thesis, no. 598, Department of Computer and Information Science, Linköping University.

Granlund, R. (2001). Web-based micro-world simulation for emergency management training. *Future Generation Computer Systems, 17*(5), 561–572.

Granlund, R. (2002). Monitoring distributed teamwork training. Linköping Studies in Science and Technology, Dissertation No. 746. Linköping, Sweden: Linköping University Electronic Press.

Granlund, R. (2003). Monitoring experiences from command and control research with the C3Fire microworld. *Cognition, Technology & Work*, *5*(3), 183–190.

Granlund, R., & Granlund, H. (2011a). GPS impact on performance, response time and communication – A review of three studies. *ISCRAM2011, 8th International Conference on Information Systems for Crisis Response and Management*. Lisbon, Portugal.

Granlund, R., & Granlund, H. (2011b). Using simulations to study the impact of GPS information in crisis response organizations. *ISCRAM2011, 8th International Conference on Information Systems for Crisis Response and Management*. Lisbon, Portugal.

Granlund, R., & Johansson, B. J. E. (2003). Monitoring distributed collaboration in the C3Fire microworld. In S. G. Schiflett, L. R. Elliott, E. Salas, & M. D. Coovert (Eds.), *Scaled worlds: Development, validation and applications*. Aldershot: Ashgate.

Granlund, R., Granlund, H., & Dahlbäck, N. (2011). Differences between students and professionals while using a GPS based GIS in an emergency response study. *14th International conference on human-computer interaction*. Orlando, FL, USA.

Granlund, R., Granlund, H., Johansson, B. J. E., & Dahlbäck, N. (2010). The effect of a geographical information system on communication in professional emergency response organizations. *ISCRAM2010, 7th international conference on information systems for crisis response and management*. Seattle, WA, USA.

Granlund, R., Johansson, B. J. E., Persson, M., Artman, H., & Mattsson, P. (2001). Exploration of methodological issues in micro-world research–Experiences from research in team decision making. *Proceedings of the first international workshop on cognitive research with microworlds*, 73–82.

Gray, W. D. (2002). Simulated task environments: The role of high-fidelity simulations, scaled worlds, synthetic environments, and laboratory tasks in basic and applied cognitive research. *Cognitive Science Quarterly*, *2*, 205–227.

Grüne-Yanoff, T., & Weirich, P. (2010). The philosophy and epistemology of simulation: A review. *Simulation & Gaming*, *41*(1), 20–50. 53470

Houghton, R. J., Baber, C., McMaster, R., Stanton, N. A., Salmon, P. M., Stewart, R., & Walker, G. H. (2006). Command and control in emergency services operations: a social network analysis. *Ergonomics*, *49*, 1204–1225.

Johansson, B. J. E. (2005). *Joint control in dynamic situations*. Linköping Studies in Arts and Science, Dissertation No. 972. Linköping, Sweden: Linköping University Electronic Press.

Johansson, B. J. E., Granlund, R., & Waern, Y. (2000). The communicative aspects of distributed dynamic decision making in the ROLF environment. *5th Conference on natural decision making*. Stockholm.

Johansson, B. J. E., Persson, M., Granlund, R., & Mattsson, P. (2003). C3Fire in command and control research. *Cognition, Technology & Work*, *5*(3), 191–196.

Johansson, B. J. E., Trnka, J., & Berggren, P. (2016). *A case study of C2 agility in the 2014 Västmanland forest fire* (No. FOI-R--4259--SE). Linköping: FOI.

Johansson, B. J. E., Trnka, J., & Granlund, R. (2007). The effect of geographical information systems on a collaborative command and control task. *Proceedings ISCRAM2007*, 191–200. Delft, The Netherlands: ISCRAM.

Johansson, B. J. E., Trnka, J., Granlund, R., & Götmar, A. (2010). The effect of a geographical information system on performance and communication of a command and control organization. *International Journal of Human- Computer Interaction*, *26*(2–3), 228–246.

Kendall, M. G., & Babington Smith, B. (1939). The problem of m rankings. *Annals of Mathematical Statistics*, *10*(3), 275–287.

Klimoski, R., & Mohammed, S. (1994). Team mental model: Construct or metaphor? *Journal of Management*, *20*(2), 403–437.

Korolja, N., & Linell, P. (1996). Episodes: coding and analyzing coherence in multiparty conversation. *Linguistics*, *34*(4), 799–832.

Langan-Fox, J., Code, S., & Langfield-Smith, K. (2000). Team mental models: Techniques, methods, and analytic approaches. *Human Factors*, *42*(2), 242–271.

Langan-Fox, J., Wirth, A., Code, S., Langfield-Smith, K., & Wirth, A. (2001). Analyzing shared and team mental models. *International Journal of Industrial Ergonomics*, *28*, 99–112.

Liu, D., Macchiarella, N. D., & Vincenzi, D. A. (2009). Simulation fidelity. In D. A. Vincenzi, J. A. Wise, M. Mouloua, & P. A. Hancock (Eds.), *Human Factors in Simulation and Training* (pp. 61–74). London: CRC Press.

McNeese, M. D., & Pfaff, M. S. (2012). Looking at macrocognition through a multimethodological lens. In E. Salas, S. M. Fiore, & M. P. Letsky (Eds.), *Theories of team cognition: Cross-disciplinary perspectives* (pp. 345–371). New York, NY: Routledge.

McNeese, M. D., Bains, P., Brewer, I., Brown, C. E., Connors, E. S., Jefferson, T., … Terrell, I. S. (2005). The NeoCITIES Simulation: Understanding the design and methodology used in a team emergency management simulation. *Proceedings of the Human Factors and Ergonomics Society 49th annual meeting*, 591–594. Santa Monica, CA.

Mohammed, S., Ferzandi, L., & Hamilton, K. (2010). Metaphor no more: A 15-year review of the team mental model construct. *Journal of Management*, *36*(4), 876–910.

Mohammed, S., Klimoski, R., & Rentsch, J. R. (2000). The measurement of team mental models: We have no shared schema. *Organizational Research Methods*, *3*(2), 123–165.

Omodei, M., Elliott, G., Walshe, M., & Wearing, A. (2005). Networked fire chief: A research and training tool that targets the human factors causes of unsafe decision making in wildfires Mary Omodei, Glenn Elliott, Matthew Walshe and Alex Wearing. In B. Butler & M. Alexander (Eds.), *Eighth international wildland fire safety summit*. Missoula, MT: The International Association of Wildland Fire.

Osinga, F. (2010). The rise of military transformation Frans Osinga. In T. Farrell, T. Terry, & F. Osinga (Eds.), *A transformation gap?: American innovations and European military change*. Stanford, CA: Stanford Security Studies.

Ouyang, M., & Wang, Z. (2015). Resilience assessment of interdependent infrastructure systems: With a focus on joint restoration modeling and analysis. *Reliability Engineering & System Safety*, *141*, 74–82.

Parush, A., & Ma, C. (2012). Team displays work, particularly with communication breakdown: Performance and situation awareness in a simulated forest fire. *Proceedings of the 56th Human Factors and Ergonomics Society annual meeting*, 383–387.

Persson, M., & Johansson, B. J. E. (2001). Creativity or diversity in command and control. In M. J. Smith, G. Salvendy, D. Harris, & R. J. Koubek (Eds.), *Usability, evaluation and interface design - cognitive engineering, intelligent agents and virtual reality (Proceedings to HCI 2001)* (Vol. 1, pp. 1508–1512). New Orleans.

Prytz, E. G., Rybing, J., Jonson, C.-O., Pettersson, A., Berggren, P., & Johansson, B. J. E. (2015). An exploratory study of a low-level shared awareness measure using mission-critical locations during an emergency exercise. *Proceedings of the 59th Human Factors and Ergonomics Society annual meeting*, 1152–1156.

Resick, C. J., Murase, T., Bedwell, W. L., Sanz, E., Jiménez, M., & DeChurch, L. A. (2010). Mental model metrics and team adaptability: A multi-facet multi-method examination. *Group Dynamics: Theory, Research, and Practice*, *14*(4), 332–349.

Salas, E., & Fiore, S. M. (2004). *Team cognition: Understanding the factors that drive process and performance*. Washington, DC: American Psychological Association.

Salas, E., Elliott, L. R., Schiflett, S. G., & Coovert, M. D. (2004). *Scaled worlds: development, validation and applications*. Aldershot: Ashgate.

Salas, E., Fiore, S. M., & Letsky, M. P. (2012). *Theories of team cognition. Cross-disciplinary perspectives*. New York, NY: Routledge.

Schiflett, S. G., Elliott, L. R., Salas, E., & Coovert, M. D. (2004). *Scaled worlds: Development, validation and applications*. Aldershot: Ashgate.

Smith-Jentsch, K. A., Campbell, G. E., Milanovich, D. M., & Reynolds, A. M. (2001). Measuring teamwork mental models to support training needs assessment, development, and evaluation: two empirical studies. *Journal of Organizational Behavior, 22*, 179–194.

Sundin, C., & Friman, H. (1998). *ROLF 2010 A joint mobile command and control concept*. Stockholm: Elanders Gotab.

Sundin, C., & Friman, H. (2000). *ROLF 2010 The way ahead and the first step: A collection of research papers*. Stockholm: Elanders Gotab.

Svenmarck, P., & Brehmer, B. (1991). D3Fire - An experimental paradigm for the study of distributed decision making. In B. Brehmer (Ed.), *Proceedings of the 3rd MOHAWC Workshop on distributed decision making* (pp. 47–77). Belgirate, Italy: Risö National Laboratory.

Swezey, R. W., & Salas, E. (1992). *Teams: Their training and performance*. Norwood, NJ: Ablex.

Tannenbaum, S. I., Mathieu, J. E., Salas, E., & Cohen, D. J. (2012). On teams: Unifying themes and the way ahead. *Industrial and Organizational Psychology, 5*(1), 56–61.

Trnka, J., & Jenvald, J. (2006). Role-playing exercise – A real-time approach to study collaborative command and control. *International Journal of Intelligent Control and Systems, 11*(4), 218–228.

Trnka, J., & Johansson, B. J. E. (2009). Collaborative command and control practice: Adaptation, self-regulation and supporting behavior. *International Journal of Information Systems for Crisis Response and Management, 1*(2), 47–67.

Wildman, J. L., Salas, E., & Scott, C. P. R. (2014). Measuring cognition in teams a cross-domain review. *Human Factors, 56*(5), 911–941.

3 The Dynamical Systems Approach to Team Cognition

Theories, Models, and Metrics

Terri A. Dunbar[1], Jamie C. Gorman[1], David A. Grimm[1], and Adam Werner[2]
Georgia Institute of Technology[1]
United States Military Academy[2]

CONTENTS

The complexity of the modern workplace has resulted in the more frequent use of teams or teams of teams to fulfill the goals and mission of the organization. Subsequently, it is important for team researchers to understand how they can best support these teams through the careful measurement and assessment of team processes and team effectiveness. To this end, in this chapter we focus on a dynamical systems approach (both theory and methods) to team processes, namely team cognition. Our goal in highlighting this approach is to provide readers with an understanding of how they can apply concepts from dynamical systems to team cognition, identify regularities in system dynamics across different work domains, and incorporate methods from dynamical systems approaches to their own research.

A dynamical system is a system (i.e., a set of interacting components; Bunge, 1979) that changes over time whose behavior is tracked to predict future states of the system (Abraham & Shaw, 1992). Informal definitions of dynamics in team cognition investigate how the construct of interest (e.g., team composition; Mathieu, Tannenbaum, Donsbach, & Alliger, 2014) changes over time but may neglect to consider multiple levels of the system, the constraints within which the team operates, or the type of explanation or mechanism posited. The type of explanation in dynamical systems theories differs from traditional cognitive theories and neuroscience theories in that rather than focusing on static explanations (e.g., knowledge structures), linear explanations (e.g., input-process-output models), or material explanations (e.g., neurons) of cognitive processes, dynamical systems theory focuses on temporal, linear and nonlinear, and nonmaterial explanations, such as changes in processes over time through forces that repel, attract, or stabilize cognitive processes (Juarrero, 1999). These types of explanations are what novices often struggle with when attempting to adopt dynamical systems theories and methods.

Systems approaches to human behavior historically originate in functionalism, an early framework from the late 19th century that sought to understand mental activities in terms of the utility of the behaviors they served (e.g., James, 1890). Functionalism stood in stark contrast to the structuralism approaches at the time, which used introspective techniques to uncover the components of the mind, often ignoring the context within which mental activities were occurring. Many 20th-century psychologists appeared inspired by functionalist approaches, including Vygotsky with activity theory (Vygotsky, 1978) and Gibson with the ecological theory of perception (Gibson, 1979). Rather than focusing on individual elements that constitute mental activities or the static representations of these activities, these functionalism-inspired theories used the person-object/environment system as the primary unit of analysis, a major theoretical transition that is still a part of dynamical systems theories today.

Although functionalism-inspired frameworks lend themselves well towards systems theorizing, historically, the methodologies behind these theories often lacked the sophisticated mathematical techniques that we associate with dynamical systems methods today. The roots of modern dynamical systems methods include mathematical techniques such as those designed to measure the degree of self-organization (i.e., where systems-level order arises from component-level interactions; Ashby, 1947) of chaotic systems (i.e., systems that produce different behaviors depending on the initial conditions of the system; Lorenz, 1963) using chaos theory-inspired methods, such as fractal analysis. Many current proponents of ecological theory also use dynamical systems methods to answer their research questions about perception and action (e.g., Coey, Kallen, Chemero, & Richardson, 2018) because these methods allow one to model the relationships between different parts of a system.

Why study team cognition as a dynamical system? A dynamical systems approach, as we will show in this chapter, offers researchers much in the form of predicting future system states and novel forms of assessment and training. Although this approach differs from what is typically studied in the behavioral sciences, we hope to illustrate that the concepts and techniques from a dynamical systems approach *can* be accessible to a wide range of researchers.

To this end, in this chapter we review concepts from dynamical systems theory, dynamical systems methods used in different work domains, and current/future directions for this approach. First, we provide a background by defining team cognition as a dynamical system and applicable concepts from dynamical systems theory. Second, we highlight four different team domains, focusing on the characteristics of the system constraints in these domains, applicable research findings, and the types of methods applied in this domain. Third, we discuss the most current work in the dynamical systems approach to team cognition: sources of variation underlying team effectiveness and real-time dynamics. Finally, at the end of the chapter, we discuss the limitations and future directions of the approach.

DYNAMICAL SYSTEMS VIEWPOINT OF TEAM COGNITION

What does team cognition mean from a dynamical systems viewpoint? From this perspective, teams *are* dynamical systems. Teams perform within a set of constraints (i.e., conditions that alter the opportunities for thoughts and actions; Juarrero, 1999), such as the goal the team is working towards, the tasks individual team members are performing, and the environmental context the team is performing within (Gorman, 2014). Within this set of constraints, team members interact with one another to work towards their goal. These goals and subgoals may change over time, depending on the constraints of the system and how the team interacts over time. The interactions between team members are important from this perspective, both in the history of interactions and the interactions as they are currently ongoing, because these interactions are what helps the team maintain and monitor their progress towards their overall goal.

Team cognition, then, from a dynamical systems viewpoint, focuses on how the interactions between team members over time are embedded within the constraints of the team's goal and task as well as the individual properties of the team members (e.g., their capabilities, limitations, and roles). This suggests that the areas of measurement for team cognition could include task and goal constraints, team interactions (verbal and nonverbal), team member knowledge and experience, and team member actions (motor behavior) or physiology (autonomic and central). We will see examples of some of these areas of measurement in the upcoming section, "Team Dynamics across Domains."

Although the methods used to measure team dynamics are similar to those from various areas of psychology or human factors, the concepts in dynamical systems differ from what is often encountered in the behavioral sciences. Traditional psychology concepts are often rooted in individual thoughts (e.g., mental models) and behaviors (e.g., individual operator's actions) rather than in processes that change over time. Investigating team cognition from a dynamical systems viewpoint requires shifting from an individualistic perspective to a systems perspective, where concepts are now focused on relationships across levels of measurement (e.g., local variability/global stability), between people (e.g., interactions between components), changes over time (e.g., phase transitions), or in response to outside disturbances (e.g., skilled behavior). To this end, we discuss some major concepts in dynamical systems theory, both the basic terminology and how the concept is applied to teams. The following concepts are important for understanding the rest of the book chapter: variability, stability, attractors, phase transitions, and skill.

Variability and Stability

Variability refers to fluctuations in patterns of behavior over time, whereas stability is its opposite, resistance to change over time. Variability and stability need not be considered a dichotomy; rather, it should be considered a continuum. When the goal of the research is to identify stable differences in behavior or cognition between experimental conditions, variability represents measurement error and stability is the more desired characteristic. However, when the goal of the research is to track cognitive processes that change over time, variability and stability both reveal important information about the characteristics of the system, such as whether the system is undergoing a change in the system state or has settled on a particular system state.

This distinction between variability and stability also depends on the timescale of measurement. For example, we can measure team cognition at different points in time, such as the team's communication during one work session or across the entire duration of time the team has worked together. At a small scale, the variations in the team's communication may appear to be highly variable. However, on a longer timescale, the variations appear less random and more stable. This overall stability occurs because the team members' communication is in service to the overall team goal, keeping the team on track to reach their goal. This phenomenon is called local variability with global stability, where behavior at the local level appears highly variable or random, but the behavior is stable and less variable at a global level (Gorman, Dunbar, Grimm, & Gipson, 2017).

Metastability refers to a stable state of the system other than the system's state of least energy (Treffner & Kelso, 1999). In a metastable state, stability is maintained through constant actions taken against the energy being pushed into the system. As a result, the output of the system in a metastable state is unpredictable. A small amount of energy might not change the system state, but a large amount of energy could push the system into a new state. For example, imagine a ball resting at the top of a hill with a slope down to either side of it. This location represents a metastable state. If the ball is strongly pushed towards either direction (i.e., energy is supplied to the system), then the ball will leave the metastable state towards a state of least energy, the bottom of the hill. If the ball is slightly nudged, then the ball may stay in the metastable state or move towards the state of least energy depending on the strength of the nudge, the steepness of the slopes, and the size of the ledge that the ball is resting on. The team's overall dynamics can be thought of as a metastable state, where communication serves to push the team into different system states (Gorman, Amazeen, & Cooke, 2010a).

Attractors and Repellers

An attractor is a system state that the system will evolve to regardless of the initial state of the system (Abraham & Shaw, 1992). A repeller, on the other hand, is the opposite—a system state that the system will not approach. Attractors and repellers are either inherently stable (outside disturbances, or perturbations, have little effect), unstable (perturbations have a large effect), or metastable (stability is being actively maintained). The influence of an attractor or repeller is evident from the time course of the system. From the initial system state, the system undergoes a transition period where the system's behavior fluctuates erratically until it either settles on a behavioral state (i.e., attractor), or settles away from a behavioral state (i.e., repeller). When predicting the future state of the system, a dynamical system will continue to gravitate towards attractors and away from repellers. In teams, attractors have been studied in motor coordination, where the primary interest was attractor formation under certain task constraints (e.g., Gorman & Crites, 2015).

Phase Transition

Phase transitions are shifts in the system from one system state to another (Haken, 1983). Phase transitions represent qualitative changes in the system state, such as the transition from a liquid to a gas. Phase transitions can be identified in the system trajectory through increased variability in the overall system's behavior. Changing the control parameter, a variable that when changed leads to a change in overall system behavior (Thelen, Ulrich, & Wolff, 1991), can induce phase transitions in the system. For example, phase transitions between different forms of infants' stepping behavior can be induced by changing the task context, such as stepping under water or on a treadmill, and detected through changes in the variability of the infants' stepping behavior itself (Thelen et al., 1991).

Skill

High-performance skill is often equated to the amount of training hours the individual or team has completed, the difficulty of acquiring expertise, and the qualitative differences between expert and novice performance (Schneider, 1985). A dynamical systems view of skill focuses less on the overall performance or expertise of the individual or team as the locus of skill and more on the pattern of behaviors the individual or team engages in and the actions they take towards outside disturbances (i.e., perturbations). The three dynamical systems characteristics of highly skilled behavior are flexible, adaptive, and resilient (Thelen & Smith, 1994). To be flexible, a team must have a repertoire of behaviors that considers the varying task, team, and environmental constraints. To be adaptive, a team must rapidly modify their behaviors in response to perturbations. To be resilient, a team must recover quickly from perturbations. In teams, skilled behavior has been examined in response to perturbations (e.g., Gorman, Cooke, & Amazeen, 2010b).

TEAM DYNAMICS ACROSS DOMAINS

Teams operate differently depending on their domain. Some teams may operate within very specific task constraints yet work flexibly across different environments. Other teams are constrained to work towards a specific goal but can perform their task in whatever way helps them reach their goal. Despite these differences, the dynamical principles underlying team cognition are similar across different work domains. In this section of the chapter, we provide an overview of the characteristics of and constraints within four work domains, highlighting the research findings and methods used to assess team dynamics in that domain. The four domains we highlight are command and control, collaborative problem solving, healthcare, and human-autonomy teaming. Table 3.1 summarizes the main constraints, findings, and methods for each domain.

TABLE 3.1
Summary of Team Dynamics across Domains

Domain	Constraints	Findings	Methods
C2	Military rules & regulations Ranks Mission goals	Experience affects novel task performance. Team members need to know who knows what information. Mixing up team composition can result in more skilled behavior.	EAST CAST
CPS	Expertise in problem type Overall goal	Macrocognition theory fits CPS team behavior. Distinct qualitative problem solving phases can be captured as phase transitions.	Lag sequential analysis Entropy Cross-wavelet coherence
Healthcare	Changing team member roles Changes in workload Emergent tasks	Events occurring during simulation are observable in the neurodynamic and communication patterns of the team.	Neurodynamic entropy Discrete recurrence analysis
HAT	Capabilities of autonomy Autonomy's interaction patterns	All-human teams and HATs perform equally well but differ in how they coordinate.	Perturbations System reorganization Entropy

Command and Control

Command and control (C2) teams are directed by a commander who executes specific orders to the team to fulfill an overall mission goal (Department of the Army, 2012). Though prevalent in other domains, C2 teams are largely associated with the military, paramilitary, and emergency response teams. C2 teams operate within very specific constraints (e.g., military rules and regulations, ranks of the commander and team members, or mission goals) with highly trained individuals under highly regimented interaction patterns. Each aspect of the system is critical to the success of the C2 team.

C2 teams can be centralized to one environment or decentralized and distributed throughout an environment or operating area. Decentralized command and control (DC2) teams are becoming more prevalent given advances in technology and the military's desire for increased mobility in combat (Heininger, 2016). Rather than being centrally located, DC2 teams flexibly and heterogeneously distribute resources across multiple environments that are tactically relevant to the overall mission (Gorman, Cooke, & Winner, 2006b). DC2 teams typically involve multiple dozens of people but could be hundreds, if not thousands, of individuals scattered throughout an operating environment. No one individual or operator is necessarily responsible for being aware of the entire DC2 environment, but instead all are responsible for their own local environment (Gorman et al., 2006b). Compared to the C2 system constraints, DC2 system constraints incorporate an additional layer of interaction complexity due to the number of people working in the system; however, the DC2 system has much greater flexibility in their environmental constraints by comparison.

Research in Command and Control

One important aspect of the C2 system is that the experience of the individual team members is vitally important to novel tasks. C2 teams who are experienced working together perform better on a novel task than teams with no experience working together (Cooke, Gorman, Duran, & Taylor, 2007). From a dynamical systems perspective, this novel task performance benefit is due to the experience team members get by constantly coordinating with one another. New interaction patterns emerge over time through repetitive coordination in the C2 problem space (Cooke, Gorman, Myers, & Duran, 2013).

C2 and DC2 systems do not need to be aware of what each team member is doing during task performance; rather, they should coordinate flexibly as needed to maintain a dynamic and collaborative situation awareness. Distributed situation awareness models (e.g., Stanton, Stewart, Harris, et al., 2006) suggest that teams should not share awareness because each team member may have a different purpose and, because of these differences, over-sharing could potentially be confusing or overwhelming to the individual team member. This implication is particularly important for DC2 teams, which can involve thousands of individuals scattered across the globe, where shared awareness is simply not feasible. Stanton and colleagues (2006) suggest instead that individuals should be aware of who has what knowledge and how that knowledge is useful to them rather than knowing everything about the situation.

Another unique finding relevant to C2 systems is that changes in team composition and retention interval can lead to improvements in the C2 system's performance (Gorman, Cooke, Pederson, et al., 2006a). Gorman et al. (2006a) examined the effects of different team compositions (intact vs. mixed) across different retention intervals (short vs. long). Mixed C2 teams exhibited more skilled behavior compared to intact C2 teams when given a longer retention interval between missions. The findings also imply that team mixing in C2 systems over longer periods of time can lead to more adaptable, better performing C2 teams. Gorman et al. (2006a) believe that this occurs because the change in teammates causes a perturbation in the team's coordination, ultimately leading to more adaptability and resilience than in intact teams.

Methods in Command and Control

Most methods for assessing C2 teams are not *dynamical* systems methods. However, systems methods such as event analysis of systematic teamwork (EAST) methodology (Stanton et al., 2006; Walker, Gibson, Stanton, et al., 2006) still approach C2 teams as systems rather than individual team members. The EAST methodology uses data from hierarchical task analyses, observations of task performance, and debriefing interviews to produce three system representations: social network (i.e., who communicates to who), task network (i.e., goals of different team members), and knowledge network (i.e., relationship between the different types of information needed for successful task performance). These networks can ultimately be used to improve the design of systems that manage these networks.

Dynamical systems methods for assessing C2 teams directly measure team member interactions in response to a changing operational environment. A process-based measure, coordinated awareness of situations by teams (CAST; Gorman et al., 2006b), assesses how team members interact to coordinate in nonroutine situations. CAST addresses unanticipated events by using roadblocks (i.e., nonroutine situations embedded in the task) to perturb the steady state of a C2 team. Teams must coordinate their perceptions and actions by interacting to overcome the roadblock. A highly skilled C2 team will successfully coordinate around these roadblocks.

Collaborative Problem-Solving

Collaborative problem solving (CPS) involves mutual coordination across people to solve a problem (Roschelle & Teasley, 1995). Although CPS is not a domain per se, CPS is often studied within particular domains where teams utilize CPS as part of their work, such as firefighting and student teams. Considering the dynamics of the CPS system, CPS teams operate within a few constraints, such as the experience of the individual team members with the problem and the problem space itself, which shape the interactions the CPS teams must take towards their overall goal of solving the problem or problem set. One theoretical approach to CPS that captures system dynamics is the macrocognition in teams theory.

Macrocognition in teams aims to determine how teams build knowledge structures while engaging in CPS and how teams accomplish their goals through coordination

(Fiore, Smith-Jentsch, Salas, Warner, & Letsky, 2010). Macrocognition is the "process of transforming internalized knowledge into externalized team knowledge through individual and team knowledge building processes" (Fiore et al., 2010. p. 204–205). Macrocognition in teams consists of five components: Individual and team knowledge building, internalized and externalized knowledge, and team problem-solving outcomes (Fiore et al., 2010). During the individual and team knowledge-building phases, individual team members process their problem space and disseminate this knowledge to their team, who translate the knowledge into actions that the team can take towards accomplishing their goal. Internalized knowledge refers to individual team member knowledge, whereas externalized knowledge refers to the relationships between knowledge and team-level knowledge. Team interactions influence team problem-solving outcomes and contribute to whether the problem is solved.

Research in Collaborative Problem Solving

Multiple studies provide support for the concept of macrocognition in teams during CPS. Hutchins and Kendall (2010) examined communication from experienced teams executing tasks such as from firefighters on 9/11. The researchers found that communication for firefighters fit within the five components of macrocognition in teams. Seeber, Maier, and Weber (2013) found similar results when examining distributed teams using collaborative software as a communication source. The distributed teams also generally followed the five components for macrocognition in teams.

Many theories in CPS, including macrocognition in teams, hypothesize that there are distinct qualitative phases that occur during successful problem solving. These qualitative phases are often measured through the team's communication. Earlier work investigating the temporality of these phases focused on how the structure of the problem impacted the immediate transitions (identified through lag-sequential analysis) between different types of problem-solving communication (Kapur, 2011). More specifically, student teams working on ill-structured problems exhibited much more complexity in their temporal patterns, shifting often between different problem-solving phases, whereas teams working on well-structured problems tended to immediately transition to the same type of problem-solving phase. Recently, Wiltshire, Butner, and Fiore (2018) discovered evidence of phase transitions between problem-solving phases. In this study, the researchers integrated dynamical systems theory with existing CPS theory by tracking the variability of different types of coded communication associated with different problem-solving phases over time. The researchers found increased variability in communication codes prior to phase transitions into new problem-solving phases. Ricca, Bowers, and Jordan (2019) applied the same analytic techniques to the communication from teams of fifth-grade students engaged in a robotics engineering design project and also found evidence of phase transitions between problem-solving phases.

Methods in Collaborative Problem Solving

CPS team dynamics are primarily studied through verbal communications (e.g., Wilshire et al., 2018), although researchers have also examined movement coordination during CPS (e.g., Wiltshire, Steffensen, & Fiore, 2019). Initial attempts investigating the temporality of CPS team dynamics used an analytic technique called lag sequential analysis (LSA), which analyzes a sequence for cross-dependencies (Kapur, 2011). More specifically, LSA compares the expected transition probabilities between observations to the actual transition probabilities, identifying the statistically significant transitions within the dataset (Bakeman & Gottman, 1997). Other techniques investigating temporal dynamics in CPS teams include identifying phase transitions in the entropy of coded team communication (Wiltshire, et al., 2018) and movement coordination across multiple timescales (Wiltshire et al., 2019).

In the CPS study identifying phase transitions in coded communication, the researchers applied a sliding window entropy technique to team communication (Wiltshire et al., 2018), where entropy is measured across a window of time (e.g., 200 seconds), shifted up one unit of time (e.g., 1 second), recalculated over the new window of time, and so on across the entire communication time series to create an entropy time series. Here, entropy refers to the level of disorder in a signal, where high entropy is highly disordered and low entropy is highly predictable (Shannon & Weaver, 1949). There was high variability in the communication entropy prior to changes in the CPS phase, indicating a phase transition occurred.

The CPS research investigating movement coordination across timescales used cross-wavelet coherence (Wiltshire et al., 2018), which measures the coherence (similar to a cross-correlation) and relative phase (i.e., the difference between the oscillations of two or more time series) between two time series across multiple frequencies and timescales (Issartel, Marin, Gaillot, Bardainne, & Cadopi, 2006). By using cross-wavelet coherence rather than only relative phase, the researchers were able to measure coordination across multiple time scales rather than a single timescale. Because this is a new area in CPS, further research is necessary to determine how well other dynamical systems concepts apply in this domain.

Healthcare

As in other domains, simulation has come to play a significant role in the training and practice of medical professionals (Aebersold & Tschannen, 2013). Indeed, a longitudinal study of nursing students (Hayden, Smiley, Alexander, et al., 2014) found that student nurses receiving traditional training (no more than 10% clinical hours in simulation) and student nurses that received enhanced simulation training (50% clinical hours in simulation) had similar levels of clinical competency, nursing knowledge, and post-training clinical competency. Moreover, reductions in adverse patient outcomes in community hospitals have been linked to simulation training (Riley, Davis, Miller, et al., 2011). Simulation training for healthcare teams often focuses on developing effective communication patterns with the ultimate goal of transferring these skills to real-world task performance. This training goal is particularly pertinent considering teamwork errors related to communication difficulties have been cited as one of the most prevalent factors contributing to medical mishaps (Sutcliffe, Lewton, & Rosenthal, 2004).

Teamwork measurement in the healthcare domain has traditionally centered on observational methods and metrics (e.g., TeamStepps). However, research on team dynamics in medical simulations has also begun to reveal how shifts in neurophysiological (Pappada, Papadimos, Lipps, et al., 2016) and communication (Gorman, Grimm, Stevens, et al., 2019) patterns can be used to identify increases in workload, emergent tasks (perturbations), and changing team member roles that can interfere with medical care. It has been argued that these dynamical systems methods for understanding shifts in teamwork in the medical domain have the potential for real-time monitoring and prediction of team interaction patterns associated with medical errors (Gorman et al., 2019).

Research in Healthcare

Stevens and colleagues (Stevens, Galloway, Halpin, & Willemsen-Dunlap, 2016b) have observed that prior to and during particularly challenging simulation training events, such as a patient seizing during surgery, the neural (EEG) dynamics of neurosurgical teams (described in the next section) enters a more disorganized state ("high entropy") compared to nominal task conditions ("low entropy"). Furthermore, they have observed that more experienced teams demonstrate this effect to a greater degree than less experienced teams (Stevens, Galloway, Gorman, et al., 2016a), which suggests that neurodynamic patterns might be more easily tied to adverse events as teams gain experience working together. Stevens and colleagues have supported the external validity of their neurodynamic research by observing similar neurodynamic patterns in high school student teams (Stevens, Galloway, Berka, & Sprang, 2009) and submarine crews (Stevens, Gorman, Amazeen, et al., 2013). More recently, Stevens, Galloway, and Dunlap (2019) have validated their neurodynamic methods developed using simulations in a live operating room (OR).

Research has begun to tie these team neurodynamics to the behavioral and speech dynamics exhibited by medical teams during challenging simulation training events. Gorman and colleagues (2019) were able to automatically detect transitions in team communication patterns that corresponded to unexpected perturbations that interfere with patient care, including fire in the OR, changes in team composition, and handoffs, during medical training simulations. In some cases these speech dynamics were correlated with neural dynamics (Willemsen-Dunlap, Halpin, Stevens, et al., 2017). However, the more general finding is that speech and neural dynamics provide complementary real-time information streams that are diagnostic of anomalies and adverse events in the OR. For example, Gorman et al. (2019) describe a medical simulation in which the communication dynamics identified a change in team membership at the start of the surgery that neurodynamics did not, but that neurodynamics identified a patient seizing event during surgery (Stevens, Galloway, Willemsen-Dunlap, et al., 2018) that communication dynamics did not. These researchers concluded that highly skilled teams engage in a continuous stream of explicit and implicit team coordination (Entin & Serfaty, 1999) in which communication dynamics address the former and neurodynamics address the latter aspect of team coordination. Currently, these methods focus on identifying team coordination anomalies related to unexpected turns in the simulation scenario and how the team reacted to them. However, these methods may also be useful for revealing natural transitions (e.g., the transition from planning to task performance; Gorman et al., 2019) during which confusion and miscommunication are also prevalent.

Taken together, these studies have revealed objective metrics of team coordination dynamics that have the potential to provide real-time feedback for medical team simulation training and potentially in the live OR. However, further research using live OR teams as well as medical simulations is needed to elucidate how feedback should be provided to enhance team coordination dynamics and learning in medical teams in real time.

Methods in Healthcare

Much of the research we have described uses a method called team neurodynamics. Neurodynamic entropy maps millisecond level recordings of electrical activity at team members' scalps using EEG onto information-based metrics of team synchrony (Stevens & Galloway, 2017). Essentially, high, medium, and low activation at the various EEG sensor sites are recorded over time for each team member, and these patterns of activation across team members constitute a symbolic set of team-member activation distributions (e.g., Team Member A – high activation/Team Member B – low activation is a different distribution than Team Member A – high/Team Member B – high). Fluctuations in these team-level symbolic distributions are analyzed over time by tracking their entropy, where high entropy corresponds to changing team distributions and low entropy corresponds to stable team distributions. Other metrics based on this method allow researchers either to model neural synchrony at the team level or to examine shared and mutual information between team member's neural patterns (Stevens et al., 2018). The resulting suite of measurements are called team neurodynamics.

The communication research cited in the healthcare domain utilized a method called discrete recurrence analysis (Gorman, Cooke, Amazeen, & Fouse, 2012a). Utilizing this method, a researcher can take any time or event series (such as a time series of speaker turn-taking events) and compute measures such as %DET (Webber & Marwan, 2015), which quantify the amount of organization in speech patterns (e.g., turn-taking patterns). The %DET measure (organization in turn-taking patterns) can be recomputed using a windowing procedure as new data come in (Gorman et al., 2019). This results in a %DET (communication organization) time series. Anomalies in this time series correspond to unusual shifts in communication patterns that can be detected in real time. This is the method Gorman et al. (2019) used to detect significant transitions in speech patterns corresponding to nonroutine medical crises in the simulated OR.

Human–Autonomy Teaming

Human-autonomy teaming (HAT) involves the coordination of human team members with autonomous, or synthetic, team members (Schulte, Donath, & Lange, 2016). Autonomy differs from automation in that autonomous systems perform entirely independently from the human operator, displaying some degree of intelligent behavior, whereas in automated systems, the system will simply perform the actions it was programmed to perform (Endsley, 2015). In HAT systems, team members are constrained not just by the individual team member capabilities, the team task, and the overall mission goal but also by the unique capabilities and interaction patterns associated with the autonomous agent.

Research in Human–Autonomy Teaming

Researchers have been investigating the dynamics of HAT in the context of remote piloted vehicle operations when there are two human team members (navigator and photographer) teamed with autonomy (pilot; Myers, Ball, Cooke, et al., 2017; Demir, Cooke, & Amazeen, 2018; Demir, McNeese, & Cooke, 2018; McNeese, Demir, Cooke, & Myers, 2018) as opposed to all-human teams. Several early experiments found that all-human and human-autonomy teams perform equally well but differ in their coordination dynamics because autonomy may not understand what it means to be a good teammate (e.g., pushing and pulling of information; back-up behaviors). In the case of HAT, the coordination dynamics rely on mutual adjustments of the human team members to facilitate human-autonomy coordination (e.g., McNeese et al., 2018).

Similar to CPS, HATs apply to a variety of domains. For instance, many of the same principles underlying HATs discussed here are present in other applications such as self-driving autonomous vehicles (Campbell, Egerstedt, How, & Murray, 2010; Lugano, 2017), urban search and rescue (Krujiff, Janíček, Keshavdas, et al., 2014), and other military applications beyond the remotely piloted vehicle operations, including unmanned ground vehicles and robotic teleoperations (Chen, Durlach, Sloan, & Bowens, 2008). Regardless of the domain, many of the potential problems that arise during HATs include brittleness (i.e., automation is unable to perform outside of designed parameters), lack of transparency (i.e., automation does not display what it is doing to the human), miscalibrated trust (i.e., human has an inappropriate level of trust in the automation), and lack of shared awareness (i.e., automation does not display the information it is using to perform the task; Shively, Lachter, Brandt, et al., 2017). Considering that these common factors arise across a variety of HAT applications, we propose that a dynamical systems approach to HATs could benefit all of these applications through the methods described here.

More recently, human–autonomy team dynamics have been viewed from the perspective of human-systems integration by investigating coordination dynamics across operators (human and autonomy) and across vehicle and control parameters of the broader remote piloted vehicle systems (Grimm, Demir, Gorman, & Cooke, 2018). This research builds on the idea that to perform effectively, the system must exhibit skilled behavior by being flexible, adaptive, and resilient. The system accomplishes this through continuously reorganizing across operator, vehicle, and control layers in response to changing task demands (i.e., destabilization due to perturbations) and/or changing system configurations (i.e., a fundamental change in the coordination task; e.g., all-human vs. human–autonomy system configurations).

Methods in Human–Autonomy Teaming

Current methods in HAT investigate skilled behavior in the system by perturbing the team and determining how the team coordinates in response to the perturbation. One way to measure the system's response to perturbations is through the length of time it takes for the system to organize, called time to system reorganization, relaxation time, or reorganization time. Time to system reorganization is calculated as the first significant peak from the time of the perturbation onset in a windowed entropy time series, described previously in the "Methods in Collaborative Problem Solving" section. Time to system reorganization can be used directly as a measure of team performance or as a predictor of system effectiveness and efficiency. For example, prior research correlated time to system reorganization with system effectiveness (i.e., a weighted composite performance score of various system parameters) and found that the configuration of the system impacted the relationship between these two variables (Gorman, Demir, Cooke, & Grimm, 2019).

CURRENT ISSUES IN TEAM DYNAMICS

Current work in the dynamical systems approach to team cognition is addressing some of the challenges from critics of this approach. For instance, dynamical systems methods have been criticized for its focus on descriptive methods such as curve fitting (Rosenbaum, 1998). Current research is now focusing on the explanatory power of dynamical systems methods, such as identifying the sources of team effectiveness. Another area of current research is on the real-time dynamics of behaviors as the team task unfolds, as this will lend itself towards practical and useful applications for real world environments.

Description vs. Explanation

Descriptive methods for team cognition aim to describe what team cognition is. In dynamical systems terms, this would be describing the dynamics of team cognition in different domains or under different constraints. Much of this research has focused on curve fitting the data to determine the underlying dynamics of the team's coordination, such as fitting communication data to a power-law distribution or an exponential distribution (e.g., Dunbar & Gorman, 2014). However, the classical curve fitting method does not identify sources of variation (e.g., specific team member behaviors) that underlie team effectiveness, which is a focus of current work.

Current work on the dynamics of team communication focuses on identifying sources of team effectiveness by using a method of post hoc filtering to identify which team members contribute significantly to team communication during perturbations. The logic behind the filtering method is that we can identify which team members contributed to the team's communication dynamics by removing their inputs post hoc and identifying how this impacts the overall dynamics or reorganizational behavior of the team. This process involves creating an initial time series consisting of a dynamical system measure such as %DET (Gorman, Hessler, Amazeen, Cooke, & Shope, 2012b) as derived from communication flow, or which speaker was speaking at a certain time throughout the time series (Grimm, Gorman, Stevens, et al., 2017). Each speaker is identified in this time series with a unique numerical identifier. It is critical that there are no speaker overlaps during this initial time series. Then a nonlinear prediction algorithm as developed by Kantz & Schreiber (1997) is used to generate another time series which measures communication reorganization, where large values indicate a high degree of anomalous team behavior and a large amount of reorganization at the team communication level. The final step before applying the filtering method would be to use average mutual information (i.e., the measure of information contained in variable *Y* about variable *X*; AMI; Abarbanel, 1996; Cover & Thomas, 2006) to generate another time series for each team member and help identify which team members may be driving team communication during perturbations.

After using this strategy to identify potential team members, team members are filtered by replacing their unique identifier in the original time series with a null value (0). Then the analysis is repeated with the hypothesized influential team member filtered out. If the overall activity of the team, as measured in the initial time series with the dynamical system measure, falls below the significance, then one can infer that the respective team member was a significant driving factor behind the communication reorganization. This finding may indicate leadership emergence because the significance of the team process dropped after the removal of the filtered team member. This process can be repeated for each team member during the filtering process. Consequently, this is strictly a post hoc filtering procedure currently. Future research could be carried out to the possibility of filtering such team members in true real-time settings.

Real-Time Dynamics

Real-time dynamics allow researchers to understand how effective teams behave at the system level. Using concepts and measurements from dynamical systems theory, methods have recently been developed that identify real-time sources of variation during team coordination. For example, in medical simulation training, entropy has been used to track a team's communication to better inform what areas of feedback instructors should target for their trainees following the training session (Grimm et al., 2017). Currently, the main benefit from the real-time method lies in team training and evaluation. However, this method could be used in other domains, for example, to identify how quickly the team responds to emergencies and how effectively the team responds to the emergency. Although many of these types of evaluations are conducted post hoc, real-time dynamics could be utilized as a method to help detect emergencies in real-life situations.

For more applied purposes, real-time analysis of team communication may allow teams to react faster to catastrophic errors by identifying anomalous team coordination patterns. For example, breakdowns in team communication are at least partially responsible for delayed responses to Hurricane Katrina (Leonard & Howitt, 2006). Real-time dynamics could help detect errors from incidents such as Hurricane Katrina much sooner than the methods that are currently available; however, software would need to be developed to facilitate real-time analysis, particularly real-time analysis of overt team member communication. In order to create this software, there are practical issues that would need to be solved first, for example, developing a method of handling several people communicating and talking at the same time. As such, software that identifies group members' communication solely by their voices may not be enough. However, technologies could be used to record vocal patterns and vibrations, but these types of technologies would need to be unobtrusive. Another alternative to overt team communication would be a covert method, such as measuring the neurophysiology of the team. Past research has used neurophysiological measures to analyze team behaviors (e.g., Stevens et al., 2016b), which could be extended to real-time analysis.

Another useful application of real-time dynamics lies in simulation training. Simulation training has been used in both healthcare (McGaghie, Issenberg, Petrusa & Scalese, 2010) and military settings (Salas, Priest, Wilson, & Burke, 2006). Real-time methods could be applied to provide a portrait of overall team coordination across the duration of a mission or during the simulated surgery. If any crises occurred during the simulations, these measures could identify any useful or harmful team-level behaviors. This information could then be used to modify the team's behavior during the simulation training.

LIMITATIONS AND FUTURE DIRECTIONS

One of the major limitations of the dynamical systems approach to team cognition is that dynamical systems theory is not a conventional mode of explanation in experimental psychology, human factors, and allied disciplines. Dynamics does not offer traditional modes of explanation, such as material explanations (e.g., neurons), but rather focuses on unseen forces that repel, attract, and stabilize cognitive processes. Hence, explanations in terms of attractors or phase transitions are less tangible than hypothesizing team knowledge structures and/or mental simulation as the fundamental mechanism of team cognition. It is the case, however, that these latter constructs operate under the constraints of dynamic interactions, and in the future, we need to do a better job educating psychologists and the public at large on how these intermediate representations are subject to the principles of dynamical organization if we are to fully explore the impact of dynamical systems approaches to team cognition.

A related limitation of the dynamical systems approach involves the concern of process vs. mechanism and what dynamical systems really explain about team cognition. Whereas systems are coordinated through observable/repeatable processes, underlying (hypothesized) mechanisms must cause those processes. The issue for the dynamical systems approach to team cognition is that if there is no theory reduction (i.e., no fundamental substance, e.g., explanations in terms of neural pathways) beyond the temporal dynamics, then what are the mechanisms? Put differently, dynamical systems theories in psychology have been criticized for drawing conclusions about psychological mechanisms simply because they carry a particular dynamical signature, such that no fundamental mechanism is posited that produces the signature (Rosenbaum, 1998). We believe, however, that this entails a confusion about what we mean by mechanism. If a mechanism must be something material like a neural pathway, then dynamical systems explanations will always be unsatisfying. However, if we move beyond materialistic modes of explanation toward nonmaterial dynamic forces as mechanisms (Juarrero, 1999), then several of the dynamical principles mentioned in this chapter might foot the bill as dynamical mechanisms of team cognition. For example, dynamical mechanisms include attractor formation and alteration, synchronization, and destabilization and adaptation of thought and behavior (Peng, Havlin, Hausdorff, et al., 1995). Are these mechanisms any less real than that neural pathways and mental representations? We owe it to the dynamical systems theory of team cognition to pursue this question.

Finally, we want to acknowledge that because the dynamical systems theory of team cognition is relatively new, there is limited evidence for its application to enhancing team effectiveness compared to more traditional shared cognition theories. In the past, training methods have been developed based on perturbation/adaptation dynamics, which was demonstrated to be as or more effective than traditional methods for training adaptive teams (e.g., Gorman et al., 2010b). Initial work of real-time perturbation detection through team communication analysis also appeared promising (e.g., Gorman et al., 2012b). However, future work should attempt to extend the practical implications of the dynamical systems approach to team cognition. As mentioned previously, we are developing real-time analysis and feedback for guiding instructor feedback to trainees following medical simulation training (e.g., Grimm et al., 2017). Ideally, this type of training tool would allow the instructor to guide trainees by incorporating feedback based on emergent team dynamics under crisis conditions. In the future, we anticipate that incorporating systems thinking into team training will benefit learners by instructing them on how team dynamics emerge and the conditions under which those dynamics can be controlled. As we have noted previously (Gorman et al., 2017), we picture this as training metacognitive processes for team dynamics or, perhaps, systems thinking from the perspective of an element within the system.

CONCLUSION

Although dynamical systems approaches to human behavior have proliferated in fields such as ecological psychology, developmental psychology, information systems, and others, team cognition researchers have been slow to adopt this approach despite the need for understanding teamwork in complex systems. The dynamical systems theory to team cognition shifts the focus to a systems-level of analysis. This means that dynamics researchers are changing their measurement from exclusively focusing on the thoughts and behaviors of individual team members to also including the coordination that occurs between team members over time within the constraints of their goal and task environment. Accordingly, the locus of explanation in this type of research has shifted from a material level to a more intangible level, perhaps making the theory and methods in this approach seem less approachable to novices.

The current state of dynamical systems research in teams has shifted from descriptive methods, describing how different concepts and methods from nonlinear dynamics also apply to human behavior, to explanatory methods that identify the sources of team effectiveness. The goal is now to develop software that can identify these sources (e.g., through machine learning algorithms trained on related training sets of data), analyze them in real time, and produce feedback on performance in a way that is accessible to a novice to dynamical systems theory. Although many research questions remain, this approach has successfully developed new conceptualizations of team cognition and performance and new methods of assessing team coordination. The future of this approach lies in its ability to assess and predict successful team performance in real time across a wide variety of domains and data sources.

REFERENCES

Abarbanel, H. (1996). *Analysis of observed chaotic data.* New York, NY: Springer

Abraham, R., & Shaw, C. D. (1992). *Dynamics: The geometry of behavior.* Boston, MA: Addison-Wesley.

Aebersold, M. & Tschannen, D. (2013). Simulation in nursing practice: The impact on patient care. *The Online Journal of Issues in Nursing, 18,* Manuscript 6.

Ashby, W. (1947). Principles of the self-organizing dynamic system. *Journal of General Psychology, 37,* 125–128.

Bakeman, R., & Gottman, J. M. (1997). *Observing interaction: An introduction to sequential analysis.* New York: Cambridge University Press.

Bunge, M. (1979). *Ontology II: A world of systems.* Netherlands: Springer.

Campbell, M., Egerstedt, M., How, J. P., & Murray, R. M. (2010). Autonomous driving in urban environments: approaches, lessons and challenges. *Philosophical Transactions of the Royal Society A: Mathematical, Physical and Engineering Sciences, 368*(1928), 4649–4672.

Chen, J. Y., Durlach, P. J., Sloan, J. A., & Bowens, L. D. (2008). Human–robot interaction in the context of simulated route reconnaissance missions. *Military Psychology, 20*(3), 135–149.

Coey, C. A., Kallen, R. W., Chemero, A., & Richardson, M. J. (2018). Exploring complexity matching and asynchrony dynamics in synchronized and syncopated task performances. *Human Movement Science, 62,* 81–104.

Cooke, N. J., Gorman, J. C., Duran, J. L., & Taylor, A. R. (2007). Team cognition in experienced command-and-control teams. *Journal of Experimental Psychology: Applied, 13*(3), 146.

Cooke, N. J., Gorman, J. C., Myers, C. W., & Duran, J. L. (2013). Interactive team cognition. *Cognitive Science, 37*(2), 255–285.Cover, T. M., & Thomas, J. A. (2006). *Elements of information theory* (2nd ed.). Hoboken, NJ: John Wiley.

Department of the Army. (2012). *Mission Command* (ADP 6-0). Washington, DC: Headquarters.

Demir, M., Cooke, N. J., & Amazeen, P. G. (2018). A conceptual model of team dynamical behaviors and performance in human-autonomy teaming. *Cognitive Systems Research, 52,* 497–507.

Demir, M., McNeese, N. J., & Cooke, N. J. (2018). The impact of perceived autonomous agents on dynamic team behaviors. *IEEE Transactions on Emerging Topics in Computational Intelligence, 2*(4), 258–267.

Dunbar, T. A. & Gorman, J. C. (2014). Fractal effects of task constraints in the self-organization of team communication. Talk presented to the *Human Factors and Ergonomics Society 58th Annual Meeting.*

Endsley, M. R. (2015). *Autonomous Horizons: System Autonomy in the Air Force - A Path to the Future* (Autonomous Horizons No. AF/ST TR 15-01). Washington DC: Department of the Air Force Headquarters of the Air Force.

Entin, E. E., & Serfaty, D. (1999). Adaptive team coordination. *Human Factors, 41*(2), 312–325.

Fiore, S. M., Smith-Jentsch, K. A., Salas, E., Warner, N., & Letsky, M. (2010). Towards an understanding of macrocognition in teams: developing and defining complex collaborative processes and products. *Theoretical Issues in Ergonomics Science, 11*(4), 250–271.

Gibson, J. (1979). *The ecological approach to visual perception.* New York, NY: Psychology Press.

Gorman, J. C. (2014). Team coordination and dynamics: Two central issues. *Current Directions in Psychological Science, 23,* 355–360.

Gorman, J. C., Amazeen, P. G., and Cooke, N. J. (2010a). Team coordination dynamics. *Nonlinear Dynamics, Psychology, and Life Sciences, 14*(3), 265–289.

Gorman, J. C., Cooke, N. J., & Amazeen, P. G. (2010b). Training adaptive teams. *Human Factors, 52*(2), 295–307.

Gorman, J. C., Cooke, N. J., Amazeen, P. G., & Fouse, S. (2012a). Measuring patterns in team interaction sequences using a discrete recurrence approach. *Human Factors, 54,* 503–517.

Gorman, J. C., Cooke, N. J., Pedersen, H. K., Winner, J., Andrews, D., & Amazeen, P. G. (2006a). Changes in team composition after a break: Building adaptive command-and-control teams. *Proceedings of the Human Factors and Ergonomics Society Annual Meeting, 50*(3), 487–491.

Gorman, J. C., Cooke, N. J., & Winner, J. L. (2006b). Measuring team situation awareness in decentralized command and control environments. *Ergonomics, 49*(12–13), 1312–1325.

Gorman, J. C., & Crites, M. J. (2015). Learning to tie well with others: Bimanual versus intermanual performance of a highly practised skill. *Ergonomics, 58*(5), 680–697.

Gorman, J. C., Demir, M., Cooke, N. J., & Grimm, D. A. (2019). Evaluating sociotechnical dynamics in a simulated remotely-piloted aircraft system: A layered dynamics approach. *Ergonomics, 62*(5), 629–643.

Gorman, J. C., Dunbar, T. A., Grimm, D., & Gipson, C. L. (2017). Understanding and modeling teams as dynamical systems. *Frontiers in Psychology.* https://doi.org/10.3389/fpsyg.2017.01053

Gorman, J. C., Grimm, D. A., Stevens, R. H., Galloway, T., Willemsen-Dunlap, A. M., & Halpin, D. J. (2019). Measuring real-time team cognition during team training. *Human Factors,* https://doi.org/10.1177/0018720819852791.Gorman, J. C., Hessler, E. E., Amazeen, P. G., Cooke, N. J., & Shope, S. M. (2012b). Dynamical analysis in real time: detecting perturbations to team communication. *Ergonomics, 55*(8), 825–839.

Grimm, D., Demir, M., Gorman, J. C., & Cooke, N. J. (2018). Systems level evaluation of resilience in human-autonomy teaming under degraded conditions. *IEEE Resilience Week Conference,* 124–130.

Grimm, D. A., Gorman, J. C., Stevens, R. H., Galloway, T. L., Willemsen-Dunlap, A. M., & Halpin, D. J. (2017). Demonstration of a method for real-time detection of anomalies in team communication. *Proceedings of the Human Factors and Ergonomics Society Annual Meeting, 61*(1), 282–286.

Haken, H. (1983). *Synergetics, an introduction: Non-equilibrium phase transitions and self-organization in physics, chemistry, and biology* (3d ed.). Berlin: Springer.

Hayden, Jennifer K.; Smiley, Richard A.; Alexander, Maryann; Kardong-Edgren, Suzan; and Jeffries, Pamela R. (2014). The NCSBN national simulation study: A longitudinal, randomized, controlled study replacing clinical hours with simulation in prelicensure nursing education. *Journal of Nursing Regulation, 5*(2), C1-S64.

Heiniger,C. (2016). *Army designing next-gen command posts.* Retrieved from https://www.army.mil/article/167807/army_designing_next_gen_command_posts.

Hutchins, S. G., & Kendall, T. (2010). The role of cognition in team collaboration during complex problem solving. In K. L. Mosier & U. M. Fischer (Eds.), *Informed by knowledge: Expert performance in complex situations* (pp. 69–89). New York, NY: Taylor & Francis.

Issartel, J., Marin, L., Gaillot, P., Bardainne, T., & Cadopi, M. (2006). A practical guide to time—frequency analysis in the study of human motor behavior: The contribution of wavelet transform. *Journal of Motor Behavior, 38*(2), 139–159.

James, W. (1890). *The principles of psychology.* London: Macmillan.

Juarrero, A. (1999). *Dynamics in action: Intentional behavior as a complex system.* Cambridge, MA: MIT Press.

Kantz, H., & Schreiber, T. (1997). Determinism and predictability. *Nonlinear time series analysis* (pp. 42–57). Cambridge, United Kingdom: Cambridge University Press.

Kruijff, G. J. M., Janíček, M., Keshavdas, S., Larochelle, B., Zender, H., Smets, N. J., ... & Sulk, M. (2014). Experience in system design for human-robot teaming in urban search and rescue. In K. Yoshida & S. Tadokoro (Eds.) *Field and Service Robotics* (pp. 111–125). Berlin Heidelberg: Springer.

Kapur, M. (2011). Temporality matters: Advancing a method for analyzing problem-solving processes in a computer-supported collaborative environment. *Computer-Supported Collaborative Learning, 6*, 39–56.

Kruijff, G. J. M., Janíček, M., Keshavdas, S., Larochelle, B., Zender, H., Smets, N. J., ... & Liu, M. (2014). Experience in system design for human-robot teaming in urban search and rescue. In K. Yoshida & S. Tadokoro (Eds.) *Field and Service Robotics* (pp. 111–125). Berlin, Heidelberg: Springer.

Leonard, H. B., & Howitt, A. M. (2006). Katrina as prelude: Preparing for and responding to Katrina-class disturbances in the United States—Testimony to U.S. Senate Committee, March 8, 2006. *Journal of Homeland Security and Emergency Management, 3*, 1–20.

Lorenz, E. N. (1963). Deterministic nonperiodic flow. *Journal of the Atmospheric Sciences, 20*(2), 130–141.

Lugano, G. (2017). Virtual assistants and self-driving cars. In *2017 15th International Conference on ITS Telecommunications (ITST)* (pp. 1–5). Warsaw, Poland: IEEE. doi: 10.1109/ITST.2017.7972192

Mathieu, J. E., Tannebaum, S. I., Donsbach, J. S., & Alliger, G. M. (2014). A review and integration of team composition models: Moving toward a dynamic and temporal framework. *Journal of Management, 40*(1), 130–160.

McGaghie, W. C., Issenberg, S. B., Petrusa, E. R., & Scalese, R. J. (2010). A critical review of simulation - based medical education research: 2003–2009. *Medical Education, 44*, 50–63.

McNeese, N. J., Demir, M., Cooke, N. J., & Myers, C. (2018). Teaming with a synthetic teammate: Insights into human-autonomy teaming. *Human Factors, 60*(2), 262–273.

Myers, C. W., Ball, J. T., Cooke, N. J., Freiman, M. D., Caisse, M., Rodgers, S. M., Demir, M., & McNeese, N. J. (2017). Autonomous intelligent agents for team training: Making the case for synthetic teammates. *IEEE Transactions on Intelligent Systems.*

Pappada, S. M., Papadimos, T. J., Lipps, J. A., Feeney, J. J., Durkee, K. T., Galster, S. M., Winfield, S. R., Pfeil, S. A., Bhandary, S. P., Castellon-Larios, K., Stoicea, N., & Moffatt-Bruce, S. D. (2016). Establishing an instrumented training environment for simulation-based training of health care providers: An initial proof of concept. *International Journal of Academic Medicine, 2*, 32–40.

Peng, C. K., Havlin, S., Hausdorff, J. M., Mietus, J. E., Stanley, H. E., & Goldberger, A. L. (1995). Fractal mechanisms and heart rate dynamics: long-range correlations and their breakdown with disease. *Journal of Electrocardiology, 28*, 59–65.

Ricca, B. P., Bowers, N., & Jordan, M. E. (2019). Seeking emergence through temporal analysis of collaborative-group discourse: A complex-systems approach. *The Journal of Experimental Education*, 1–17. https://doi.org/10.1080/00220973.2019.1628691

Riley, W., Davis, S., Miller, K., Hansen, H., Sainfort, F., & Sweet, R. (2011). Didactic and simulation nontechnical skills team training to improve perinatal patient outcomes in a community hospital. *Joint Commission Journal on Quality and Patient Safety / Joint Commission Resources, 31*, 357–364.

Roschelle, J., & Teasley, S. D. (1995). The construction of shared knowledge in collaborative problem solving. In C. O'Malley (Ed.) *Computer Supported Collaborative Learning* (pp. 69–97). Berlin, Heidelberg: Springer.

Rosenbaum, D. A. (1998). Is dynamical systems modeling just curve fitting? *Motor Control, 2,* 101–104.

Salas, E., Priest, H. A., Wilson, K. A., & Burke, C. S. (2006). Scenario-based training: Improving military mission performance and adaptability. In A. B. Adler, C. A. Castro, & T. W. Britt (Eds.), *Operational Stress. Military life: The psychology of serving in peace and combat: Operational stress* (pp. 32–53). Westport, CT: Praeger Security International.

Schneider, W. (1985). Training high-performance skills: Fallacies and guidelines. *Human Factors, 27*(3), 285–300.

Schulte, A., Donath, D., & Lange, D. S. (2016). Design patterns for human-cognitive agent teaming. In D. Harris (Ed.), *Engineering Psychology and Cognitive Ergonomics* (pp. 231–243). Cham, Switzerland: Springer International Publishing.

Seeber, I., Maier, R., & Weber, B. (2013). Macrocognition in collaboration: Analyzing processes of team knowledge building with CoPrA. *Group Decision and Negotiation, 22*(5), 915–942.

Shannon, C., & Weaver, W. (1949). *The mathematical theory of communication.* Urbana: University of Illinois Press.

Shively, R. J., Lachter, J., Brandt, S. L., Matessa, M., Battiste, V., & Johnson, W. W. (2017). Why human-autonomy teaming? In *International Conference on Applied Human Factors and Ergonomics* (pp. 3–11). Los Angeles, CA: Springer.

Stanton, N. A., Stewart, R., Harris, D., Houghton, R. J., Baber, C., McMaster, R., ... & Linsell, M. (2006). Distributed situation awareness in dynamic systems: theoretical development and application of an ergonomics methodology. *Ergonomics, 49*(12–13), 1288–1311.

Stevens, R. H. & Galloway, T. (2017). Are neurodynamic organizations a fundamental property of teamwork? *Frontiers in Psychology, 8,* 644.

Stevens, R. H., Galloway, T., Berka, C., & Sprang, M. (2009). Neurophysiologic collaboration patterns during team problem solving. *Proceedings of the Human Factors and Ergonomics Society, 53*(12), 804–808.

Stevens, R., H., Galloway, T., Gorman, J., Willemsen-Dunlap, A., & Halpin, D. (2016a). Toward objective measures of team dynamics during healthcare simulation training. *Proceedings of the International Symposium on Human Factors and Ergonomics in Health Care, 5,* 50–54.

Stevens, R., Galloway, T., Halpin, D., & Willemsen-Dunlap, A. (2016b). Healthcare teams neurodynamically reorganize when resolving uncertainty. *Entropy, 18*(12), 427.

Stevens, R. H., Galloway, T., Willemsen-Dunlap, A., Gorman, J. C., Halpin, D., & Grimm, D. A. (2018). Making sense of team information. *Proceedings of the Human Factors and Ergonomics Society, 62*(1), 114–118.

Stevens, R. H., Gorman, J. C., Amazeen, P. G., Likens, A., Galloway, T. (2013). The organizational neurodynamics of teams. *Nonlinear Dynamics, Psychology, & Life Sciences, 17,* 67–86.

Sutcliffe, K. M., Lewton, E., & Rosenthal, M. E. (2004). Communication failures: An insidious contributor to medical mishaps. *Academic Medicine, 79,* 186–194.

Thelen, E., & Smith, L. B. (1994). *A dynamic systems approach to the development of cognition and action.* Cambridge, MA: MIT Press.

Thelen, E., Ulrich, B. D., & Wolff, P. H. (1991). Hidden skills: A dynamic systems analysis of treadmill stepping during the first year. *Monographs of the Society for Research in Child Development, 56*(1), 1–103.

Treffner, P. J., & Kelso, J. A. S. (1999). Dynamic encounters: Long memory during functional stabilization. *Ecological Psychology, 11,* 103–137.

Vygotsky, L. S. (1978). *Mind in society: The development of higher psychological processes.* Cambridge, UK: Harvard University.

Walker, G. H., Gibson, H., Stanton, N. A., Baber, C., Salmon, P., & Green, D. (2006). Event analysis of systemic teamwork (EAST): a novel integration of ergonomics methods to analyse C4i activity. *Ergonomics, 49*(12–13), 1345–1369.

Webber, C. L. & Marwan, N. (Eds.) (2015). *Recurrence quantification analysis: Theory and best practices.* New York: Springer.

Willemsen-Dunlap, A., Halpin, D. Stevens, R.H., Galloway, T.L., Gorman, J. (2017) Identifying neural and speech correlates of uncertainty during healthcare simulation training. Talk presented at the *March, 2017 HFES-Health Care International meeting*, New Orleans, LA.

Wiltshire, T. J., Butner, J. E., & Fiore, S. M. (2018). Problem - solving phase transitions during team collaboration. *Cognitive Science, 42*(1), 129–167.

Wiltshire, T. J., Steffensen, S. V., & Fiore, S. M. (2019). Multiscale movement coordination dynamics in collaborative team problem solving. *Applied Ergonomics, 79*, 143–151.

4 Distributed Cognition in Self-Organizing Teams

Neelam Naikar

CONTENTS

INTRODUCTION

Cognitive work in sociotechnical systems has individual, social, material, and cultural dimensions. The theory of distributed cognition (Hollan, Hutchins, & Kirsh, 2000; Hutchins, 1995; Norman, 1991) explains how computational tasks or processes are distributed across these dimensions of work practice. However, this theory, and its associated modeling tools (e.g., Furniss, Masci, Curzon, Mayer, & Blandford, 2015; Sellberg & Lindblom, 2014; Stanton, 2014), though important developments in the study of cognitive work practice, gives insufficient emphasis to the self-organizing behaviors of these systems, which is vital for effective performance. The concept of self-organization in sociotechnical systems (Naikar & Elix, 2019) recognizes that computational work processes are not only distributed across actors and artifacts, but are also self-organizing, such that the distribution and content of the computations may change unpredictably with the spontaneous behaviors of actors, even over very short timescales. Consequently, models of work practice must accommodate not just the distributed but also the spontaneous, emergent characteristics of computational task performance, so that we can design systems that properly account for manifestations of such cognitive phenomena in actual workplaces. As a result, workers will be better supported in dealing with the challenges of cognitive work in unstable, uncertain, and unpredictable environments.

This chapter considers the nature of distributed cognition in self-organizing teams. This frame of reference was motivated by the observation that, in sociotechnical workplaces, new or different organizational structures emerge from individual, interacting actors' spontaneous behaviors, in ways that migrate toward becoming fitted to the demands of a continually evolving task environment, so that these systems are self-organizing. This chapter therefore begins by illustrating the nature of individual and organizational adaptation in sociotechnical workplaces, and by describing the concept of self-organization as it manifests in these settings. It then considers the nature of distributed cognition in sociotechnical workplaces, showing how observations from Hutchins's (1990, 1991, 1995) classic field studies support the argument that the computational processes of work are not only distributed but are also self-organizing. Following that, the chapter examines the implications of the self-organization phenomenon for modeling and designing computational work in these systems, again using examples from Hutchins's case studies to illustrate the arguments. This chapter therefore leads to an understanding of how complex cognitive work can be supported through design to promote the inherent capacity of sociotechnical systems for self-organization, a phenomenon that is essential for dealing with instability, uncertainty, and unpredictability in the task conditions.

ADAPTATION IN SOCIOTECHNICAL WORKPLACES

Sociotechnical workplaces are systems with strong psychological, social, cultural, and technological attributes (see also Vicente, 1999). In such workplaces, which include hospitals, air traffic control centers, naval vessels, and emergency management organizations, human workers bring both psychological and social dimensions to the task performance, and the work is carried out in context of a substantive cultural and technological environment. In some cases, such as petrochemical refineries, the technological dimension may be more prominent, whereas in other cases, such as stock market trading, the psychological, social, and cultural dimensions may be more apparent. Nevertheless, in all cases, all of these facets of complex cognitive work must be accounted for in system design. Designs that fail to consider any one or more of these dimensions may compromise performance, sometimes with disastrous consequences, as demonstrated by a number of recent, high-profile accidents, such as the fatal crashes of two Boeing 737 MAX 8 aircraft in Indonesia and Ethiopia (National Transportation Safety Board, 2019) and the Fukushima Daiichi nuclear power plant disaster (Director General, 2015).

Sociotechnical workplaces are necessarily distributed systems. The scale and demands of the work are such that it cannot be undertaken by any one individual alone. The work is therefore distributed across multiple actors and artifacts, such that it may span mental, social, physical, and temporal spaces, as Hutchins (1995) describes in depth in his classic text on distributed cognition.

The challenges of cognitive work in sociotechnical systems arises in large part from the instability, uncertainty, and unpredictability of the task environment (Naikar & Brady, 2019; Vicente, 1999; Woods, 1988). Workers must operate in highly dynamic conditions, so the problems, demands, or pressures they are faced with may change constantly with the evolving situation. As an example, the threats posed by wildfires to emergency management workers may shift constantly depending on the weather conditions, such as the amount of rainfall and wind directions, and the habitation and infrastructure in the affected areas. Workers must also operate with considerable uncertainty. In naval work, for instance, it is difficult to establish the presence of obstacles or adversaries in a ship's underwater environment with complete certainty because of imperfections in the onboard sensors. Finally, workers must also contend with significant unpredictability in the work requirements. That is, they must deal with events that cannot be foreseen or specified fully a priori by analysts or designers (e.g., Leveson, 1995; Perrow, 1984; Rasmussen, 1969; Reason, 1990; Vicente, 1999). Examples of such events include a new type of military threat (Herzog, 2011; Reich, Weinstein, Wild, & Cabanlong, 2010), an unanticipated reaction of a patient to an anesthetic during a surgical procedure (Hoppe & Popham, 2007), or an unforeseen string of supplier collapses in the aftermath of a natural disaster (Park, Hong, & Roh, 2013).

Field studies show that, given conditions of instability, uncertainty, and unpredictability in the task environment, workers adjust both their individual behaviors and collective structures in line with the evolving demands (Naikar & Brady, 2019; Naikar & Elix, 2016, 2019). Such observations have been documented in a variety of settings including emergency management (e.g., Bigley & Roberts, 2001; Lundberg & Rankin, 2014), military (e.g., Hutchins, 1990, 1991, 1995; Rochlin, La Porte, & Roberts, 1987), commercial aviation (e.g., Hutchins & Klausen, 1998), law enforcement (e.g., Linde, 1988), transport (e.g., Heath & Luff, 1998; Luff & Heath, 2000), and healthcare (e.g., Bogdanovic, Perry, Guggenheim, & Manser, 2015; Klein, Ziegert, Knight, & Xiao, 2006).

As an example, in a field study of emergency management workers, specifically teams of firefighters, Bigley and Roberts (2001) observed considerable flexibility in workers' use of tools, rules, and routines. When a fire truck arrives at the scene of an incident, for instance, firefighters may have no choice but to improvise with the tools available on the truck to handle the emergency. In addition, workers may find it necessary to breach standard operating procedures or rules. In one case, firefighters deliberately chose the strategy of "opposing hose streams" to deal with an emergency, despite the fact that this strategy is prohibited by procedure, as one group can push the fire into another. Finally, firefighters often tailor their standard routines, such as those for "hose laying" or "ladder throwing," to local contingencies.

Field studies also provide clear evidence for changes in the work structure or organization. A compelling case is provided by Rochlin et al. (1987), who conducted detailed observations of naval aircraft carriers. Their findings showed that the formal organizational structure of these carriers is strongly hierarchical and defined by clear chains of command and means of enforcing authority. However, during complex operations, the work organization shifts—without a priori planning, external intervention, or centralized coordination—to configurations that may be described as informal in that they are not officially documented. These informal configurations, which are flat and distributed, are not defined by any simple mapping between people and roles. Instead, the mapping changes spontaneously with the local circumstances.

SELF-ORGANIZATION IN SOCIOTECHNICAL WORKPLACES

The preceding case studies show that adaptations in sociotechnical systems can be observed at two levels: in the behaviors of individual actors and the structures of multiple actors. The concept of self-organization (e.g., Haken, 1988; Heylighen, 2001; Hofkirchner, 1998) is important in this context because it provides a plausible and parsimonious framework for understanding why adaptations in both actors' individual behaviors and collective structures are necessary, and how such adaptations can be achieved spontaneously, continuously, and relatively seamlessly (Naikar, 2018; Naikar & Brady, 2019; Naikar & Elix, 2019; Naikar, Elix, Dâgge, & Caldwell, 2017). Moreover, in that the adaptations signify continuous, spontaneous change in computational work processes, the self-organization concept contributes to the understanding of cognitive task performance in distributed systems.

Specifically, the concept of self-organization (e.g., Haken, 1988; Heylighen, 2001; Hofkirchner, 1998), when analyzed in the context of sociotechnical workplaces, suggests that a system's formal structure bounds individual actors' degrees of freedom for action in ways that are suited to particular circumstances, usually those that are routine or familiar. Therefore, when new or different conditions are encountered, the formal structure may constrain the system's response in ways that are unsuitable or ineffective (Figure 4.1). However, in responding to the changed conditions, the spontaneous actions of individual, interacting actors may result in changes in the work organization. Consequently, when a new or different structure emerges that is better suited to the present circumstances, it will stabilize, such that it will continue to constrain and enable the behaviors of individual actors in ways that are fitted to the local conditions. However, when the situation changes again in some fundamental respect, the spontaneous behaviors of individual, interacting actors will result in further structural changes, so that the system is self-organizing (Naikar, 2018; Naikar & Brady, 2019; Naikar & Elix, 2019; Naikar et al., 2017).

Further, the concept of self-organization in sociotechnical workplaces recognizes that, under conditions of instability, uncertainty, and unpredictability, actors move away from formal structures or procedures toward the intrinsic constraints of the sociotechnical system as the principal governing mechanism for their conduct (Figure 4.1). These constraints, which are boundaries on behavior that must be respected by actors for a system to perform effectively, still afford actors many degrees of freedom for action (Rasmussen, Pejtersen, & Goodstein, 1994; Vicente, 1999). For example, organizational values and resources place limits on actors' behaviors, but still afford actors many possibilities for action. Therefore, given a commitment to these fundamental constraints on behavior, rather than to formal specifications, workers can safely and productively adjust their actions to the local conditions, such that new structures may emerge from their spontaneous behaviors. This phenomenon of self-organization is integral to the process of a system adapting to its environment.

As Naikar and Elix (2019) discuss, the processes of self-organization in sociotechnical workplaces are not without challenges and may not be regarded as perfect or flawless (e.g., Lundberg & Rankin, 2014), at least when assessed against idealized measures or benchmarks. However, many sociotechnical systems are described as high-reliability organizations (La Porte, 1996; Rochlin, 1993; Weick & Sutcliffe, 2001) because of their capacity to balance safety and productivity goals successfully in the face of considerable instability, uncertainty, and unpredictability. Therefore, the processes of self-organization can be relatively seamless, particularly in well-established systems, and the fact is they are necessary. Alternative strategies, such as a priori planning, centralized coordination, or external intervention, are rarely

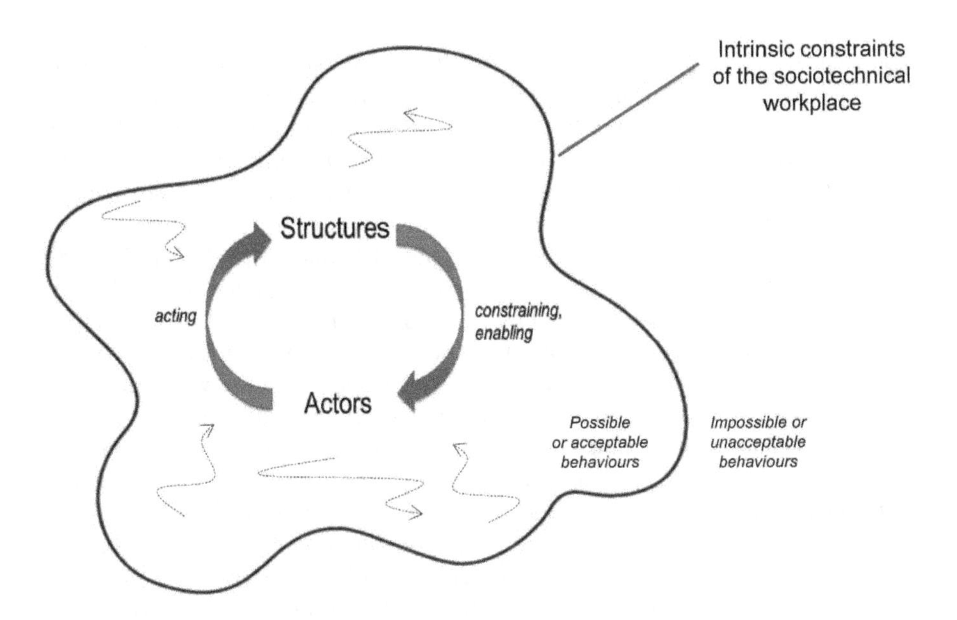

FIGURE 4.1 The phenomenon of self-organization in sociotechnical workplaces: Given the opportunities for action, afforded by the intrinsic constraints of the physical, social, and cultural work environment, new or different organizational structures may emerge from actors' spontaneous behaviors. The emergent structures constrain and enable actors' behaviors in ways that are fitted to the local conditions, such that the system is self-organizing.

Note: The dashed lines signify possible or acceptable behavioral trajectories, which are innumerable.

feasible under conditions of instability and ambiguity. Instead, spontaneous behaviors and emergent work structures are required to resolve—in time and in situ—the "proper, immediate balance" (Rochlin et al., 1987, pp. 83–84) between a system's safety and productivity objectives.

DISTRIBUTED COGNITION IN SOCIOTECHNICAL WORKPLACES

The theory of distributed cognition (Hollan et al., 2000; Hutchins, 1995; Norman, 1991) focuses on human cognition in its natural habitat or in naturally occurring contexts. This theory suggests that, in sociotechnical workplaces, it does not make sense to draw the boundaries of cognitive processes around individuals. Rather, the boundaries must extend beyond individuals to their social, physical, and temporal spaces to account for interactions between people and their environments. Thus, this theory recognizes that in actual workplaces—or in the wild—cognitive accomplishments are necessarily a product of the sociotechnical system, resulting from interactions between mental, social, material, and cultural structures.

The theory of distributed cognition is compelling and has led to significant advancements in the understanding of the computational processes of work in sociotechnical systems (e.g., Furniss et al., 2015; Sellberg & Lindblom, 2014; Stanton,

2014). However, the concept of self-organization suggests that it is important to appreciate that these processes are not only distributed, but are also self-organizing. Given the self-organizing behavior of distributed cognitive systems, the distributions of computational work, as well as its content, may change in ways that are closely fitted to evolving circumstances, which cannot be specified fully a priori. Therefore, the phenomenon of self-organization has significant implications for modeling and designing distributed cognitive systems.

In this section, this argument is illustrated and elaborated with Hutchins's (1990, 1991, 1995) classic studies of the computational task of navigation on naval ships. Although his works contain many observations that may be used to substantiate this argument, the theory of distributed cognition does not directly address the self-organizing characteristics of distributed cognitive work in complex workplaces.

First, Hutchins's (1990, 1991, 1995) field studies may be used to support the claim that the distributions of computational processes across human workers in socio-technical systems may vary with their spontaneous behaviors. In the navigation task that Hutchins (1995) describes, it is formally the job of the two bearing takers, who are located on the ship's left and right wings, to identify nominated landmarks in the surrounding area and to measure and report their bearings. In addition, it is the bearing timer-recorder's responsibility to time and record the reported bearings in a log, and the plotter's responsibility to plot those bearings onto a navigation chart. However, Hutchins (1990) observed this formal structure being violated on a number of different occasions. In one incident, the bearing timer-recorder, who is stationed in the pilothouse, assisted the bearing taker on the ship's right wing with identifying a landmark over the telephone. In another incident, the plotter, who is also stationed in the pilothouse, walked out onto the ship's right wing to point out a landmark to the bearing taker. Furthermore, in a third incident, the bearing timer-recorder performed the first step of the plotting job when the plotter was called away from the chart table for a consultation with the ship's captain. These observations suggest that the distributions of cognitive processes across workers may be reconfigured continually in the course of their spontaneous interactions, as they "negotiate" their responsibilities on the job and "participate" in each other's work (Luff & Heath, 2000).

Hutchins's (1990, 1991, 1995) field studies may also be used to make the case that it is not just the distributions of cognitive activities across human workers that may vary, but also the distributions of computational processes across material artifacts. In one incident observed by Hutchins (1991), the ship's gyrocompass failed when it lost its power supply, so that the true bearing of landmarks could no longer be measured directly with this instrument but, instead, had to be computed arithmetically from the relative bearings. This meant that to navigate the ship safely through restricted waters, being as it was in a narrow harbor on its return to port, new computational tasks to establish the ship's position in space and an efficient structure for organizing these activities across the team's internal (i.e., mental) and external resources were required. Initially, as the navigation team improvised with different processes for the computational task, such as performing the arithmetic mentally or with a hand-held electronic calculator, there was no consistent pattern in the distribution of computational processes across actors and artifacts. In fact, 30 lines of position for the ship were plotted before the distribution of the cognitive work stabilized. Hutchins (1991, p. 23) commented that "the social structure (division of labor) seems to have emerged from the interactions among the participants without any explicit planning." This emergent social distribution, which involved the plotter and the bearing timer-recorder calculating the true bearings of landmarks with the aid of an electronic calculator, in place of the bearing takers measuring the true bearings directly with the gyrocompass, was fitted to the local conditions in that it allowed the individuals in the team to perform the new computational work in a manner that was cognitively efficient for each of them given the artifacts available at the time.

Third, observations from Hutchins's (1990, 1991, 1995) field studies may be used to support the argument that the distributions of computational processes across actors and artifacts may vary in ways that are not fully specifiable a priori. As an example, in the incident from Hutchins's (1991) study when the ship's gyrocompass failed, the bearing timer-recorder seemed to prefer performing the calculation for obtaining the true bearing of landmarks from their relative bearings by adding the mathematical terms in the order of the availability of the data, despite the fact that the plotter was encouraging him to perform the computation in another, arguably more meaningful, order. This "conflict" was resolved temporarily, prior to the stabilized structure being reached, by the bearing timer-recorder adopting the order the plotter desired when performing the computation in interaction with him, but not otherwise. These observations suggest that computational strategies that were cognitively efficient for the plotter were not cognitively efficient for the bearing timer-recorder. More generally, the distributions that emerge at any point in time depend on specific details of the local situation, including individual differences between human participants, so that it would be impossible to specify in advance a distribution to cover every circumstance.

Fourth, Hutchins's (1990, 1991, 1995) field studies may be used to make the point that it is not just the distributions of the computations that may vary, but also the computations that are required in the first place. In the incident in Hutchins's (1991) study when the ship's gyrocompass failed, actors were suddenly faced with the task of computing the true bearing of landmarks from the relative bearings, rather than measuring the true bearings directly with the gyrocompass. In other words, the computational tasks that were required when the gyrocompass was functional were different from when it was not. Thus, both the content and distribution of the computational processes may vary.

Finally, Hutchins's (1990, 1991, 1995) case studies may be used to demonstrate that the distribution and content of the computational work varies with actors' spontaneous acts on the affordances of the environment. For example, in the incident when the ship's gyrocompass failed, the computational processes exhibited depended on whether the plotter chose to perform the arithmetic mentally; whether the plotter decided to perform the computation with the aid of a handheld calculator, which he had to obtain by walking over to the charthouse; or whether the plotter elected to share the task with the bearing timer-recorder, when he realized that he could not keep up with the timing requirements for the task even with the aid of an electronic calculator. These observations suggest that the work environment affords different computational possibilities for the job, and the content and distribution of the cognitive work varies depending on which affordances the actors spontaneously act on for the job.

In summary, the concept of self-organization in sociotechnical systems suggests that, given the inherent instability and ambiguity of the task conditions, the computational processes of work are not only distributed across actors and artifacts, but are also self-organizing. In responding to the local conditions, actors' spontaneous acts on the affordances of the environment produce variations in the distribution and content of the computational work, such that the system's behavior migrates toward becoming fitted to the demands of the evolving circumstances. This type of self-organizing behavior is evident in a variety of distributed cognitive systems, including in commercial and military aviation, healthcare, transport, law enforcement, and emergency management, as discussed in more depth elsewhere (Naikar & Elix, 2019). The capacity of distributed cognitive systems for self-organization—or for spontaneous change in the computational processes—is fundamental to their viability in task conditions with high levels of instability, uncertainty, and unpredictability.

MODELS FOR WORK ANALYSIS AND DESIGN

The concept of self-organization presented in this chapter recognizes that, in sociotechnical systems, computational work processes change in ways that become closely fitted to evolving circumstances, which cannot be specified fully a priori. The spontaneous behaviors of individual, interacting actors, and the emergence of new structural forms or organizations from these behaviors, make it possible for a system to adapt to the changing demands of the work environment, and thus to "survive in the wild." This phenomenon of self-organization has distinct implications for modeling and designing distributed cognitive systems or, in other words, for modeling the work requirements of these systems and developing designs for interfaces, teams, and training systems, for instance, that support the work requirements effectively.

Models for work analysis and design may be viewed as having normative, descriptive, or formative orientations (Rasmussen, 1997; Vicente, 1999). In this section, the capacity of these three types of approaches to support the analysis and design of distributed cognitive systems, which are self-organizing, is examined. As in the preceding section, the following arguments are substantiated with observations from Hutchins's (1990, 1991, 1995) field studies of the navigation task on naval vessels.

First, a key implication of the self-organization phenomenon is that normative models, which formalize or prescribe the processes of work required under specific conditions, are inadequate for distributed cognitive systems. This view may be supported by Hutchins's (1995) observations of the formal procedures for the navigation task, which he reproduces in his classic text. His findings show that, although these procedures are detailed, they are only nominal processes for ideal conditions. Consequently, the procedures are routinely violated during everyday operations (Hutchins, 1990, 1995). Furthermore, his studies suggest that it would be impossible to specify a distribution of computational processes to cover every single circumstance. As a case in point, there was no specified procedure for how the navigation task should be carried out when the ship's gyrocompass failed (Hutchins, 1991). Normative models, therefore, are likely to be incomplete and inflexible, leading to designs that are limited to supporting computational processes in situations that can be pre-specified or anticipated.

Second, the self-organization phenomenon highlights that descriptive approaches, which characterize or describe the processes of work observed in particular situations, are also unsuitable for distributed cognitive systems. In his classic text, Hutchins (1995) presents a detailed "activity score," which describes how computational processes for position fixing (i.e., establishing the ship's location in space) are distributed, or coordinated temporally, across actors and artifacts on the vessel. However, this activity score is likely only relevant for a single instance of position fixing—that which Hutchins observed in creating the record. The computational processes in other instances, even very similar ones, are likely to be different. Certainly, Hutchins's activity score is not relevant to the case when the ship's gyrocompass failed. Moreover, activity scores cannot be produced for instances that have not been observed, and it is impossible to observe the full range of instances, especially those that have not yet occurred. Consequently, descriptive models of computational work are also likely to be incomplete and inflexible, leading to designs that are limited to supporting computational processes that have or can be observed. Such designs, like those produced by normative models, may inhibit, rather than foster, the spontaneity needed in the workplace for dealing with instability and ambiguity in the task requirements. Such arguments are also applicable to tools developed more recently for modeling distributed cognition, such as Event Analysis of Systemic Teamwork (EAST; Stanton, 2014) and Distributed Cognition for Teamwork (DiCoT; Furniss et al., 2015).

Compared with normative and descriptive approaches, formative models are more appropriate for distributed cognitive systems, given the intent to accommodate their self-organizing characteristics. A well-established framework that falls in this category is cognitive work analysis (CWA; Rasmussen et al., 1994; Vicente, 1999). This framework provides a strong starting point for modeling and designing distributed cognitive systems because it focuses attention on the possibilities for computational work, rather than on formalized or observed computational processes. Designs based on such models can accommodate adaptations in the computational processes to the local conditions, even those that have not been observed or anticipated.

The CWA framework is particularly powerful in the approach it offers for modeling the constraints, or affordances, of the work environment. Specifically, work domain analysis (Naikar, 2013; Rasmussen et al., 1994; Vicente, 1999), the first dimension of the framework, models the constraints of the physical, social, and cultural environment at five levels of abstraction, relating to the system's purposes, values and priorities, functions, processes, and objects. As indicated above in the section "Self-Organization in Sociotechnical Workplaces", although these constraints place limits on actors' behaviors, they also afford actors may degrees of freedom for action. Moreover, as these constraints are event-independent, the affordances are relevant to a wide range of situations, including those that have not been observed or anticipated. Further, activity analysis and strategies analysis, the second and third dimensions of CWA, provide increasingly detailed views of the constraints in terms that relate to the possible activities and strategies in recurring classes of situation.

One problem with the standard CWA framework (Rasmussen et al., 1994; Vicente, 1999), however, is the approach it takes for modeling the distributions of work across actors (Naikar & Elix, 2016, 2019). Specifically, in social organization and cooperation analysis, the fourth dimension of the framework, the distributions are limited to recurring classes of situation and even further to organizational structures that have been observed or are judged to be reasonable in these situations. As a result, the standard framework does not account for the range of work organization possibilities, especially those that may emerge in novel or unforeseen circumstances.

Recently, however, a new modeling tool has been developed for social organization and cooperation analysis, namely the diagram of work organization possibilities (Naikar & Elix, 2016). This tool models the behavioral opportunities of individual actors and the structural possibilities of multiple actors, as afforded by the constraints of the system, in a single, integrated representation—in a way that is compatible with the phenomenon of self-organization in sociotechnical workplaces (Naikar & Elix, 2019). Therefore, this tool has the potential to support the analysis and design of distributed cognitive systems, which are self-organizing.

The basic form of the diagram of work organization possibilities is shown in Figure 4.2. In this representation, the actors signify agents, whether human or artificial, that are capable of goal-directed behaviors. The behavioral opportunities of individual actors are demarcated by sets of work demands, or constraints, from the first three CWA dimensions. The terms "constraints" and "work demands" are used interchangeably in CWA because the constraints place demands on actors by defining boundaries that must be respected in their actions for effective performance (Vicente, 1999). Accordingly, for each actor, the work demands collectively demarcate a field of opportunities for safe and productive behavior. The structural possibilities of multiple actors, then, emerge from the behavioral opportunities of individual, interacting actors (cf. Figure 4.1).

As a simplified example, Figure 4.2 shows that the behavioral spaces of both Actors A and C, though not B, are delimited by work demand 1. Therefore, some of the structural forms that can emerge from these actors' spontaneous behaviors are: Actor A is engaged in behaviors accommodated by work demand 1, while Actors B and C are occupied in other behaviors; Actor C is involved in behaviors accommodated by work demand 1, while Actors A and B are engaged in other behaviors; or both Actors A and C are involved in behaviors accommodated by work demand 1, while Actor B is occupied in other behaviors. Which structural possibility emerges at any point in time depends on the details of the circumstances, including individual differences between human workers, which cannot be predicted in full a priori.

Figure 4.3 presents a diagram of work organization possibilities for a future airborne system for maritime surveillance (Elix & Naikar, 2019). This diagram shows, for instance, that actors on the flight deck as well as actors located in the cabin of the aircraft, specifically at two observer stations and six tactical workstations, are afforded behaviors for satisfying the constraints of navigation. Therefore, any one or more of these actors may be occupied in such behaviors at any point in time, so that the emergent structural forms may vary with the circumstances. Another example is that only the actors on the flight deck and at the observer stations in the cabin are afforded behaviors for sighting or observing targets out of a window. However, at

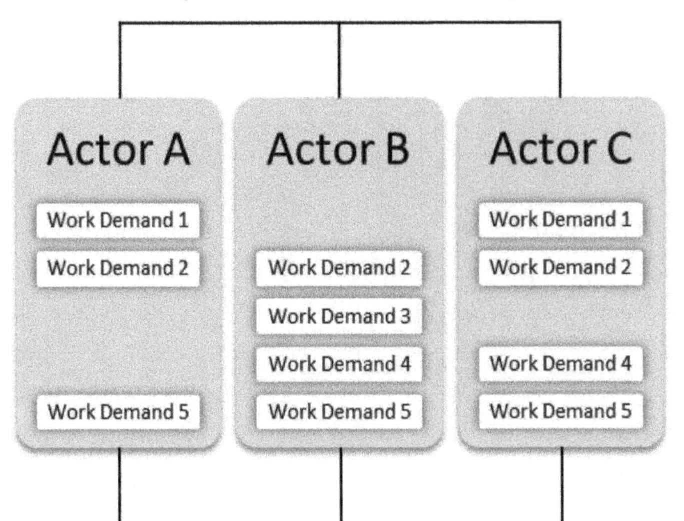

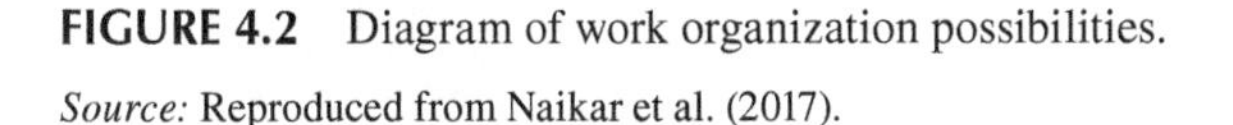

FIGURE 4.2 Diagram of work organization possibilities.

Source: Reproduced from Naikar et al. (2017).

any point in time, any one or more of the two flight deck actors and two observer station actors, who are normally located in these positions, may be occupied in these behaviors. Furthermore, if there is an electrical failure, so that some of the sensors available to the six workstation actors can no longer be used for detecting, tracking, localizing, or identifying targets, any one or more of these actors might relocate to the stations with a window to increase the chances of finding the target. The diagram of work organization possibilities therefore accommodates a range of behavioral and structural possibilities for adaptation, which are relevant to a wide range of situations, including novel or unanticipated events.

The processes for constructing a diagram of work organization possibilities have been described in depth elsewhere (Elix & Naikar, 2019; Naikar & Elix, 2016). However, generally the processes involve applying a set of criteria to the work demands from the earlier CWA dimensions to identify the *limits* on the distribution of the work demands across actors, or the organizational constraints. The set of criteria, which includes compliance with policies or regulations, safety and reliability, access to information/controls, and feasible coordination, competencies, and workload, govern shifts in work organization dynamically in sociotechnical systems (Rasmussen et al., 1994; Vicente, 1999). Therefore, in applying these criteria to construct a model of work organization possibilities, the aim is to rule out behaviors and structures that cannot be manifested in situ, regardless of the circumstances, and to support the possible behaviors and structures in design.

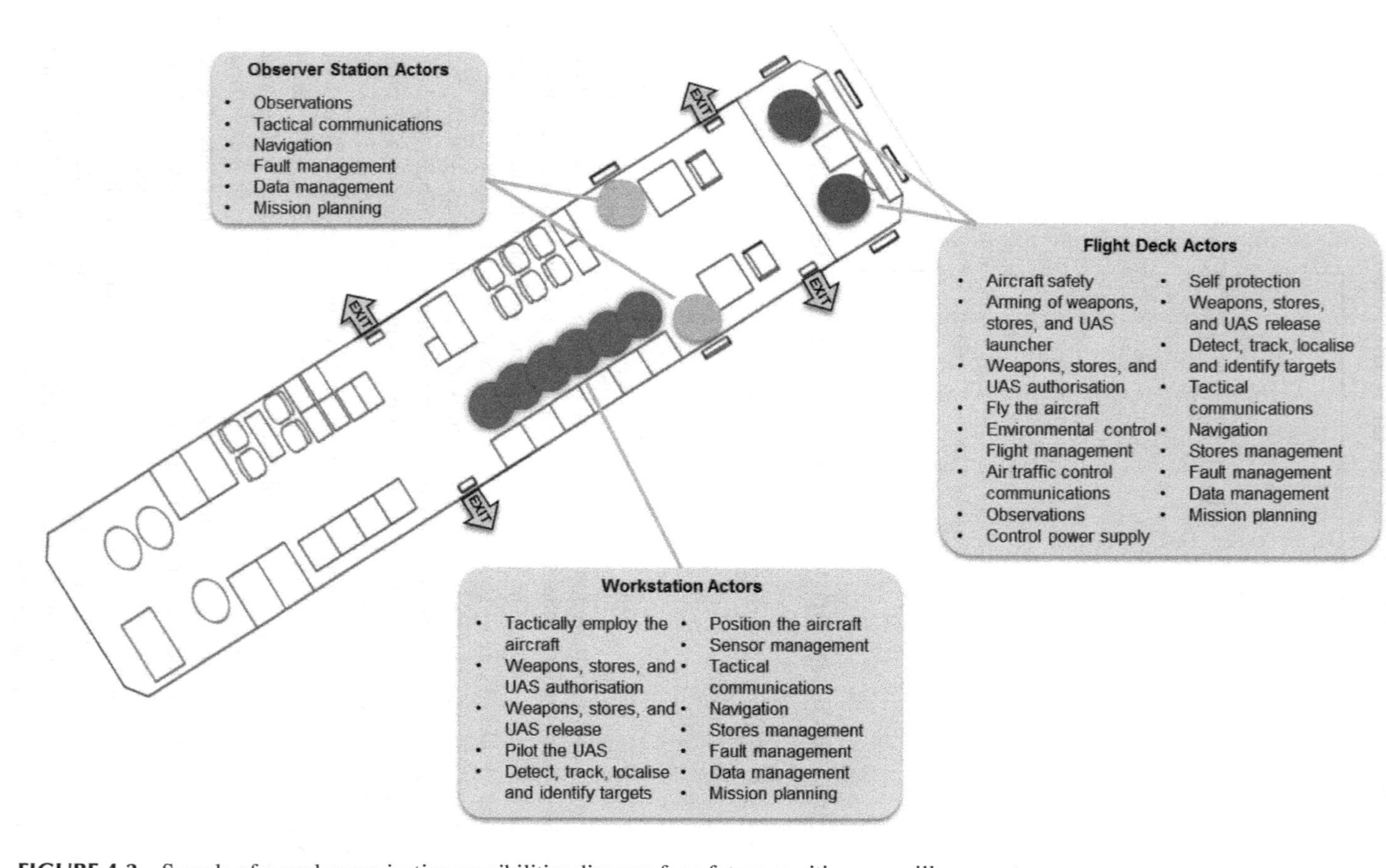

FIGURE 4.3 Sample of a work organization possibilities diagram for a future maritime surveillance system.

Source: Reproduced from Elix and Naikar (2019).

In the context of a discussion on distributed cognitive systems, it is important to emphasize that the diagram encompasses the computational possibilities for the work given both the actors and artifacts in the system. As mentioned, the actors in the diagram may represent human workers or artificial agents that are capable of goal-directed behaviors. Therefore, some cognitive artifacts, or "physical objects made by humans for the purpose of aiding, enhancing, or improving cognition" (Hutchins, 1999, p. 126), may be represented as actors. As Hutchins (1999) recognizes, computers are a special class of cognitive artifact that "mimic certain aspects of human cognitive function" (p. 127). Cognitive artifacts that are not capable of goal-directed behaviors are incorporated in the diagram as objects or resources in the work environment that define the behavioral spaces of actors. Therefore, depending on whether actors choose to act on the affordances of such artifacts or not, the actors' cognitive trajectories will be different. Accordingly, the computational structures for the work will vary. (Although physical objects or resources were included in the analysis of the future maritime surveillance system discussed earlier in this section, they are not evident in the diagram shown in Figure 4.3 because, as Elix and Naikar (2019) explain, the constraints were represented at a higher level of abstraction for practical reasons in this case.)

To return to Hutchins's (1990, 1991, 1995) field studies of the navigation task on naval vessels, let us assume that Actors A, B, and C in the basic form of the work organization possibilities diagram represent the positions of the bearing timer-recorder, bearing taker, and plotter, respectively (Figure 4.4). Further, let us superimpose some of the cognitive artifacts referenced in this study, specifically a gyrocompass (G), paper and pencil (P), and hand-held electronic calculator (C), onto the behavioral spaces of these actors. This representation shows that the gyrocompass is only available to an actor in the position or location of the bearing taker, whereas the paper and pencil and electronic calculator are accessible to actors in any of the three positions. This means that only an actor in the position of the bearing taker can measure the true bearings of landmarks directly with the gyrocompass. However, in the incident when the ship's gyrocompass failed, any combination of these three actors could have been involved in computing the true bearings of landmarks from their relative bearings using the paper and pencil or hand-held calculator, given the afforded behavioral spaces of these actors. In Hutchins's (1991) study, the stabilized social structure involved the bearing timer-recorder and the plotter sharing this computational task, with the aid of an electronic calculator, and this possibility for the distribution of computational work is accommodated in the diagram. Notably, this possibility was not accounted for in the formal procedures of the ship or documented in an activity score a priori.

By modeling the computational work possibilities, given the actors and artifacts in the system, the diagram of work organization possibilities provides a means for designing distributed cognitive systems. Designs based on this diagram, whether of interfaces, teams, training systems, or workspace layouts, will not be limited to prespecified computational processes, such as the formal procedures for the task of navigation on naval vessels (Hutchins, 1995). In addition, designs based on this diagram will not be limited to computational processes that have already been observed, such as those documented in the activity score for position fixing presented by Hutchins

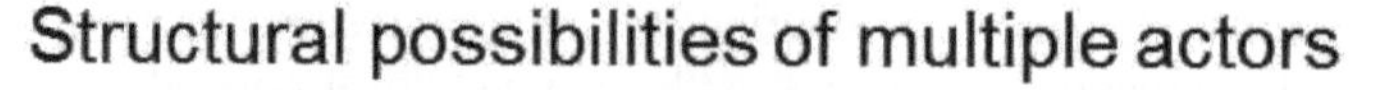

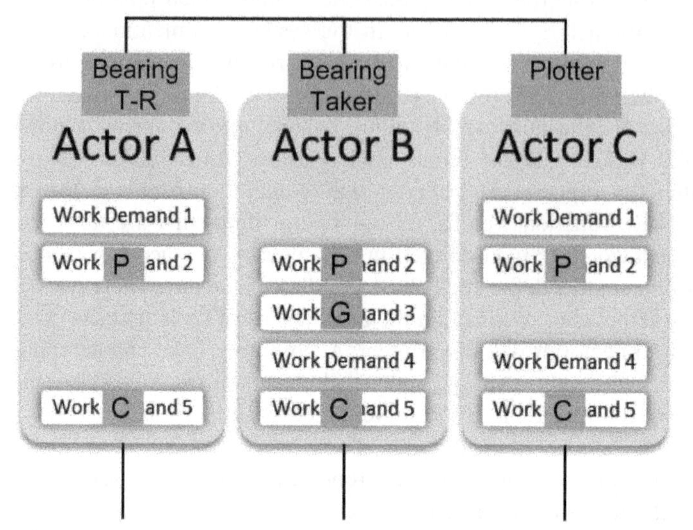

FIGURE 4.4 The work organization possibilities diagram encompasses the computational possibilities for work.

(1995). Instead, designs based on this diagram will have the potential to accommodate the range of possibilities for computational work, regardless of the circumstances. Consequently, actors may be better supported in responding spontaneously to the local conditions, whether these are defined by small variations in routine situations, dramatic changes in circumstances, or workers' individual preferences. As a result, the system's inherent capacity for self-organization, a phenomenon that is essential for dealing with instability and ambiguity in the task environment, may be preserved.

For example, Elix and Naikar (2019) present a case study of the future airborne maritime surveillance system discussed earlier in this section that demonstrates how the diagram (Figure 4.3) led to a team design that is integrated with the training and career progression pathway of the crew in a way that maximizes the system's behavioral and structural possibilities for adaptation, specifically in relation to a novel operational concept involving the control of an uninhabited aerial system (UAS) from the manned aircraft. In this approach, the learning requirements of the flight deck actors, observer station actors, and workstation actors are based on their spaces of behavioral opportunities, as defined by the intrinsic constraints of the system, rather than on pre-specified roles or responsibilities. Consequently, their actual roles and responsibilities on any occasion emerge from their bounded spaces of possibilities for action.

Designs based on the diagram of work organization possibilities, then, may differ significantly from designs produced with conventional approaches. The diagram for the future maritime surveillance system (Figure 4.3) identified that all six workstation actors had the capacity for piloting the UAS and detecting, tracking, localizing, and identifying targets with its sensor. Therefore, an integrated system design was produced, which combined the team, training, and career progression elements in a way that enabled all six workstation actors to operate the UAS. As a result, the design accommodates a range of possibilities for managing the instability, uncertainty, and unpredictability of the tasking environment, such as shifting responsibility for the UAS or combining some of the roles to allow for a dedicated UAS operator if necessary.

In contrast, normative and descriptive approaches are likely to have identified the best crew member for the job in view of specific circumstances. Specifically, normative approaches are likely to have identified the best crew member for dealing with known or anticipated situations, whereas descriptive approaches are likely to have identified the best crew member for the job in situations that have been observed. However, in sociotechnical workplaces, events are rarely reproduced exactly and novel or unforeseen events are possible (Hoffman & Woods, 2011; Rasmussen et al., 1994; Vicente, 1999). Therefore, normative and descriptive approaches are likely to result in designs that are limited in changed or novel circumstances.

Indeed, many field studies show that workers often act outside the boundaries of their professional roles when necessary (e.g., Bogdanovic et al., 2015; Lundberg & Rankin, 2014; Rochlin et al., 1987). For example, in Hutchins's studies (1990, 1991, 1995), the plotter and the bearing timer-recorder were observed to act outside the formal definitions of their roles following the loss of the gyrocompass as well as in more routine situations. However, despite the existing body of empirical evidence, conventional design practices persist in utilizing relatively inflexible forms of social organization, which are rarely manifested during actual operations, in the system development process. As a result, workers may not be well supported by the system design in their improvisations, which are necessary for survival in the wild.

For instance, Lundberg and Rankin (2014) found that role improvisations by workers in crisis response teams, which were not accounted for in training, meant that jobs were sometimes performed less effectively or efficiently than specialists performing the same work. Moreover, Klein, Wiggins, and Deal (2008) observed that the Three Mile Island accident highlighted that nuclear power plant control rooms presented information to workers in ways that sometimes interfered with their ability to understand the situation and adapt to the circumstances. Clearly, then, designs based on inflexible work practices are not only inefficient, but they may also be dangerous, particularly when the task circumstances stretch the capacity of workers to keep the system afloat (pardon the pun!).

In some systems, relatively robust designs may evolve through the bottom-up practices of workers, as opposed to resulting from the top-down practices of designers. For example, in considering the division of labor among members of the navigation team on naval vessels, Hutchins (1990) observes that "the progress of various team members through the career cycle of navigation practitioners produces an overlapping distribution of expertise" (p. 191) that allows team members to take responsibility for all parts of the process to which they can contribute—not just for their own jobs—which enables the system to adapt to changing circumstances. This observation may explain, for instance, how the bearing timer-recorder was able to perform the first step of the plotting job, when the plotter was called away from the chart table for a consultation with the ship's captain. In addition, this observation may explain how the plotter and bearing timer-recorder were able to establish the true bearing of landmarks from the relative bearings when the gyrocompass failed, despite the fact that establishing the true bearings of landmarks is formally the responsibility of the bearing takers.

In contrast, in the case of designing future systems, or designing upgrades to existing systems, robust work practices wouldn't already been evolved, at least not for any new or revolutionary components. Instead, robust work practices must be envisaged and created by designers. For example, in relation to the future maritime surveillance system, all of the six workstation actors are afforded behaviors for navigation, such that they could assist the flight deck actors with these activities. However, although such capabilities may be provided by the software at all of the workstations by default, as it is inexpensive to do so, and the crew may take advantage of these opportunities, these possibilities must be accounted for in the team, interface, and training design of the system to support adaptation effectively. By deliberately seeking to enable adaptation within a set of boundary conditions on safe and productive performance, the diagram provides a means for systematically designing for constrained flexibility in the workplace.

DISCUSSION

This chapter has considered the nature of distributed cognition in self-organizing teams. It has been observed that, while the theory of distributed cognition (Hollan et al., 2000; Hutchins, 1995; Norman, 1991) has led to significant advancements in the understanding of computational work processes in sociotechnical systems (e.g., Furniss et al., 2015; Sellberg & Lindblom, 2014; Stanton, 2014), it is important to appreciate that these processes are not only distributed, but are also self-organizing. Given the system's intrinsic constraints, or affordances, new computational structures may emerge from actors' spontaneous behaviors, and constrain and enable their behaviors, in a way that is fitted to the local conditions. This phenomenon is essential to the system's viability under conditions of instability, uncertainty, and unpredictability. Therefore, the concept of self-organization in sociotechnical workplaces was put forward in this chapter as a framework for characterizing the self-organizing behavior of distributed cognitive systems.

This chapter has also proposed an approach for modeling and designing distributed cognitive systems. Specifically, it has been suggested that the diagram of work organization possibilities (Naikar & Elix, 2016), a tool recently developed as an addition to CWA, is suitable for this purpose. This tool, which has a formative orientation, is aligned with the phenomenon of self-organization in sociotechnical workplaces (Naikar & Elix, 2019; Naikar et al., 2017). Therefore, designs based on this framework have the potential to support the self-organizing behavior of distributed cognitive systems.

It should be acknowledged explicitly that Hutchins's (1990, 1991, 1995) work does not disregard the importance of human adaptation for successful performance in complex environments. His studies of the navigation task on naval vessels report a number of incidents in which adaptations are not only described but are regarded as important for system robustness. However, the theory of distributed cognition naturally places greater emphasis on the distributed nature of computational work processes, rather than on their self-organizing characteristics. Accordingly, the methods that Hutchins (1995) utilizes for mapping distributions of computational processes across actors and artifacts, namely the activity score, cannot readily accommodate the self-organizing behaviors of distributed cognitive systems.

Future research on distributed cognition should focus on elaborating its self-organizing properties, so that more robust accounts of how complex cognitive work is accomplished in sociotechnical workplaces can be developed. For example, research that examines the sociocognitive mechanisms by which self-organization occurs in distributed cognitive systems would be beneficial. In addition, future research should be concerned with methods for modeling and designing distributed cognitive systems. The diagram of work organization possibilities, which is aligned with the phenomenon of self-organization in sociotechnical systems, provides a potential starting point. However, further to some initial studies (Elix & Naikar, 2019; Naikar & Elix, 2016), other proof-of-concept demonstrations and analytical and empirical validation of the diagram, whether in laboratory, field, or industrial settings, are needed.

In conclusion, this chapter has examined some of the characteristics of self-organizing behaviors in distributed cognitive systems. It has been suggested that cognitive theories and methods used in the analysis and design of sociotechnical workplaces must be able to account for these behaviors. By developing a more complete account of how cognitive work is actually accomplished in complex settings, designs may be created that better support actors in dealing with instability, uncertainty, and unpredictability, and therefore in surviving in the wild.

ACKNOWLEDGMENTS

I am grateful to Mike McNeese and Mica R. Endsley for their thoughtful comments on a draft of this chapter, which were very helpful to me in revising it for publication.

REFERENCES

Bigley, G. A., & Roberts, K. H. (2001). The incident command system: High-reliability organizing for complex and volatile task environments. *Academy of Management Journal*, *44*(6), 1281–1299.

Bogdanovic, J., Perry, J., Guggenheim, M., & Manser, T. (2015). Adaptive coordination in surgical teams: An interview study. *BMC Health Services Research*, *15*, 128–139.

Director General. (2015). *The Fukishima Daichii accident*. Vienna, Austria: International Atomic Energy Agency.

Elix, B., & Naikar, N. (2020). Designing for adaptation in workers' individual behaviors and collective structures with cognitive work analysis: Case study of the diagram of work organization possibilities. *Human Factors*. https://doi.org/10.1177/0018720819893510

Furniss, D., Masci, P., Curzon, P., Mayer, A., & Blandford, A. (2015). Exploring medical device design and use through layers of Distributed Cognition: How a glucometer is coupled with its context. *Journal of Biomedical Informatics*, *53*, 330–341.

Haken, H. (1988). *Information and self-organization: A macroscopic approach to complex systems*. Berlin, Heidelberg, New York: Springer.
Heath, C., & Luff, P. (1998). Convergent activities: Line control and passenger information on the London underground. In Y. Engeström & D. Middleton (Eds.), *Cognition and communication at work* (pp. 96–129). Cambridge: Cambridge University Press.
Herzog, S. (2011). Revisiting the Estonian cyber attacks: Digital threats and multinational responses. *Journal of Strategic Security*, *4*(2), 49–60.
Heylighen, F. (2001). The science of self-organization and adaptivity. *The Encyclopedia of Life Support Systems*, *5*(3), 253–280.
Hoffman, R. R., & Woods, D. D. (2011). Beyond Simon's slice: Five fundamental trades-offs that bound the performance of macrocognitive work systems. *IEEE Intelligent Systems*, *26*(6), 67–71.
Hofkirchner, W. (1998). Emergence and the logic of explanation: An argument for the unity of science. *Acta Polytechnica Scandinavica: Mathematics, Computing and Management in Engineering Series*, *91*, 23–30.
Hollan, J., Hutchins, E., & Kirsh, D. (2000). Distributed cognition: Toward a new foundation for human-computer interaction research. *ACM Transactions on Computer-Human Interaction*, *7*(2), 174–196.
Hoppe, J., & Popham, P. (2007). Complete failure of spinal anaesthesia in obstetrics. *International Journal of Obstetric Anesthesia*, *16*(3), 250–255.
Hutchins, E. L. (1990). The technology of team navigation. In J. Galagher, R. E. Kraut, & C. Egido (Eds.), *Intellectual teamwork: Social and technological foundations of cooperative work* (pp. 191–221). Hillsdale, NJ: Lawrence Erlbaum Associates.
Hutchins, E. L. (1991). Organizing work by adaptation. *Organization Science*, *2*(1), 14–39.
Hutchins, E. L. (1995). *Cognition in the wild*. Cambridge, MA: The MIT Press.
Hutchins, E. L. (1999). Cognitive artifacts. In R. A. Wilson & F. C. Keil (Eds.), *The MIT encyclopedia of the cognitive sciences* (pp. 126–128). Cambridge, MA: The MIT Press.
Hutchins, E. L., & Klausen, T. (1998). Distributed cognition in an airline cockpit. In Y. Engeström & D. Middleton (Eds.), *Cognition and communication at work* (pp. 15–34). Cambridge: Cambridge University Press.
Klein, G., Wiggins, S., & Deal, S. (2008). Cognitive systems engineering: The hype and the hope. *IT systems perspective*, March, 95–97.
Klein, K. J., Ziegert, J. C., Knight, A. P., & Xiao, Y. (2006). Dynamic delegation: Shared, hierarchical, and deindividualized leadership in extreme action teams. *Administrative Science Quarterly*, *51*, 590–621.
La Porte, T. R. (1996). High reliability organizations: Unlikely, demanding and at risk. *Journal of Contingencies and Crisis Management*, *4*(2), 60–71.
Leveson, N. G. (1995). *Safeware: System safety and computers*. Reading, MA: Addison-Wesley.
Linde, C. (1988). Who's in charge here? Cooperative work and authority negotiation in police helicopter missions. In *Proceedings of the second annual ACM conference on computer supported collaborative work* (pp. 52–64). Portland, Oregon: ACM Press.
Luff, P., & Heath, C. (2000). The collaborative production of computer commands in command and control. *International Journal of Human-Computer Studies*, *52*, 669–699.
Lundberg, J., & Rankin, A. (2014). Resilience and vulnerability of small flexible crisis response teams: Implications for training and preparation. *Cognition, Technology and Work*, *16*, 143–155.
Naikar, N. (2013). *Work domain analysis: Concepts, guidelines, and cases*. Boca Raton, FL: CRC Press.

Naikar, N. (2018). Human-automation interaction in self-organizing sociotechnical systems. *Journal of Cognitive Engineering and Decision Making*, *12*(1), 62–66. https://doi.org/10.1177/1555343417731223

Naikar, N., & Brady, A. (2019). Cognitive systems engineering: Expertise in sociotechnical systems. In P. Ward, J. M. Schraagen, J. Gore, & E. Roth (Eds.), *The Oxford handbook of expertise: Research & application*. Oxford: Oxford University Press.

Naikar, N., & Elix, B. (2016). Integrated system design: Promoting the capacity of sociotechnical systems for adaptation through extensions of cognitive work analysis. *Frontiers in Psychology*, *7*(962), 44–64. http://journal.frontiersin.org/article/10.3389/fpsyg.2016.00962/full

Naikar, N., & Elix, B. (2019). Designing for self-organisation in sociotechnical systems: Resilience engineering, cognitive work analysis, and the diagram of work organisation possibilities. *Cognition, Technology & Work*. https://doi.org/10.1007/s10111-019-00595-y

Naikar, N., Elix, B., Dâgge, C., & Caldwell, T. (2017). Designing for self-organisation with the diagram of work organisation possibilities. In J. Gore & P. Ward (Eds.), *Proceedings of the 13th international conference on naturalistic decision making* (pp. 159–166). Bath: University of Bath. Retrieved from www.eventsforce.net/uob/media/uploaded/EVUOB/event_2/GoreWard_NDM13Proceedings_2017.pdf

National Transportation Safety Board. (2019). *Assumptions used in the safety assessment process and the effects of multiple alerts and indications on pilot performance*. Washington, DC: National Transportation Safety Board.

Norman, D. A. (1991). Cognitive artifacts. In J. M. Carroll (Ed.), *Designing interaction: Psychology at the human-computer interface* (pp. 17–38). Cambridge: Cambridge University Press.

Park, Y., Hong, P., & Roh, J. J. (2013). Supply chain lessons from the catastrophic natural disaster in Japan. *Business Horizons*, *56*(1), 75–85.

Perrow, C. (1984). *Normal accidents: Living with high risk technologies*. New York, NY: Basic Books.

Rasmussen, J. (1969). *Man-machine communication in the light of accident records* (Report No. S-1–69). Roskilde, Denmark: Danish Atomic Energy Commission, Research Establishment Risø.

Rasmussen, J. (1997). Merging paradigms: Decision making, management, and cognitive control. In R. Flin, E. Salas, M. Strub, & L. Martin (Eds.), *Decision making under stress: Emerging themes and applications* (pp. 67–81). Aldershot: Ashgate.

Rasmussen, J., Pejtersen, A. M., & Goodstein, L. P. (1994). *Cognitive systems engineering*. New York, NY: John Wiley & Sons.

Reason, J. (1990). *Human error*. Cambridge: Cambridge University Press.

Reich, P. C., Weinstein, S., Wild, C., & Cabanlong, A. S. (2010). Cyber warfare: A review of theories, law, policies, actual incidents—and the dilemma of anonymity. *European Journal of Law and Technology*, *1*(2), 1–58.

Rochlin, G. I. (1993). Defining "high reliability" organisations in practice: A taxonomic prologue. In K. Roberts (Ed.), *New challenges to understanding organizations*. New York: Macmillan.

Rochlin, G. I., La Porte, T. R., & Roberts, K. H. (1987). The self-designing high-reliability organization: Aircraft carrier flight operations at sea. *Naval War College Review*, *40*(4), 76–90.

Sellberg, C., & Lindblom, J. (2014). Comparing methods for workplace studies: A theoretical and empirical analysis. *Cognition, Technology & Work*, *16*, 467–486.

Stanton, N. A. (2014). Representing distributed cognition in complex systems: How a submarine returns to periscope depth. *Ergonomics*, *57*(3), 403–418.

Vicente, K. J. (1999). *Cognitive work analysis: Toward safe, productive, and healthy computer-based work*. Mahwah, NJ: Lawrence Erlbaum Associates.

Weick, K., & Sutcliffe, K. M. (2001). *Managing the unexpected: Assuring high performance in an age of complexity*. San Francisco, CA: Jossey Bass.

Woods, D. D. (1988). Coping with complexity: The psychology of human behaviour in complex systems. In L. P. Goodstein, H. B. Andersen, & S. E. Olsen (Eds.), *Tasks, errors, and mental models: A festschrift to celebrate the 60th birthday of professor Jens Rasmussen* (pp. 128–148). London: Taylor & Francis.

5 Unobtrusive Measurement of Team Cognition

A Review and Event-Based Approach to Measurement Design

Salar Khaleghzadegan, Sadaf Kazi, and Michael A. Rosen

CONTENTS

INTRODUCTION

In today's complex and dynamic work environment, organizations look to teams to address problems effectively and efficiently. Team-based work structures drive safety, innovation, productivity, and other important organizational outcomes across industries. One key dimension of teamwork at the forefront of understanding team performance across these criteria and work settings is collective or team cognition, a multi-level phenomenon underlying how teams process information, make decisions, plan, learn, and adapt. There are many perspectives on these general phenomenon ranging from those that focus primarily on how individual cognition (i.e., knowledge or cognitive processing) is shared or distributed across space, time, and team members (e.g., DeChurch & Mesmer-Magnus, 2010), to those taking a broader perspective to understand cognitive processing as a property of the sociotechnical system as a whole, inclusive of humans, technological agents, information systems, and artifacts in the physical environment (Hollan, Hutchins, & Kirsh, 2000; Zhang & Norman, 1994). These person-focused and whole-system-focused approaches have been integrated in different ways (e.g., Fiore et al., 2010) over the years. This chapter focuses primarily on unobtrusive measurement of team process, and therefore is most closely aligned with the view of team social interaction processes constituting a critically important aspect of distributed cognition (Cooke, Gorman, Myers, & Duran, 2013). However, these methods can be extended to understand broader aspects of distributed cognition such as artifact use in team settings (e.g., Li, Yao, Pan, Johannaman et al., 2016).

While team cognition is important for all teams, it is especially vital for teams that operate in dynamic and complex work settings that contain high degrees of stress, workload, and severity of consequences for performance lapses. The measurement of team cognition is therefore critical to both researchers seeking to build and refine better theories of team cognition, and to practitioners seeking to develop interventions to compose, train, or support team cognition in vivo. Over the years, a variety of methods have been developed and refined to measure different aspects of team cognition (Wildman, Salas, & Scott, 2014). However, new technologies and analytic tools provide the opportunity to create a new generation of team cognition measurement methods. Advances in wearable and environmental sensors, natural language processing, video analysis, and high dimensional time series analysis promise to deliver scalable and meaningful data collection around social interactions in field and lab settings (Yarkoni, 2012). Potential advantages of these emerging technologies include less labor-intensive data acquisition over longer durations of time, ultimately allowing for the measurement of team cognition in actual task environments rather than simulated conditions. While there are clearly open methodological issues to resolve (Chaffin et al., 2017), technical capabilities and evidence are maturing quickly.

In this chapter, we review the state of the science in applying unobtrusive measurement approaches to team cognition. To that end, we address three goals. First, we briefly review key concepts and traditional measurement strategies for team cognition. Second, we present a framework of unobtrusive measurement features (i.e., patterns in communication, physiological, and activity monitoring data) which can be used as indictors of team cognitive processing. Third, we detail a method for designing unobtrusive team cognition measurement systems within training and assessment simulations. This approach builds from an event-based approach to training (EBAT) methods of concurrent simulation and measurement system development (Fowlkes et al., 1997). Using this approach, training objectives and desired competencies drive the design of scenario events delivered through simulation. Observation of targeted responses demonstrate the presence or absence of competencies tested in the scenario and feedback can help improve performance for that desired competency.

TEAM COGNITION: CONCEPT AND MEASUREMENT

Key Concepts in Team Cognition

Team cognition encompasses a variety of concepts, including team mental models (TMM), transactive memory systems (TMS), and team situation awareness (TSA). These different terms conceptualize team cognition as inputs (e.g., TMMs), processes (e.g., interactive team cognition; Cooke et al., 2013), and emergent states (e.g., TSA; Gorman, Cooke, & Winner, 2006). TMMs describe the content and structures of individual team member's knowledge (Mohammed, Ferzandi, & Hamilton, 2010). Expanding previous research on TMMs, temporal TMMs incorporate time-related dimensions of teamwork which had been lacking in previous TMM frameworks because of the critical role that time can play on taskwork and performance (Mohammed, Hamilton, Tesler, Mancuso, & McNeese, 2015). These knowledge structures may relate to teamwork (e.g., roles and responsibilities, team interdependencies, model of interaction, etc.) or task work (e.g., nature of team task, tools and technology, etc.; Cooke, Kiekel, & Helm, 2001; Mathieu, Heffner, Goodwin, Salas, & Cannon-Bowers, 2000; Mohammed et al., 2010). The concept of TMS also concerns knowledge, but focuses on its distribution between team members, and on processes through which knowledge is organized (Wegner, 1987).

As opposed to the knowledge-based perspective of team cognition, Cooke and colleagues (2008, 2013) adopt a more dynamic view of team cognition that focuses on processes through which team members interact. TSA is also a dynamic construct of team cognition because it refers to the collection of an individual team member's perception and comprehension of the current state of the environment and projection of its state in the future (Bolstad & Endsley, 2003; Endsley, 1988). The use of scaled world simulations in research has been instrumental in conducting empirical research in concepts like TSA because of the ability to study real-world problems in realistic settings within a controlled lab environment (McNeese, McNeese, Endsley, Reep, & Forster, 2017). Many of the scaled world simulations that are used today in studies of different aspects of teamwork constructs stem from the NeoCITIES simulation, which is an interactive simulation environment for emergency crisis management (Hamilton et al., 2010; Hellar & McNeese, 2010). These simulations are used today in various team environments to study different variables like TSA, including their use in surgical teams (Hazlehurst, McMullen, & Gorman, 2007) and cybersecurity teams (Mancuso & McNeese, 2013).

Concepts in team cognition differ in their unit level of analysis (e.g., measurement of individual- vs. team-level data), as well as how frequently team cognition is measured. Some concepts within team cognition consider team cognition as the aggregate (e.g., sum, similarity, dispersion, fit) of the knowledge of individual team members at any given point in time. For example, TMMs are typically measured by assessing knowledge structures of individual team members at a single time point, and then comparing similarities between members' knowledge structures and accuracy level of knowledge structures (Mohammed et al., 2010). Similar, or shared, mental models can indicate that team members have similar descriptions, predictions, and explanation of events (Cannon-Bowers, Salas, & Converse, 1993). Accuracy of knowledge structures is assessed by comparing TMMs with objective evaluations of team performance (Edwards, Day, Arthur, & Bell, 2006). The research connecting shared mental models to team performance is mixed. Whereas some studies have found task work accuracy to be associated with better team performance (Cooke et al., 2001), others have found that teamwork similarity was more important than task work similarity for team performance (Webber, Chen, Payne, Marsh, & Zaccaro, 2000).

Measures relying on simple aggregations of individual team member cognitions may not be appropriate to make inferences about team cognition because they assume that interaction between team members is linear and that all members contribute equally across tasks (Webber et al., 2000). Therefore, Cooke and colleagues propose a more dynamic view of team cognition that is focused on evaluating team interactions across multiple points in time. This view considers team cognition as an emergent phenomenon that is adaptive to and shaped by interactions between team members as they accomplish task goals. Measures based in the interactive team cognition view focus on holistic, team-level communication that encapsulates dynamic cognitive processing as the locus of measurement.

Traditional Methods of Team Cognition Measurement

Similar to other domains of research in psychology, team cognition has been traditionally measured through self-reported knowledge and perceptions (e.g., surveys, interviews) and behavioral observations. Interviews and focus groups are used to elicit the structure, content, and organization of knowledge. In addition, these methods are also valuable in assessing recollections of interactive team processes. There is great diversity in methods to assess self-reported individual knowledge and perceptions about the team. These include surveys (e.g., transactive memory; Lewis, 2003), knowledge tests (Smith-Jentsch, Mathieu, & Kraiger, 2005), relatedness ratings (Fisher, Bell, Dierdorff, & Belohlav, 2012), card sorting (Smith-Jentsch, Campbell, Milanovich, & Reynolds, 2001), concept mapping (Marks, Sabella, Burke, & Zaccaro, 2002), and in-task probes (Bolstad & Endsley, 2003). Audio, video, and task logs and observations are valuable sources of capturing real-time team interactive processes. Data from all these sources are most commonly analyzed through aggregation, although researchers also use content analysis, analysis of patterns (e.g., recurrence, dominance, etc.), computational scaling, and holistic consensus (Wildman et al., 2014).

Decisions about methods of data collection should be theoretically guided by the team cognition construct under consideration and the overarching research question. However, it is equally important to be cognizant about the feasibility of data collection from the desired sample. Wildman et al. (2014) propose a variety of criteria to guide researchers in choosing appropriate methods of collecting data on team cognition. These include determinations about knowledge vs. interaction processes contained in the data, the underlying dynamic structure of the phenomenon, etc. In addition, Wildman and colleagues also evaluated methods of measuring team cognition in terms of their resource requirements for data collection and storage and the level of obtrusiveness to participant work. Audio and video observations may require low researcher burden during data collection and are relatively unobtrusive, but require significant resources during storage, processing, and analysis. Methods relying on self-report may be well suited to capturing knowledge about teamwork structures. However, they may be obtrusive and interrupt work, thus changing the nature of work.

EMERGING METHODS FOR ASSESSING TEAM PERFORMANCE

In this section, we provide an overview of emerging methods used to study group and team performance that could be applied to the measurement of team cognition. These methods are unobtrusive in nature and can complement the traditional methods reviewed in the previous section. These methods include team physiological dynamics, location and activity sensing, and linguistic and paralinguistic features of communication. In Table 5.1, we present a brief overview of data streams, data measures, and representative findings of these methods. We expand on these measures below.

TEAM PHYSIOLOGICAL DYNAMICS

Team physiological dynamics (TPD) refers to the continuous assessment of patterns of physiological arousal, synchrony, and organization of individual team members during team performance episodes (Lewis, 2003). TPD has been studied on diverse populations, including submarine operators (Gorman et al., 2016), healthcare teams (Stevens, Galloway, Halpin, & Willemsen-Dunlap, 2016), and flight deck pilot crews (Toppi et al., 2016), in addition to college students (Chanel, Kivikangas, & Ravaja, 2012). This body of research has enabled the study of important variables in team cognition.

Gorman and colleagues (Gorman et al., 2016) investigated neurophysiological and communication patterns of submarine operators in a simulation. Teams of novice and experienced submarine operator teams were measured prior to the task during briefing, during the task, and after the task during debriefing. Neurophysiological activity was measured through neurodynamic entropy (a measure of the change in neurophysiological distribution of a team) with a measure of low entropy representing a smaller change in team neurophysiological distribution (i.e. relatively fixed team mental state) and a measure of high entropy representing a higher change in team neurophysiological distribution (i.e. more flexible team mental state). In addition, communication content was analyzed using latent semantic analysis, a computational tool that measures the degree to which communication is synchronous or asynchronous based on the level of similarity between words that co-occur in dialog (Landauer, Foltz, & Laham, 1998; Dong, 2005). Teams of experienced submarine operators were more flexible during the task than novice teams, as shown by higher neurodynamic entropy in experienced teams during the scenario. This flexibility allows teams to function effectively and efficiently without becoming too rigid or deterministic in their operation (Stevens & Galloway, 2014). In addition, changes in communication patterns occurred before changes in neurophysiological patterns in experienced teams. This suggests that experience level and communication patterns can influence TPD.

TABLE 5.1
Unobtrusive Methods for Measuring Team Cognition

	Physiology	Activity Tracing	Linguistic Communication	Paralinguistic Communication
Data Streams	• Cardiac system • Electrodermal system (EDA) • Facial electromyography • Electroencephalography (EEG)	• Wearable and environmental sensors • Accelerometer • Use of the byproducts of interaction captured through information systems used for collaboration (e.g., email, paging)	• Lexical analysis/dictionary-based methods • Supervised learning • Generative language modeling	• Vocal Features • Communication flow • Gesture and posture • Facial expression and gaze behavior
Data Measures	• Interbeat interval, beats per minute, respiratory sinus arrhythmia, etc. • Tonic/phasic activity • Degree of physiological synchrony in a team • Degree of stability in patterns of brain activation across team members over time	• Patterns of interaction • Level of activity • Physical movements	• Frequency of word category use • Linguistic style matching (LSM) • Speech act/functional coding counts • Anticipation ratio • Semantic content and similarity	• Speech energy • Pitch • Rate/tempo • Speech dominance of members • Frequency of gesture and posture associated with emotional states • Degree of postural sway • Synchrony in gaze behavior
Representative Findings	• Experienced submarine operator teams show more diverse neural physiological organization during team performance which may indicate greater cognitive flexibility in the team (Gorman et al., 2016) • Higher level of cardiovascular arousal across team resulted in worse team performance (Walker, Muth, Switzer, & Rosopa, 2013)	• Higher perception of mental exertion in nurses associated with greater time spent in high activity non-service areas (Rosen et al., 2018) • Close proximity of team members is associated with fewer emails exchanged between them (Olguín et al., 2019)	• LSM predicts team cohesion and performance outcomes (Gonzales, Hancock, & Pennebaker, 2010) • High-performing teams have a higher anticipation ratio (Gontar et al., 2017)	• Egalitarian turn taking positively predicts team task outcomes or collective intelligence (Woolley et al., 2010) • Synchrony in postural sway negatively predicts team cohesion (Strang et al., 2014)

TPD has also been used as an indicator of cognitive readiness. Walker et al. (2013) measured cardiovascular activity when participants were engaged in a process control simulation. The simulation involved a team of two participants monitoring five tanks. Each participant was responsible for simultaneously monitoring levels of three parameters on two tanks; the fifth tank was jointly monitored. Walker and colleagues manipulated the difficulty level of the task by changing the frequency with which individual and team tanks had to be monitored. Results showed that higher levels of cardiovascular arousal across the team resulted in worse team performance, and that team cardiovascular activity accounted for 10% of variance in team error on the process control task. Overall, however, team performance was better predicted by individual team members' cardiovascular activity.

The study of TPD offers a new window into team cognition. TPD can be used to study variables across the IMO framework of studying teams (Kazi et al., 2019). Developments in relatively unobtrusive sensors capable of tracking physiological states can help accelerate study in TPD. This is a promising time to establish guidelines in the use of sensor technology in capturing TPD. These guidelines can help direct not only appropriate data collection and analysis methods, but also guide the linkage between team cognition and team member physiology.

Location and Activity Sensing

Radiofrequency identification (RFID) and infrared sensors are used in creating low-cost sensors to capture location and activity data. Rosen et al. (2018) investigated nursing workload in the intensive care unit by using sensors to capture location, activity, and speech patterns, and focus groups and perceptions of physical and mental exertion. Data from the sensors, focus groups, and surveys was interpreted in the context of patient census during each shift and patient acuity. Rosen and colleagues found that perceptions of mental and physical workload were differentially affected by individual and unit-level activity. Higher perceptions of mental exertion were predicted by noise levels in patient rooms, greater time spent in high-activity non-service areas, high-intensity speech that occurred close together with periods of sparser speech activity, and a high number of patients on insulin drip. On the other hand, higher perceived physical exertion was predicted by high levels of noise in areas containing patient supplies, unpredictable physical movements that occurred in close succession, greater time spent in nursing areas and more time spent speaking outside work areas, a negative interaction between noise in service areas and higher activity in patient rooms, and mean speech level at nursing stations and average patient load.

Proximity sensors can also be used to investigate team communication and interaction. Olguín and colleagues (2009) investigated patterns of face-to-face and email communication for 22 office workers though activity tracking sensors and email interactions. They found that being in close proximity with team members was associated with fewer emails exchanged with those members. Interestingly, a study of interaction of individuals in an organization found that changing the office to an open layout resulted in a significant reduction in face-to-face interaction and an increase in electronic communication (Bernstein & Turban, 2018).

Linguistic Features of Communication

Team communication is an essential process for all types of organizations (Marlow, Lacerenza, Paoletti, Burke, & Salas, 2018). The words that are used to communicate information comprise of the linguistic aspects of communication. The most common method for measuring linguistic communication is using self-report and content analysis of communication. Content analysis has traditionally been done by the manual coding of transcripts in order to categorize different measures of communication (Brauner, Boos, & Kolbe, 2018). In the following section, we discuss emerging methods of linguistic analysis that are done unobtrusively and do not depend on burdensome manual coding of interaction. These methods include dictionary-based lexical analysis, supervised learning, and generative (or unsupervised learning) methods (Grimmer & Stewart, 2013).

Lexical or *dictionary-based methods* require the use of pre-defined lists of words and phrases that are associated with known constructs of interest. The degree to which a specific construct is present in a team's communication depends on the rate at which these key words or phrases appear in the team's communication. Tausczik and Pennebaker (2010) review Linguistic Inquiry and Word Count (LIWC), a well-validated example of this approach. LIWC is a text analysis tool that counts words and phrases in psychologically meaningful ways in order to provide cues to different constructs such as thought processes, intentions, motivations, and emotional states (Tausczik & Pennebaker, 2010). While LIWC has been a popular tool among researchers interested in this domain, there are a multitude of novel tools available, especially for sentiment analysis which quantifies the emotional valence and intensity in communication content (Gilbert & Hutto, 2014). Dictionary-based methods such as LIWC provide measures of domain-independent language use including verb tense or pronoun use. The main advantage to such a technique is its broad applicability across teams and tasks done by teams. One disadvantage of this technique is that it does not capture any domain-specific or technical content that is used in language. A laboratory study of linguistic style matching (LSM), which is the degree to which members in a team mimic one another's use of function words or speech rate, positively predicts cooperation, group cohesion, and team performance (Gonzales, Hancock, & Pennebaker, 2010). In addition, lexical analysis has shown predictive validity for teams in complex, real-world tasks. The patterns of general language use, such as the use of first person plural, was found to be correlated with error rates amongst commercial aircrews (Sexton & Helmreich, 2000). Similarly, there is evidence that both general patterns of language use and task domain-specific communication are related to performance outcomes for spaceflight teams (Fischer, McDonnell, & Orasanu, 2007).

Supervised learning methods use a range of machine learning techniques to automate the coding of text (Evans & Aceves, 2016). These machine learning algorithms initially require the input of a training set of documents that have been previously coded by humans as well as a validation process using an additional set of documents that were not initially included in the training set. Supervised learning methods are most appropriate when there is a well-defined coding scheme and it is feasible to code a subset, although not the complete set, of documents. The advantage of these supervised learning methods is their potential to reduce the burden that is normally felt during manual coding of interaction.

Generative (or *unsupervised learning*) *methods* are latent variable modeling techniques that are applied in natural language processing (NLP) and machine learning tasks. These approaches work using previously unclassified documents. As an individual generates a speech or a document, they have in mind certain ideas, such as the subject that they are writing or speaking about. The idea(s) that a person has in mind may influence the likelihood of words chosen to help describe that idea (Landauer et al., 1998). Generative modeling identifies a set of topics that describe a set of documents, such as latent variables produced from a set of documents or speech acts. One such generative modeling technique is latent semantic analysis (LSA), which infers expected relationships between contextual usages of words in a discourse. Team performance outcomes was predicted with a reasonable degree of accuracy (correlation of $r = .63$) using LSA algorithms (Gorman, Foltz, Kiekel, Martin, & Cooke, 2003; Martin & Foltz, 2004). In addition, LSA revealed higher semantic similarity among high-performing teams (Gorman et al., 2016).

These methods each have the potential to contribute to team interaction measurement systems in different ways. Dictionary-based methods are popular because they are highly generalizable and relatively easy to use, but fail to capture context-specific characteristics of speech. Supervised learning methods need pre-existing categories, large training sets of coded documents, and a complex validation process, but they can have the potential to expand the throughput of a human coding team. Generative models such as topic modeling and LSA do not require initial coding like supervised learning methods, but are atheoretical and model diagnostics remain underdeveloped (Chuang, Gupta, Manning, & Heer, 2013). There is a rich community of researchers actively working on further developing these methods, so they are bound to grow in power and ease of use.

Paralinguistic Features of Communication

Paralinguistic aspects of communication are currently not frequently researched *within work team settings*. However, there is an extensive literature that team researchers can draw from in order to better measure, understand, and improve team communication (Gatica-Perez, 2009). Measuring paralinguistic features of communication include approaches for capturing vocal features, communication flow, gesture and posture, and facial expressions and gaze behavior.

Vocal features of potential interest to investigators that measure team communication include pitch, tempo, and energy (Vinciarelli, Pantic, & Bourlard, 2009). These communication features are validated as indicators of personality, perceptions of speaker competence, and persuasiveness (Schuller et al., 2015). Prosody, which includes different attributes of speech delivery such as stress and intonation, and voice quality distinguish between the most and least dominant group members (Charfuelan, Schröder, & Steiner, 2010). Principal components analysis (PCA) and support vector machines (SVM) have been used to identify vocal features associated with other team factors such as the speaker role (Charfuelan & Schröder, 2011). SVM and naive classifiers were used alongside human annotation and coding in order to estimate cohesion using nonverbal communication behavior (Hung & Gatica-Perez, 2010). Maximum overlap of speaking rate was found to be significantly higher in cohesive teams, where there was active participation among team members. Furthermore, nonverbal cues like prosody, speaking activity, and variation in energy (i.e., loudness heard by ear) and visual nonverbal features like head activity predict emergent leadership (Sanchez-Cortes, Aran, Mast, & Gatica-Perez, 2012).

Communication flow includes the temporal dynamics of communication acts. This is possibly the most extensively researched paralinguistic aspect of communication within work team settings. Communication flow involves the overall assessments of turn-taking behavior (e.g., speech dominance of individual members compared to the egalitarian sharing of turns; Woolley, Chabris, Pentland, Hashmi, & Malone, 2010), the degree of stability in turn taking behavior (Gontar, Fischer, & Bengler, 2017), and the occurrence of specific patterns of interaction (Tschan, 2002). Unobtrusive measures of communication flow can be derived from a variety of data streams, including audio recordings, automated speech detection systems (which identify segments of paralinguistic communication from microphones but do not actually record the content of communication in order to protect privacy in work settings), and activity traces (e.g., email, paging, phone, chats, and access logs in shared information systems). An example of measuring communication flow is an experimental paradigm that used a push-button-to-talk technique which allowed for tracking conversational flow (Gorman, Hessler, Amazeen, Cooke, & Shope, 2012). Gorman and colleagues showed that information from the communication device, which looked at who was selecting to talk to whom and when, provided important information about team interaction. Another study also examined different aspects of communication flow such as pauses between individual turns, pauses between floor exchanges, turn lengths, and overlapping speech, among others, using supervised machine learning techniques (Hung & Gatica-Perez, 2010). There were a number of significant findings for identifying highly cohesive teams. For example, it was found that total pause time represents how actively attentive members of the team are to one another. In addition, total overlap in communication was expected to be negatively correlated with cohesion. However, it turns out that overlap is actually a feature of highly cohesive teams because it indicates good rapport when team members are able to finish one another's sentences.

Gesture and posture includes the communication of emphasis and intent, as well as the emotional states of team members (Vinciarelli et al., 2009). Data for gesture and posture data can be generated through multiple methods, such as the analysis of video data or equipping individuals with sensors to capture position, movement, and muscle activation. There have been some studies which link gesture and posture within teams to specific outcomes or other constructs of interest. First, synchrony in postural sway negatively predicts team cohesion (Strang, Funke, Russell, Dukes, & Middendorf, 2014). In addition, it was found that authoritarian leaders are likely to move their arms more frequently than considerate leaders, while considerate leaders are more likely to imitate posture changes and head nods of their team members (Feese, Arnrich, Troster, Meyer, & Jonas, 2012). Mimicry is generally considered an indicator of group cohesion. These paralinguistic features of communication provide another opportunity to expand team communication measurement toolbox.

Facial expression and *gaze behavior* are powerful for their potential to communicate the attentional focus and emotional states within the team. Facial expressions can be coded using video data or measures of facial muscle activity captured by electromyography (EMG). Gaze behavior can be captured through eye-tracking systems or inferred from the position of the head using video data. Interestingly, synchrony of facial expressions between members in a team measured via video analysis or facial muscle activity is positively predictive of team cohesion (Mønster, Håkonsson, Eskildsen, & Wallot, 2016) and task performance outcomes (Chikersal, Tomprou, Kim, Woolley, & Dabbish, 2017). In addition, teams with low levels of synchrony in facial muscle activity were found to be more likely to adapt their strategies across different performance episodes (Mønster et al., 2016). In a study of airline crews, synchrony in gaze behavior, as an indicator of shared attentional focus, was predictive of performance outcomes (Gontar & Mulligan, 2016). In another study of student air traffic control teams, investigation of individual and team situation awareness (SA) suggested that eye tracking had the potential to be used to measure individual and team SA by using the co-occurrence of information seeking and acquisition to predict specific aspects of simulation (Hauland, 2008).

EVENT-BASED METHODS FOR UNOBTRUSIVE MEASUREMENT OF TEAM COGNITION

One of the main benefits of the emerging methods detailed above is high-volume, high-frequency data collection. However, this comes with challenges in analysis and interpretation, as this data is information rich, but often missing the context needed to generate understanding and meaningful insights into team performance (Rosen, Dietz, & Kazi, 2018). In field settings, this can be addressed to some degree by looking at patterns across sensor streams, or by targeting the use of traditional observational or self-report measures at identified critical or otherwise sampled time periods (Beal, 2015). However, the controlled but realistic environment of simulations provides more options for rigorous assessment. In this section, we briefly introduce event-based methods, discuss their value as an approach to unobtrusive measures, and provide example applications.

Simulation has been used extensively to both train and assess teams across a range of industries and research communities (Salas, Rosen, Held, & Weissmuller, 2009). The event-based approach to assessment is based on the premise that training or measurement objectives can drive the design scenario events that represent opportunities for team members to demonstrate and raters or trainers to observe behaviors targeted for assessment and development (Fowlkes, Dwyer, Oser, & Salas, 1998). A scenario event can be anything under the control of the scenario designers including shifting task demands, manipulation of information resources, or changes to team composition (Rosen et al., 2008). From a test development perspective, events represent the items that team members respond to with their interactions. Traditionally, event-based measurement links behavioral checklists to critical task situations experienced by the team as a way to focus rater attention on key moments. We argue here that creating critical events within scenarios also provides an anchor for the development of unobtrusive measures of team cognition by adding needed context to interpret the large volume of data generated by these sensor systems. As the field of unobtrusive measurement advances, there likely will be some indicators or markers of the level or quality of different team cognition constructs that are valid across team and task types. It is also likely that there will be unique indicators within different contextual constraints. This is similar to the notion of team and task *specific* and team and task *generic* competencies (Cannon-Bowers, Tannenbaum, Salas, & Volpe, 1995). Event-based methods provides an organizing framework for thinking about how different features of unobtrusive measurement may relate to team cognition constructs under different conditions. In Figure 5.1, we provide a brief overview of this framework and follow with more detailed examples below.

We introduce example event types and associated unobtrusive markers below. These are based on empirical findings reviewed above, but speculative in their connection to specific team cognition constructs as these relationships need to be further tested. First, the *leadership role structure* within the team can be manipulated as a scenario event and provide opportunities to measure TMM development. For example, scenarios can include conditions where formal leadership roles move from one team member to another (e.g., moving from one phase of work to another; induction of a patient led by an anesthesia provider in a surgical procedure to the actual procedure led by a surgeon), or where a novel situation is presented to the team without a clear leadership structure, yet an emergent leader is necessary. While equality of turn taking appears to be a general marker of effective team information processing (Woolley et al., 2010), teams exhibiting a shift in speech dominance such that the new or emergent leader has a higher proportion of speaking time may have more clearly recognized and adapted to the situation (Chaffin et al., 2017). This pattern of behavior could be indicative of better TMM of role structure within the team. Second, *task workload demands* within a simulation are frequently manipulated as critical events. Various makers of individual physiology are used to assess workload and the degree of synchrony in team physiological dynamics in response to shifts in task workload demand may be a valid marker of TSA, indicating shared perceptions of and similar reactions to new demands. Third, *team member turnover* or the addition of new team members (real or confederates) provides another opportunity to observe the team's development of a shared mental model of communication norms by capturing linguistic markers of style and their convergence or lack thereof. Fourth, *disruptions to communication structures* within the teams is common in military simulations (e.g., a team where all members could directly communicate shifts to a hub and spoke communication structure with a simulated communication channel outage). Measures of team member activity following this type of event (e.g., changes to use of available synchronous or asynchronous communication channels) is another opportunity to measure the TSA component of team adaptation (Burke, Stagl, Salas, Pierce, & Kendall, 2006).

In sum, the control over the environment afford by simulation-based assessment provides opportunities to focus unobtrusive measures, align them with traditional observational approaches, and create needed context for interpretation. As the evidence base grows, the repository of team and task general and team and task specific unobtrusive markers can be refined and serve as a valuable resource for practitioners and researchers.

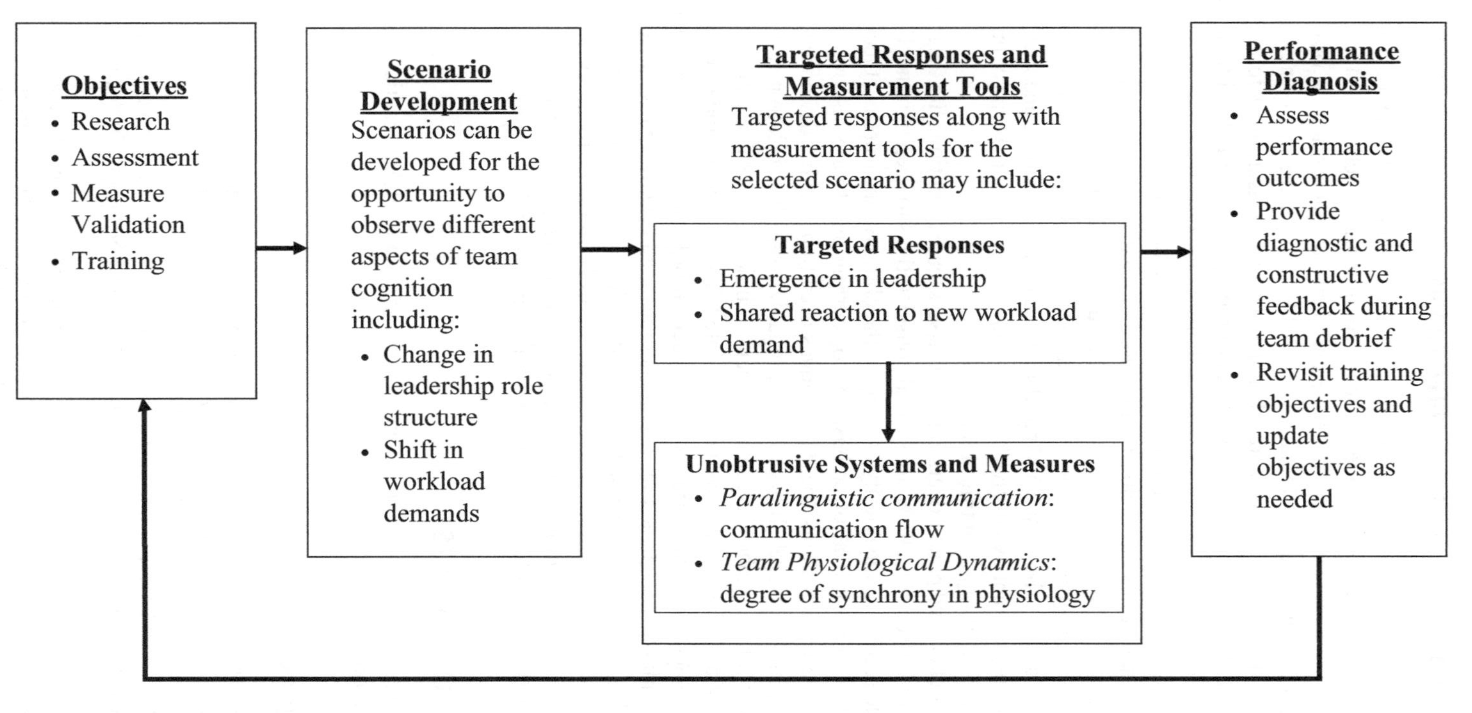

FIGURE 5.1 Organizing Framework for Event-Based Unobtrusive Measurement of Team Cognition.

CONCLUSION AND FUTURE DIRECTIONS

Team cognition plays a crucial role in team effectiveness and performance (DeChurch & Mesmer-Magnus, 2010). Decades of team cognition research has led to a multitude of measurement strategies that have enriched our understanding of different team cognition constructs (Wildman et al., 2014). The emergence of new technological tools such as wearables and other unobtrusive measurement tools provide us the opportunity to elevate our understanding of the science behind team cognition. These technological advances make it feasible for researchers to analyze data streams larger than ever before due to reduction in the burden for capturing this data. In addition, these tools can help provide more context to understanding the quality of cognition within teams, especially when used in conjunction with traditional measures. The four different categories of data streams reviewed in this chapter (i.e., physiological activity, linguistic and paralinguistic communication, and location and activity sensing) are not exhaustive but rather representative of the type of promising research that is developed in this arena.

For researchers and practitioners hoping to assess team cognition with the measurement tools discussed here, there are several considerations to keep in mind. First, it is important to recognize the novelty of many of the obtrusive measurement techniques. While there is a great deal of enthusiasm for current and future discoveries made possible by these innovations, these techniques need to be further developed with an emphasis on improving reliability. Second, organizations that seek to measure team cognition should use unobtrusive measurement techniques alongside more traditional measurement techniques that have been validated. Third, further studies of unobtrusive measurement in team cognition are needed in order to validate these methods for team cognition constructs. Improvements in team and cognitive task analysis that are better linked to unobtrusive methods should help by allowing us to know specific markers to look for in event-based approaches. This is undoubtedly an exciting time for researchers and practitioners as the presence of innovations has sparked a new era of discovery that will only strengthen our understanding in the coming years.

ACKNOWLEDGMENTS

This work was partially funded by a grant from the National Aeronautics and Space Administration (Grant # NNX17AB55G; PI: Rosen).

REFERENCES

Beal, D. J. (2015). ESM 2.0: State of the art and future potential of experience sampling methods in organizational research. *Annual Review of Organizational Psychology and Organizational Behavior*, *2*(1), 383–407.

Bernstein, E. S., & Turban, S. (2018). The impact of the "open" workspace on human collaboration. *Philosophical Transactions of the Royal Society B: Biological Sciences*, *373*(1753), 20170239.

Bolstad, C. A., & Endsley, M. R. (2003). Measuring shared and team situation awareness in the army's future objective force. In *Proceedings of the human factors and ergonomics society annual meeting* (Vol. 47, pp. 369–373). Los Angeles, CA: Sage Publications.

Brauner, E., Boos, M., & Kolbe, M. (2018). *The Cambridge handbook of group interaction analysis*. Cambridge: Cambridge University Press.

Burke, C. S., Stagl, K. C., Salas, E., Pierce, L., & Kendall, D. (2006). Understanding team adaptation: A conceptual analysis and model. *Journal of Applied Psychology, 91*(6), 1189.

Cannon-Bowers, J. A., & Salas, E., & Converse, S. (1993). Shared mental models in expert team decision making. In N. J. Castellan (Eds.), *Individual and group decision making: Current issues* (p. 221). Hilldale, NJ: Erlbaum.

Cannon-Bowers, J. A., Tannenbaum, S. I., Salas, E., & Volpe, C. E. (1995). Defining competencies and establishing team training requirements. *Team Effectiveness and Decision Making in Organizations, 333*, 380.

Chaffin, D., Heidl, R., Hollenbeck, J. R., Howe, M., Yu, A., Voorhees, C., & Calantone, R. (2017). The promise and perils of wearable sensors in organizational research. *Organizational Research Methods, 20*(1), 3–31.

Chanel, G., Kivikangas, J. M., & Ravaja, N. (2012). Physiological compliance for social gaming analysis: Cooperative versus competitive play☆." *Interacting with Computers, 24*(4), 306–316.

Charfuelan, M., & Schröder, M. (2011). Investigating the prosody and voice quality of social signals in scenario meetings. In *International conference on affective computing and intelligent interaction* (pp. 46–56). London: Springer.

Charfuelan, M., Schröder, M., & Steiner, I. (2010). Prosody and voice quality of vocal social signals: The case of dominance in scenario meetings. In *Eleventh annual conference of the international speech communication association*. Makuhari, Japan. Retrieved from https://www.isca-speech.org/archive/interspeech_2010/i10_2558.html

Chikersal, P., Tomprou, M., Kim, Y. J., Woolley, A. W., & Dabbish, L. (2017). Deep structures of collaboration: Physiological correlates of collective intelligence and group satisfaction. In *CSCW* (pp. 873–888), Portland, OR: ACM.

Chuang, J., Gupta, S., Manning, C., & Heer, J. (2013). Topic model diagnostics: Assessing domain relevance via topical alignment. In *Proceedings of the 30th International conference on machine learning* (pp. 612–620). Atlanta, GA: JMLR-W & CP.

Cooke, N. J., Gorman, J. C., Myers, C. W., & Duran, J. L. (2013). Interactive team cognition. *Cognitive Science, 37*(2), 255–285.

Cooke, N. J., Gorman, J. C., & Rowe, L. J. (2008). An ecological perspective on team cognition. In *Team effectiveness in complex organizations* (pp. 191–216). Abingdon: Routledge.

Cooke, N. J., Kiekel, P. A., & Helm, E. E. (2001). Measuring team knowledge during skill acquisition of a complex task. *International Journal of Cognitive Ergonomics, 5*(3), 297–315.

DeChurch, L. A., & Mesmer-Magnus, J. R. (2010). The cognitive underpinnings of effective teamwork: A meta-analysis. *Journal of Applied Psychology, 95*(1), 32.

Dong, A. (2005). The latent semantic approach to studying design team communication. *Design Studies, 26*(5), 445–461.

Edwards, B. D., Day, E. A., Arthur Jr., W., & Bell, S. T. (2006). Relationships among team ability composition, team mental models, and team performance. *Journal of Applied Psychology, 91*(3), 727.

Endsley, M. R. (1988). Design and evaluation for situation awareness enhancement. In *Proceedings of the human factors society annual meeting* (Vol. 32, pp. 97–101). Los Angeles, CA: Sage Publications.

Evans, J. A., & Aceves, P. (2016). Machine translation: Mining text for social theory. *Annual Review of Sociology, 42*.

Feese, S., Arnrich, B., Troster, G., Meyer, B., & Jonas, K. (2012). Quantifying behavioral mimicry by automatic detection of nonverbal cues from body motion. In *Privacy, Security, Risk and Trust (PASSAT), 2012 International Conference on and 2012 International Confernece on Social Computing (SocialCom)* (pp. 520–525). Piscataway, NJ: IEEE.

Fiore, S. M., Rosen, M. A., Smith-Jentsch, K. A., Salas, E., Letsky, M., & Warner, N. (2010). Toward an understanding of macrocognition in teams: Predicting processes in complex collaborative contexts. *Human Factors*, *52*(2), 203–224.

Fischer, U., McDonnell, L., & Orasanu, J. (2007). Linguistic correlates of team performance: Toward a tool for monitoring team functioning during space missions. *Aviation, Space, and Environmental Medicine*, *78*(5), B95.

Fisher, D. M., Bell, S. T., Dierdorff, E. C., & Belohlav, J. A. (2012). Facet personality and surface-level diversity as team mental model antecedents: Implications for implicit coordination. *Journal of Applied Psychology*, *97*(4), 825.

Fowlkes, J., Dwyer, D. J., Oser, R. L., & Salas, E. (1998). Event-based approach to training (EBAT). *The International Journal of Aviation Psychology*, *8*(3), 209–221.

Gatica-Perez, D. (2009). Automatic nonverbal analysis of social interaction in small groups: A review. *Image and Vision Computing*, *27*(12), 1775–1787.

Gilbert, C. J., & Hutto, E. (2014). Vader: A parsimonious rule-based model for sentiment analysis of social media text. In *Eighth International Conference on Weblogs and Social Media (ICWSM-14)*. Retrieved April 20, 2016 from http://comp.social.gatech.edu/papers/icwsm14.vader.hutto.pdf

Gontar, P., Fischer, U., & Bengler, K. (2017). Methods to evaluate pilots' cockpit communication: Cross-recurrence analyses vs. speech act—based analyses. *Journal of Cognitive Engineering and Decision Making*, *11*(4), 337–352.

Gontar, P., & Mulligan, J. B. (2016). Cross recurrence analysis as a measure of pilots' visual behaviour. In A. Droog, M. Schwartz, & R. Schmidt (Eds.), *Proceedings of the 32nd conference of the European association for aviation psychology*. Groningen, NL.

Gonzales, A. L., Hancock, J. T., & Pennebaker, J. W. (2010). Language style matching as a predictor of social dynamics in small groups. *Communication Research*, *37*(1), 3–19.

Gorman, J. C., Cooke, N. J., & Winner, J. L. (2006). Measuring team situation awareness in decentralized command and control environments. *Ergonomics*, *49*(12–13), 1312–1325.

Gorman, J. C., Foltz, P. W., Kiekel, P. A., Martin, M. J., & Cooke, N. J. (2003). Evaluation of latent semantic analysis-based measures of team communications content. In *Proceedings of the Human Factors and Ergonomics Society Annual Meeting* (Vol. 47, pp. 424–428). Los Angeles, CA: Sage Publications.

Gorman, J. C., Hessler, E. E., Amazeen, P. G., Cooke, N. J., & Shope, S. M. (2012). Dynamical analysis in real time: Detecting perturbations to team communication. *Ergonomics*, *55*(8), 825–839.

Gorman, J. C., Martin, M. J., Dunbar, T. A., Stevens, R. H., Galloway, T. L., Amazeen, P. G., & Likens, A. D. (2016). Cross-level effects between neurophysiology and communication during team training. *Human Factors*, *58*(1), 181–199.

Grimmer, J., & Stewart, B. M. (2013). Text as data: The promise and pitfalls of automatic content analysis methods for political texts. *Political Analysis*, *21*(3), 267–297.

Hamilton, K., Mancuso, V., Minotra, D., Hoult, R., Mohammed, S., Parr, A., . . . McNeese, M. (2010). Using the NeoCITIES 3.1 simulation to study and measure team cognition. In *Proceedings of the human factors and ergonomics society annual meeting* (Vol. 54, no. 4, pp. 433–437). Los Angeles, CA: Sage Publications.

Hauland, G. (2008). Measuring individual and team situation awareness during planning tasks in training of en route air traffic control. *The International Journal of Aviation Psychology*, *18*(3), 290–304.

Hazlehurst, B., McMullen, C. K., & Gorman, P. N. (2007). Distributed cognition in the heart room: How situation awareness arises from coordinated communications during cardiac surgery. *Journal of Biomedical Informatics*, *40*(5), 539–551.

Hellar, D. B., & McNeese, M. (2010). NeoCITIES: A simulated command and control task environment for experimental research. In *Proceedings of the human factors and ergonomics society annual meeting* (Vol. 54, no. 13, pp. 1027–1031). Los Angeles, CA: Sage Publications.

Hollan, J., Hutchins, E., & Kirsh, D. (2000). Distributed cognition: Toward a new foundation for human-computer interaction research. *ACM Transactions on Computer-Human Interaction (TOCHI)*, *7*(2), 174–196.

Hung, H., & Gatica-Perez, D. (2010). Estimating cohesion in small groups using audio-visual nonverbal behavior. *IEEE Transactions on Multimedia*, *12*(6), 563–575.

Kazi, S., Khaleghzadegan, S., Dinh, J. V., Shelhamer, M. J., Sapirstein, A., Goeddel, L. A., . . . Rosen, M. A. (2019). Team physiological dynamics: A critical review. *Human Factors*, 0018720819874160.

Landauer, T. K., Foltz, P. W., & Laham, D. (1998). An introduction to latent semantic analysis. *Discourse Processes*, *25*(2–3), 259–284.

Lewis, K. (2003). Measuring transactive memory systems in the field: Scale development and validation. *Journal of Applied Psychology*, *88*(4), 587.

Li, X., Yao, D., Pan, X., Johannaman, J., Yang, J., Webman, R., . . . Burd, R. S. (2016). Activity recognition for medical teamwork based on passive RFID. In *2016 IEEE international conference on RFID (RFID)* (pp. 1–9). Orlando, FL: IEEE.

Mancuso, V. F., & McNeese, M. (2013). TeamNETS: Scaled world simulation for distributed cyber teams. In *International conference on human-computer interaction* (pp. 509–513). Berlin, Heidelberg: Springer.

Marks, M. A., Sabella, M. J., Burke, C. S., & Zaccaro, S. J. (2002). The impact of cross-training on team effectiveness. *Journal of Applied Psychology*, *87*(1), 3.

Marlow, S. L., Lacerenza, C. N., Paoletti, J., Burke, C. S., & Salas, E. (2018). Does team communication represent a one-size-fits-all approach?: A meta-analysis of team communication and performance. *Organizational Behavior and Human Decision Processes*, *144*, 145–170.

Martin, M. J., & Foltz, P. W. (2004). Automated team discourse annotation and performance prediction using LSA. In *Proceedings of HLT-NAACL 2004: Short papers* (pp. 97–100). Boston, MA: Association for Computational Linguistics.

Mathieu, J. E., Heffner, T. S., Goodwin, G. F., Salas, E., & Cannon-Bowers, J. A. (2000). The influence of shared mental models on team process and performance. *Journal of Applied Psychology*, *85*(2), 273.

Mohammed, S., Ferzandi, L., & Hamilton, K. (2010). Metaphor no more: A 15-year review of the team mental model construct. *Journal of Management*, *36*(4), 876–910.

Mohammed, S., Hamilton, K., Tesler, R., Mancuso, V., & McNeese, M. (2015). Time for temporal team mental models: Expanding beyond "what" and "how" to incorporate "when". *European Journal of Work and Organizational Psychology*, *24*(5), 693–709.

Mønster, D., Håkonsson, D. D., Eskildsen, J. K., & Wallot, S. (2016). Physiological evidence of interpersonal dynamics in a cooperative production task. *Physiology & Behavior*, *156*, 24–34.

McNeese, M., McNeese, N. J., Endsley, T., Reep, J., & Forster, P. (2017). Simulating team cognition in complex systems: Practical considerations for researchers. In *Advances in neuroergonomics and cognitive engineering* (pp. 255–267). Cham: Springer.

Olguín, D. O., Waber, B. N., Kim, T., Mohan, A., Ara, K., & Pentland, A. (2009). Sensible organizations: Technology and methodology for automatically measuring organizational behavior. *IEEE Transactions on Systems, Man, and Cybernetics, Part B (Cybernetics)*, *39*(1), 43–55.

Rosen, M. A., Dietz, A. S., & Kazi, S. (2018). Beyond coding interaction. In *The Cambridge handbook of group interaction analysis* (Cambridge handbooks in psychology, pp. 142–162). Cambridge: Cambridge University Press. https://doi.org/10.1017/9781316286302.009.

Rosen, M. A., Dietz, A. S., Lee, N., Wang, I.-J., Markowitz, J., Wyskiel, R. M., . . . Gurses, A. P. (2018). Sensor-based measurement of critical care nursing workload: Unobtrusive measures of nursing activity complement traditional task and patient level indicators of workload to predict perceived exertion. *PLoS One*, *13*(10), e0204819.

Rosen, M. A., Salas, E., Wilson, K. A., King, H. B., Salisbury, M., Augenstein, J. S., . . . Birnbach, D. J. (2008). Measuring team performance in simulation-based training: Adopting best practices for healthcare. *Simulation in Healthcare*, *3*(1), 33–41.

Salas, E., Rosen, M. A., Held, J. D., & Weissmuller, J. J. (2009). Performance measurement in simulation-based training: A review and best practices. *Simulation & Gaming*, *40*(3), 328–376.

Sanchez-Cortes, D., Aran, O., Mast, M. S., & Gatica-Perez, D. (2012). A nonverbal behavior approach to identify emergent leaders in small groups. *IEEE Transactions on Multimedia*, *14*(3), 816–832.

Schuller, B., Steidl, S., Batliner, A., Nöth, E., Vinciarelli, A., Burkhardt, F., . . . Bocklet, T. (2015). A survey on perceived speaker traits: Personality, likability, pathology, and the first challenge. *Computer Speech & Language*, *29*(1), 100–131.

Sexton, J. B., & Helmreich, R. L. (2000). Analyzing cockpit communications: The links between language, performance, error, and workload. *Human Performance in Extreme Environments*, *5*(1), 63–68.

Smith-Jentsch, K. A., Campbell, G. E., Milanovich, D. M., & Reynolds, A. M. (2001). Measuring teamwork mental models to support training needs assessment, development, and evaluation: Two empirical studies. *Journal of Organizational Behavior: The International Journal of Industrial, Occupational and Organizational Psychology and Behavior*, *22*(2), 179–194.

Smith-Jentsch, K. A., Mathieu, J. E., & Kraiger, K. (2005). Investigating linear and interactive effects of shared mental models on safety and efficiency in a field setting. *Journal of Applied Psychology*, *90*(3), 523.

Stevens, R. H., & Galloway, T. L. (2014). Toward a quantitative description of the neurodynamic organizations of teams. *Social Neuroscience*, *9*(2), 160–173.

Stevens, R. H., Galloway, T. L., Halpin, D., & Willemsen-Dunlap, A. (2016). Healthcare teams neurodynamically reorganize when resolving uncertainty. *Entropy*, *18*(12), 427.

Strang, A. J., Funke, G. J., Russell, S. M., Dukes, A. W., & Middendorf, M. S. (2014). Physio-behavioral coupling in a cooperative team task: Contributors and relations. *Journal of Experimental Psychology: Human Perception and Performance*, *40*(1), 145.

Tausczik, Y. R., & Pennebaker, J. W. (2010). The psychological meaning of words: LIWC and computerized text analysis methods. *Journal of Language and Social Psychology*, *29*(1), 24–54.

Toppi, J., Borghini, G., Petti, M., He, E. J., Giusti, V. D., He, B., . . . Babiloni, F. (2016). Investigating cooperative behavior in ecological settings: An EEG hyperscanning study. *PLoS One*, *11*(4), e0154236.

Tschan, F. (2002). Ideal cycles of communication (or Cognitions) in triads, dyads, and individuals. *Small Group Research*, *33*(6), 615–643.

Vinciarelli, A., Pantic, M., & Bourlard, H. (2009). Social signal processing: Survey of an emerging domain. *Image and Vision Computing*, *27*(12), 1743–1759.

Walker, A. D., Muth, E. R., Switzer III, F. S., & Rosopa, P. J. (2013). Predicting team performance in a dynamic environment: A team psychophysiological approach to measuring cognitive readiness. *Journal of Cognitive Engineering and Decision Making*, *7*(1), 69–82.

Webber, S. S., Chen, G., Payne, S. C., Marsh, S. M., & Zaccaro, S. J. (2000). Enhancing team mental model measurement with performance appraisal practices. *Organizational Research Methods*, *3*(4), 307–322.

Wegner, D. M. (1987). Transactive memory: A contemporary analysis of the group mind. In *Theories of group behavior* (pp. 185–208). London: Springer.

Wildman, J. L., Salas, E., & Scott, C. P. R. (2014). Measuring cognition in teams: A cross-domain review. *Human Factors*, *56*(5), 911–941.

Woolley, A. W., Chabris, C. F., Pentland, A., Hashmi, N., & Malone, T. W. (2010). Evidence for a collective intelligence factor in the performance of human groups. *Science*, *330*(6004), 686–688.

Yarkoni, T. (2012). Psychoinformatics: New horizons at the interface of the psychological and computing sciences. *Current Directions in Psychological Science*, *21*(6), 391–397.

Zhang, J., & Norman, D. A. (1994). Representations in distributed cognitive tasks. *Cognitive Science*, *18*(1), 87–122.

6 A Method for Rigorously Assessing Causal Mental Models to Support Distributed Team Cognition

Jill L. Drury, Mark S. Pfaff, and Gary L. Klein

CONTENTS

INTRODUCTION

The concept of team cognition has evolved as the global culture of teamwork has continuously changed. A variety of non-traditional distributed teaming situations now occur frequently, such as:

- A cross-organizational team comprised of individuals scattered across time zones who never meet face-to-face and have different organizational reporting structures
- A team including Amazon Mechanical Turk workers as temporary colleagues from the "gig" economy, at least some of whom are unfamiliar with

corporate cultural norms, and others having a variety of experiences with corporate cultures
- Assistance on tasks rendered by pseudonym-identified contributors working from unknown locations who post helpful comments, observations, or answers to questions in response to social media blog entries describing a work challenge

Such situations have called into question the classic definitions of teams: "a distinguishable set of two or more people who interact, dynamically, interdependently, and adaptively toward a common and valued goal/objective/mission" (Salas, Dickinson, Converse, & Tannenbaum, 1992, p. 4). The primary sticking point pertains to the concept of a "common and valued goal." While all individuals have goals when they engage in activities (even if the goal is to avoid being bored), people who work together may not have the same goals. For example, the pseudonym-identified contributor described in the third situation above has the goal of demonstrating his or her expertise and garnering "likes" and other forms of reputational currency in an online environment, whereas the person being helped has the goal of resolving a specific business challenge.

Another issue regarding the definition of teaming arises from the nature of team members' interdependence—specifically, the level, constancy, and/or predictability of their interdependence. In traditional teaming arrangements, there is an expectation that members will engage with each other with at least somewhat predictable and/or pre-negotiated dependencies of each other's work processes and products. In a social media environment, unpredictable interactions and dependencies may occur with individuals due to different, perhaps previously unknown, people choosing to engage in addressing different challenges based on the relevance of each work task to the potential respondents' expertise and interests.

Some would argue that the lack of common goals and variable interdependence necessarily mean that the set of people do not constitute a team, but instead are members of a group (a set of individuals who coordinate their efforts), an online distributed group, or a collective (a large, undifferentiated set of people with diverse membership characteristics). However, given the fluidity of interactivity and the potential emergence over time of common goals from online social interaction, the definitions of teams and teamwork are softening (McNeese, 2019). Besides crowdsourcing and other forms of social online communication, the current culture of the gig economy, rapid change, and technological innovation may require stretching the teaming definitions that have been applied historically. It is now quite common, in fact, to speak even of "human-machine teaming" despite the fact that machines do not have the sentience and individual volition that are normally deemed necessary to form and adapt goals. Thus, when we speak of "distributed teams" in the context of this chapter we are taking a broad view of team characteristics, knowing that some may disagree with our inclusive stance. This broadened definition of distributed teams means that the conception of a teaming or shared mental model among team members may need to be quite different than our traditional notion of it. In turn, the ways we model and measure these kinds of teaming entities may have to change or be addressed in new ways. Accordingly, this chapter offers a new way to model and measure distributed team cognition.

The broadening of distributed team characteristics and configurations means that team members' viewpoints, backgrounds, motivations, and loci of attention are almost certainly becoming more diverse. In the aggregate, these characteristics may result in team members belonging to different cognitive cultures (Sieck, Rasmussen, & Smart, 2010). This broadening increases the need to make salient components of distributed team cognition more transparent, so that team members' diverse conceptualizations of the situation and possible decision choices can be readily perceived and thus more fully considered (Endsley, Reep, McNeese, & Forster, 2015). These conceptualizations of options and their potential outcomes can be described as mental models, most often defined as a "mechanism whereby humans generate descriptions of system purpose and form, explanations of system functioning and observed system states, and predictions of future system states" (Rouse & Morris, 1986, p. 360).

To reinforce the aspect of mental models that leads to predictions of future states, we refer to them as causal mental models. "A causal mental model is a network of ideas that constitute people's explanations for how things work. These models influence people's behavior, judgments and real-world decision making" (Gentner & Stevens, 1983, cited in Sieck et al., 2010). This description emphasizes the importance of capturing and including in mental models the causation chains that lead to state transitions. A causal model can be contrasted with a non-causal declarative model that captures the features of an object or lists facts about a situation.

The commonly occurring non-collocation of team members has a deleterious effect on forming shared causal mental models: "Team distribution has its largest impact on the development and utilization of shared mental models because of the team opacity [decreased awareness of team member interactions] arising from distributed interaction" (Fiore, Salas, Cuevas, & Bowers, 2003, p. 346). Achieving synchrony among team members is difficult when causal mental models among team members differ and those differences are not revealed. For example, two people might agree that the last ten years have been the hottest in recorded history, but if they disagree on why that is happening, they will disagree on how to address it. Shared causal mental models typically derive from continual articulation of events as situations evolve, including negotiating the meaning and causation of what is being shared. This negotiation is simply more difficult to achieve when team members are distributed because of diminished opportunities for establishing common understanding, including the inability to use deictic references and visual cues to establish whether conversation partners are experiencing confusion (McNeese, 2019). Having shared causal mental models, or at least an articulated understanding of divergences among causal mental models held by team members, is important because it can lead to higher team performance (Mathieu, Heffner, Goodwin, Salas, & Cannon-Bowers, 2000).

The literature around diverse workforces also notes that "diversity is a key driver of innovation and is a critical component of being successful on a global scale" (Forbes Insights Team, 2011, p. 3). The act of identifying that diversity and then discussing differing viewpoints so that the resulting shared causal mental model is different from anyone's original model is important to achieving better work outcomes (Gallo, 2018). Even when it is not possible to develop a single causal mental model, having awareness of different views of causation held by team members could enable decision makers to avoid taking actions that would be catastrophic if any of the known alternative viewpoints hold true. Further, maintaining awareness of diverse viewpoints makes it possible for team members to be alert for information that confirms one viewpoint versus another. Bringing this information back to the team could result in merging causal mental models over time.

The centrality of causal mental models as a mechanism to describe and operationalize team cognition is broadly accepted, despite the diversity in frameworks and theories pertaining to distributed team cognition (e.g., Hutchins, 1995; Fiore et al., 2003; Letsky, Warner, Fiore, Rosen, & Salas, 2007; Jonker, van Riemsdijk, & Vermeulen, 2010; Hinsz & Ladbury, 2012; Perry, 2017). It is challenging, however, to extract causal mental models from people's heads and compare them in a fashion that is both rigorous and accessible to people who are not modeling experts. The Descriptive to Executable SIMulation modeling methodology (DESIM; Pfaff, Drury, & Klein, 2015) provides a means to describe and assess causal mental models qualitatively and quantitatively, and therefore can allow team members to explicitly discuss their different understandings of the ways in which they believe the situation and their potential courses of action will evolve.

The DESIM methodology captures mental models in the form of influence diagrams that allow us to assess and quantify the relationships in models, and which ultimately allow us to execute models as simulations. The first part of DESIM is based upon Sieck's work in cognitive network analysis, which focuses on "gathering, analyzing, and representing the relevant cultural concepts, beliefs and values that drive decisions" (Sieck et al., 2010, p. 237). Our contribution is that we have extended a method by Osei-Bryson for individually quantifying the relationships in the model (Osei-Bryson, 2004) to a flexible crowd-sourcing technique to quantify the strengths of the relationships in the model. Crowd-sourcing enables modelers to quickly tap into a variety of knowledgeable subject matter experts. This technique is flexible because it handles incomplete data, if necessary, from each expert, thus enabling modelers to split up the information-provision load among the experts. Finally, we apply our own algorithm to execute the quantified model as a simulation. DESIM's unique contribution is integrating these separate steps into an efficient crowd-sourced process and developing prototype tools to streamline the elicitation, validation, and simulation of causal mental models.

The contribution of this chapter is a presentation of a rigorous method to expose differences in causal mental models held by distributed team members such that they can be resolved or otherwise accounted for. In the following sections, we review the different types of knowledge from the standpoint of their abilities to provide the raw data for causal mental models. Next we present DESIM, followed by an example of using DESIM to highlight its support to distributed team cognition. This chapter concludes by identifying the relationship between DESIM and the theoretical construct from distributed team cognition known as a distributed coordination space (Fiore et al., 2003).

TASK KNOWLEDGE, TEAM KNOWLEDGE, AND TEMPORAL KNOWLEDGE

Usually causal mental models are developed from task-based, team-based, and temporal knowledge (Mohammed, Hamilton, Tesler, Mancuso, & McNeese, 2015). Task-based knowledge pertains to an understanding of what needs to be done, including a shared terminology so communications about tasks can occur successfully. Team-based knowledge refers to understanding who team members are, whether they are available to engage in joint work at any given time, which team member knows what, and who is responsible for what parts of the joint task flow. We suggest that this list of team-based knowledge can be augmented by an awareness of the differences in causal mental models held by team members. Temporal knowledge refers to a detailed understanding of who is taking (or should take) what action at any given time, and how to pursue interactions with team members at any given moment. Team members may have good shared comprehension of the task (task-based knowledge) and their team members' skills and responsibilities (team-based knowledge), but if they have poor knowledge about the intricacies of time-based dependencies (temporal knowledge), their performance may suffer (Mohammed et al., 2015). These three knowledge types are described in more detail below.

Task Knowledge

An in-depth knowledge of the facts of the environment in which the task is situated is most often known as situation awareness (Endsley, 1988). Situation awareness is defined as the perception of an environment within a volume of time and space, plus the comprehension of that environment to the degree that enables the projection of what will happen in the near future (Endsley, 1988). Situation awareness originated as a concept to describe an individual's knowledge of an environment, but it has been expanded to encompass and describe the knowledge held by members of a collaborating team that is working towards a joint goal (e.g., Endsley, 1995). Several approaches to team situation awareness have been proposed (see She & Li, 2017 for a review). For example, one approach consists of the aggregation of the facts that individuals know (Endsley, 1995), and another focuses on the content of the team's factual knowledge plus the understanding of which knowledge content is shared (Espinosa & Clark, 2014).

While having good situation awareness is a necessary prerequisite for creating a causal mental model of the environment, making decisions and taking actions in any given situation, it is not sufficient to completely support the decision-making process (Pfaff et al., 2013). Option awareness is also needed: the knowledge of the available courses of action and the relative desirability of one choice versus another (Drury, Klein, Pfaff, & More, 2009). To paraphrase, option awareness is the understanding of the options available and plausible outcomes of choosing one option versus another. Still, people who possess the same situation awareness and have been given the same type of option awareness support, may nevertheless assess options differently if they hold different causal mental models.

A concept for team-based option awareness (Klein, Drury, Pfaff, & More, 2010; Liu, Moon, Pfaff, Drury, & Klein, 2011) relies on identifying synergistic joint courses of action. Simply collaborating over individual decisions will not achieve collaborative option awareness because jointly executing even the most robust individual options may not yield the most robust joint option (Klein et al., 2010). Collaborative decision making under these conditions is complex not only due to the difficulty of forecasting the impact of this synergy, but also the need to achieve a common causal mental model among the decision-making participants.

Team Knowledge

Team-based knowledge refers to an understanding of what team members are doing, whether they are present or absent, and their goals and intentions for future task-related actions. Knowing about team members' activities is more challenging when team members are distributed across locations and potentially time zones, adding asynchronicity to the mix. A process for generating team-based knowledge is described as "macrocognition in teams," which Fiore, Smith-Jentsch, Salas, Warner, and Letsky (2010) defined as "the process of transforming internalised team knowledge into externalised team knowledge through individual and team knowledge building processes" (p. 258). Knowledge of one's collaborators' states is described as having awareness of what team members are doing, have done, or are intending to do (Dourish & Bellotti, 1992; Drury & Williams, 2002; Gutwin & Greenberg, 2002; Gaver et al., 1992; Gross, Stary, & Totter, 2005). This type of awareness is essential for synchronizing activities and avoiding duplicating work or leaving important tasks undone and primarily serves as a means to promote effortless coordination (Gross, 2013). This type of knowledge supplements and augments users' causal mental models of the situation to the degree that team related information affects the available options and the plausible outcomes of choosing one option versus another.

It is difficult to assess the degree to which a common understanding among team members has been attained. Espinosa and Clark (2014) developed a network model that shows which elements of the situation are known by which team members. While such networks illustrate the degree to which there is common knowledge of the facts of the situation (that is, situation awareness), they do not provide an assessment of whether team members share common causal models that can lead to a shared option awareness.

Temporal Knowledge

Team members have sufficient temporal knowledge when they are aware of the time-based dependencies of their assigned tasks, the moment-by-moment progress towards task completion, and the point when they need to execute their tasks to respect those dependencies (Mohammed et al., 2015). Such knowledge is enriched by a shared causal model because it can inform why the temporal dependencies exist and what to do to repair a situation in which the task sequence is disrupted.

CAPTURING AND COMPARING MENTAL MODELS USING DESIM

Overview

The DESIM process elicits and transforms descriptive causal models into executable computer simulation models based on information obtained from multiple experts in a subject area. A computer user interface backed by computational algorithms produces quantitative values for the strengths of causal relationships between variables in the descriptive models, resulting in unbiased distributions of estimated values for each relationship and enabling the models to be computationally processed. The result is a quantitative depiction of causal mental models, shown as influence diagrams, plus an improved depiction of potential outcomes known as a decision space visualization (Pfaff et al., 2013). Decision space visualizations present the relationships among options, actions, or variables that can be used to analyze a focus question and support decision making. Figure 6.1 depicts the DESIM process.

The advantages of this process include capturing perceptions that are difficult to quantify and collecting data in an engaging, participatory approach that is accessible to domain experts unfamiliar with modeling processes (Özesmi & Özesmi, 2004). Participatory modeling (Prell et al., 2007; McNeese, Zaff, Citera, Brown, & Whitaker, 1995) provides domain experts with more control over how the model is constructed than situations in which the model development occurs solely via the interpretation by the analyst of the expert's answers to interview questions. Involving domain experts as co-constructors of the models is critical to developing accurate models quickly. Using multiple experts in participatory modeling ensures that the

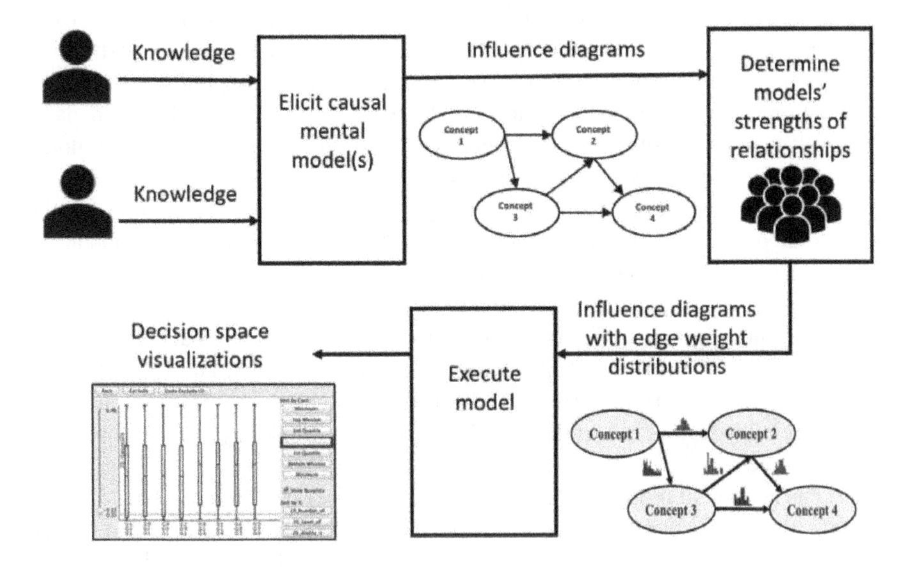

FIGURE 6.1 Overview of the DESIM process.

Note: Boxes indicate processes and arrows indicate inputs to and/or outputs from the processes.

resulting model(s) are not idiosyncratic of a single domain expert (Vennix, 1999). Finally, this method allows for detailed comparisons of causal models developed with different domain experts, enabling the differences to be highlighted and probed. For example, multiple sessions may be held with domain experts to negotiate developing a composite model, or they may "agree to disagree," resulting in preserving multiple models that represent the different viewpoints.

The rest of this section describes the methods for eliciting one or more cognitive models of a problem, representing the models interactively, eliciting values for the models from many experts, and analyzing and viewing the data resulting from executing the models.

Eliciting Causal Mental Models and Representing Them Interactively

The DESIM process starts by eliciting one or more domain expert's causal mental model via interviews that explore a focal question such as "What is causing employee dissatisfaction?" Most often, multiple domain experts are interviewed to understand different and potentially conflicting perspectives on the problem. Experts are asked questions requesting that they identify model components, which are factors that experts believe influence the outcomes of the focus question; links between components; and the dynamic and functional relationships among the components. This process is facilitated by an analyst who displays concept mapping software to develop one causal mental model per expert.

The causal mental models are represented on a computer as a concept map or influence diagram: a graph of nodes connected by edges. A node in a descriptive model is a variable that represents a concept such as an action, option, or policy that has a continuous or discrete range of values. Multiple tools exist for visually depicting and computationally representing mental models in this manner. CMapTools (Cañas et al., 2004), for example, provides a graphical interface for constructing and editing cognitive models and provides machine-readable output for use by other computational tools. Similarly, MentalModeler (Gray, Gray, Cox, & Henly-Shepard, 2013) was designed to support a fuzzy cognitive modeling process, including model elicitation, graphical representation, and exploratory simulation.

The validation process begins by displaying a graphical representation of the newly created model to the domain expert, who checks it for completeness and accuracy (Sieck et al., 2010). We have had good results with using a mental modeling tool interactively during the interview session to turn responses to questions immediately into nodes and edges (see Figure 6.2 for an example). This approach enables the analyst to obtain model-related information, construct a model, and receive initial validation and/or real-time corrections from the expert during the same meeting, usually one to two hours in length depending upon the complexity of the subject area being explored.

The causal mental models created for each expert may be similar or may diverge. The divergence may consist of additional factors (nodes) that were included by one or more experts. Alternatively, the same set of nodes may be connected by experts in different ways (that is, the sets of edges are dissimilar), signifying differences in beliefs regarding the relationships among the same factors. It is also possible that

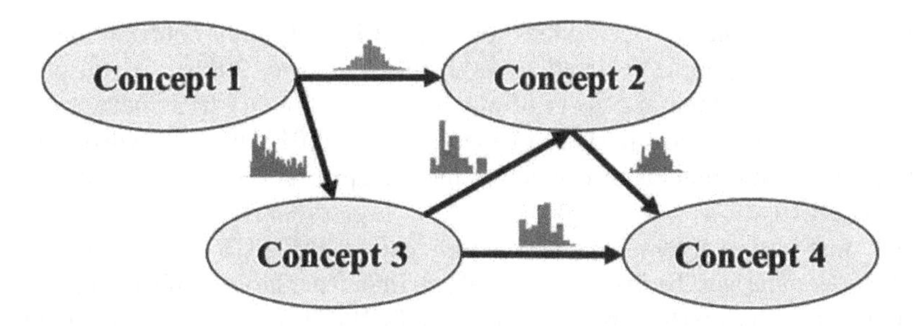

FIGURE 6.2 Example mental model illustrating causal relationships between concepts (arrows) and distributions of edge weights elicited from domain experts (histograms).

the direction of the edges differs from one expert's model to another, indicating a disagreement in which factors cause other factors to occur (a chicken-and-egg dilemma). Analysts combine multiple models based on concepts they hold in common and invite the experts to review the resulting model for completeness and accuracy. However, it is also illuminating to have more than one model describing the experts' mental models and apply the remainder of the DESIM process to each one and compare the results.

Assigning Edge Weights

Once a model is elicited from one or more domain experts and represented structurally, the DESIM computer program processes it in parts to obtain edge weights with the range from −1 to +1. An edge weight quantifies a causal association or relationship between the two or more nodes that are connected by the edge, similar to that described by Perusich and McNeese (2006) and McNeese, Rentsch, and Perusich (2000). The sign of an edge weight denotes a direction of correlation between nodes, and the magnitude of an edge weight denotes the strength of the causal relationship between the nodes. While a static value for an edge weight could be elicited from a single expert (but may be unreliable), a more trustworthy distribution of values for the edge weights can be determined through appropriately querying multiple experts. In a departure from classic fuzzy cognitive maps, an algorithm determines how the feedback defines a distribution of edge weights for each edge.

While the interviewed experts are able to give the sign (+ or −) of a causal relationship, they are less able to give an accurate estimate of the magnitude (Osei-Bryson, 2004). Because subjective point estimates are unreliable, another method is necessary to produce accurate edge weights. This is achieved through a systematic set of pairwise comparisons of the connected node pairs in the model, for which an expert rates the comparative strength of two relationships. The choice is whether relationship $X_1 \rightarrow X_2$ is stronger than $X_3 \rightarrow X_4$, and by how much, repeated for as many pairwise comparisons of edges in the model necessary to produce a complete set of edge weights.

DESIM uses "expert sourcing" (crowdsourcing among domain experts) to quantify the relationships in the descriptive causal model. Crowdsourcing is a process of obtaining services, ideas, or content by soliciting contributions from a large group of people referred to as a crowd (Howe, 2006). Crowdsourcing combines the incremental efforts of numerous contributors to achieve a greater result in a relatively short period of time. Lin et al. (2012) used crowdsourcing to understand mental models of privacy in mobile applications, but did not create a model explicitly.

Our web-based automated survey tool called IMPACT (Interactive Model PAirwise Comparison Tool) can obtain information from large numbers of people in a population with expertise about the problem. IMPACT takes the machine-readable model produced in the preceding steps and generates a set of pairwise comparisons that are presented in sequence to the experts. First, a single relationship $X_1 \rightarrow X_2$ is presented graphically to the user with the question "Do you agree with this relationship?" The two choices are "Agree: An increase on the left causes and increase on the right," and "Disagree: An increase on the left does not cause an increase on the right."

After the respondent has agreed with at least two relationships in the model, these relationships ($A = X_1 \rightarrow X_2$ and $B = X_3 \rightarrow X_1$) are presented with the question "Which relationship is stronger?" with the choices "A is stronger than B," "B is stronger than A," or "A is the same as B." If either of the first two choices are selected, the respondent is additionally asked "How much stronger?" and is presented with a slider ranging from "A is much stronger than B" to "A is the same as B." After answering, the respondent then proceeds to the next comparison. When the respondent disagrees with a given relationship, it is given a weight of zero and eliminated from all future pairwise comparisons given to that expert.

After a respondent completes all of the essential pairwise comparisons, IMPACT analyzes the results of the pairwise comparisons using a form of the analytic hierarchy process (AHP; Saaty, 1990) modified to accommodate incomplete sets of comparisons (Harker, 1987). From this analysis, it calculates the complete set of edge weights and as a final calibration step it asks the respondent to provide an absolute weight for the relationship with the strongest edge. A respondent who rates one relationship stronger than the rest does not necessarily believe that it is a very strong relationship, but without this calibration step, that strongest edge would be rated near the top of the scale. Rescaling each respondent's edge weights accordingly more accurately captures their true beliefs and enables more accurate comparisons between experts and across models.

The model validation begun earlier continues by computing the internal consistency of pairwise responses (creating a consistency ratio), and by examining the level of support from respondents for each relationship as derived from the computed edge weights (see Pfaff, Klein, & Egeth, 2017, for more validation details).

The sets of edge weights for all respondents are aggregated and used to populate the original model with distributions of edge weights for each relationship in the model. Using these distributions of weights, multiple simulation model processing runs can be performed to assess how the values in the distributions for each variable affect the ranges of outcomes for each possible decision option to address the focal question. These processing runs generate one or more outcome depictions using an iterative fuzzy cognitive modeling (FCM) method (Kosko, 1986). In this method, initial node values and edge weights can be varied for each processing run to create the distribution of outcomes. Team members' analyses of the resulting distribution of outcomes provides a more comprehensive understanding of the tradeoffs and tipping points than a single aggregated mean estimate regarding how various variables impact the focal question (Pfaff et al., 2013).

Examining Edge-Weight Differences and Viewing Results

The edge weights can be interpreted not only as a respondent's estimate of the strength of the causal relationship between two nodes, but also his or her level of agreement with the idea that the two nodes are related at all. Calculated weights that are at or very near zero indicate broad disagreement with a proposed relationship. While there may have been agreement within the small population that was interviewed by analysts to produce the initial model, the larger population accessed via the IMPACT survey may reveal differences in beliefs regarding the model's structure.

Even subpopulations that agree completely on the structure of models may assign different values to edge weights. For example, there may be bi-modal or multi-modal distributions of edge weights among subpopulations based on differences in belief regarding the relative strengths of relationships between factors represented by the model's nodes. Each relationship for which disagreement exists can form the basis for further probing in follow-up interviews or focus groups.

The model can be executed across each of the sets of edge weights elicited from the subject matter experts. To score the outcomes, one or more nodes in the model, such as cost or mission effectiveness, are selected as evaluative criteria. Because the fuzzy cognitive modeling method used in DESIM is computationally lightweight, we can additionally examine a wide variety of plausible conditions, such as environmental factors beyond the control of decision makers, under which an option might be executed. An example of these environmental factors is weather, which, for certain domains such as military command and control or emergency response, may affect the team's performance when executing an option. (For example, performance in a case that assumes dry and gusty conditions may be very different from another case that assumes a drenching downpour with limited visibility.) The result is a range of outcomes computed for each decision option, generated by running each set of edge weights crossed with a set of plausible values for each of the variables in the environment.

The outcomes are displayed as a range of possible results for each option—a decision space visualization, such as can be seen in Figure 6.3. Team members can apply their beliefs and judgment regarding the distribution of outcomes that would be acceptable and can dig into the parameters of each data point to determine tradeoffs and tipping points. Further, they can use decision space visualizations to facilitate team discussions regarding risks and costs (or other evaluative criteria).

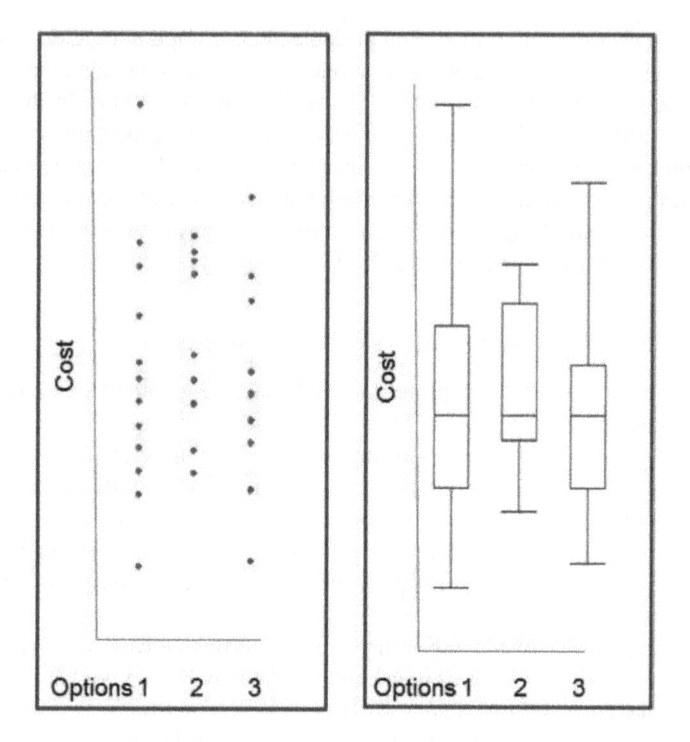

FIGURE 6.3 Two visualizations of the same example decision space, using a scatter plot and box plots.

Note: The box plots summarize the maximum, minimum, mean, and highest and lowest values of the middle 50% of the cases for each option, which are scored based on cost in this situation. Note that the three options have almost the same mean values, but the shapes of their distributions are very different. If only means are reported rather than showing their entire distributions, much of the nuance of the results for each option is lost.

EXAMPLE USE OF DESIM

DESIM was used in a real-world application after an organization's Likert-scale survey about employee satisfaction revealed unexpected results that required further examination. DESIM was chosen because of its potential to uncover the causal relationships staff members believed affect various aspects of job satisfaction and to reveal the differences in opinions among subpopulations.

To start, 50 employees participated in hour-long semi-structured interviews. Analysis resulted in seven causal models being created using MentalModeler (Gray et al., 2013) regarding the subjects of compensation, connections among staff, getting good work, management practices, perceived value, conflict between organizational groups (called divisions), and promotions. Models were validated with employees.

These seven models contained a total of 80 nodes and 98 edges, leading to a total of 199 pairwise comparisons. Because the time needed to complete all 199 comparisons would be too long to expect of an individual, employees participating

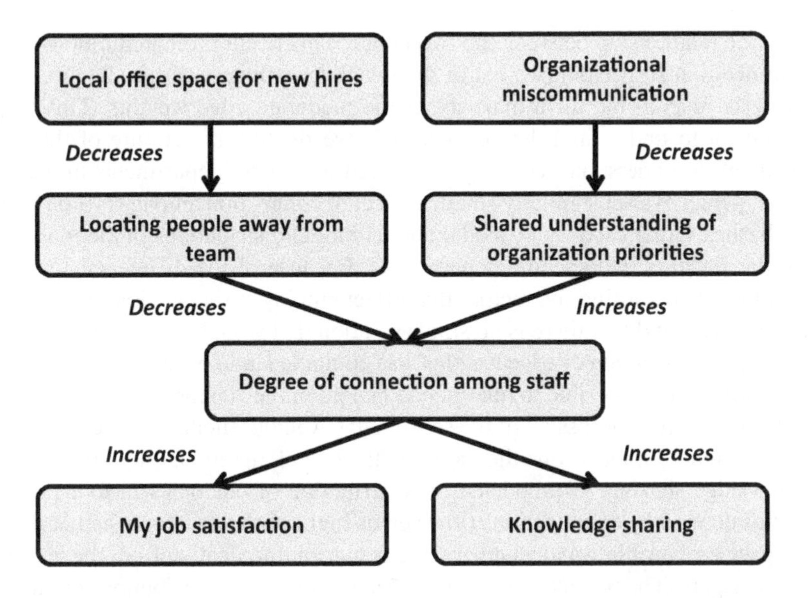

FIGURE 6.4 Influence diagram describing a causal mental model for the topic "Connections Among Staff," part of an investigation into the factors that affect job satisfaction and knowledge sharing.

Source: Adapted from Pfaff et al., 2017.

in the crowdsourcing phase were asked to complete the comparisons for three randomly selected models, with the remaining model comparisons being optional if the employee wished to continue.

A total of 232 employees used IMPACT to create over 10,000 individual edge weights. After computing the consistency ratio (CR), the "compensation" model was revealed as having particularly inconsistent results, meaning that the model required further review and clarification before its results could be trusted. In contrast, the "connections among staff" model was shown to have extremely consistent results (see Figure 6.4). Within this model, however, there were disagreements regarding the strengths of relationships.

For example, roughly one-third of respondents explicitly disagreed with the proposition formed by the node-edge-node tuple of "Shared understanding of division priorities increases the degree of connection among staff." An additional approximately one-third gave the relationship a weight of less than 0.05, which indicates a very small effect. The remainder believed there was at least a moderately strong relationship, leading to a multi-modal response pattern.

Another example of disagreement regarding edge weights pertains to the proposition "Degree of connection among staff increases knowledge sharing." This proposition received a lot of support overall, with a quarter of the respondents rating this edge strongest in the model and only one respondent explicitly disagreeing with it. A bi-modal distribution, however, revealed two subpopulations: one believing that connection among staff strongly increases knowledge sharing and one that sees only a moderate relationship between the two nodes. This result prompted follow-up that determined that staff from two out of the seven departments within the total population were responsible for a majority of the moderate edge weights. This helped management to understand the necessity of investigating the culture of those two departments to understand why they, compared with other departments in the same division, perceive a substantially smaller effect of connection among staff on knowledge sharing. Other cases of bi-modal and tri-modal distributions in the models led to similar insights into specific departments. Because of the differences in causal mental models regarding the factors that affect employee satisfaction, management came to understand that there is no single solution that would satisfy everyone.

Once the distribution of edge weights was computed, and by assigning initial values to nodes, it was possible to manipulate factors in the "connections among staff" model and execute it across all of the sets of edge weights elicited from employees to see their potential impacts on other nodes in the model, in this case looking at effects on knowledge sharing. We modeled the distribution of outcomes in four potential scenarios defined by combinations (low versus high) of two factors: whether there is local space for new hires and whether there is miscommunication from the organization's managers. The box plots in Figure 6.5 summarize the distributions of outcome

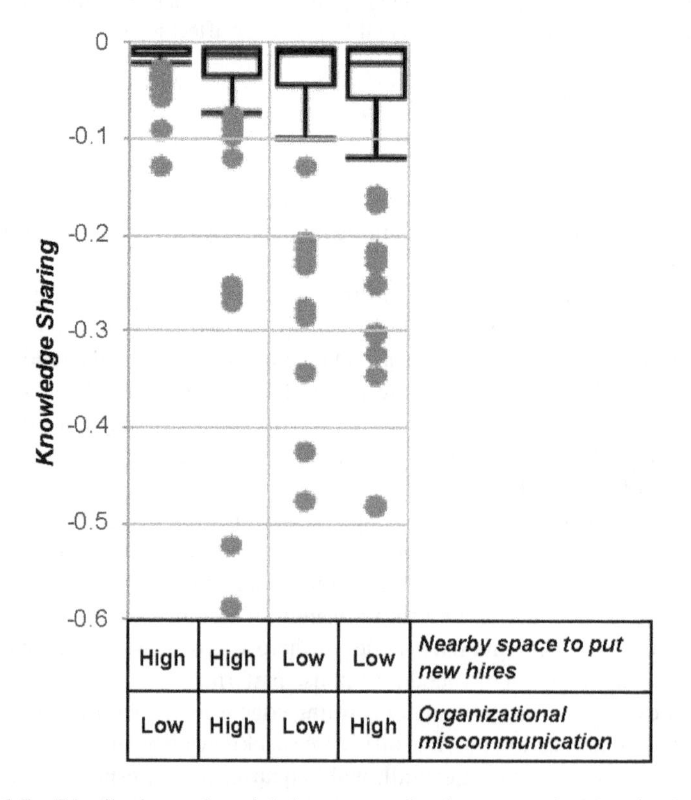

FIGURE 6.5 Distributions of modeled outcomes for four scenarios that imply different courses of action.

Source: Adapted from Pfaff et al., 2017.

values ordered by decreasing mean value, with the outlier points indicating more extreme values. A large fraction of the possible outcomes of the first two scenarios suggest little to no effects of either factor on knowledge sharing, but many of the possible outcomes were more negative for the last two scenarios. These results indicate that non-collocation of department members has a stronger negative effect on knowledge sharing than communications from management, shown through the clearly more negative outcomes of having little local office space, regardless of the level of organizational miscommunication. Without this type of visual assistance, people are usually incapable of exploring the implications of their mental models. Using the visualizations, we worked with department management to understand the results so that they could determine data-supported courses of action for organizational improvement.

IMPLICATIONS FOR SUPPORT TO DISTRIBUTED TEAM COGNITION

The foregoing explained how DESIM supports distributed teamwork by identifying the similarities and differences of team members' causal mental models and by providing visualizations that depict the range of possible outcomes for the analyzed options (such as in Figures 6.3 and 6.5). But how should DESIM be situated in relationship to theories that describe how distributed teamwork takes place? To investigate this question, we examined DESIM's relationship to a theoretical construct known as a distributed coordination space (Fiore et al., 2003). A distributed coordination space is "a theoretical framework designed to elucidate the many issues surrounding distributed team performance, emphasizing how work characteristics associated with such teams may alter both the processes and the products emerging from distributed interaction" (Fiore et al., 2003, p. 340).

DESIM and the Coordination Space Framework

The distributed coordination space framework consists of three phases of team-based information sharing to achieve coordination. In doing so, it highlights issues that are most characteristic of distributed team performance during each phase and asserts that this team-based information sharing may modify the processes and products resulting from the team's interaction. As described in this section, DESIM models can enhance the coordination space framework in each of the framework's phases: pre-process coordination, in-process coordination, and post-process coordination.

Pre-process coordination for information management is defined as distributing targeted (relevant) information to specific team members during pre-briefs or executive summaries prior to actual interaction (Fiore et al., 2003). By capturing the mental models at or prior to the pre-process phase, there is a baseline from which to compare later activations of long-term memory, which may be influenced by the context and therefore evolve over time. Memories will be biased by the context of what one does in the moment, and so having a shared mental model created prior to the in-process stage can provide a more stable basis for team interaction during the in-process stage.

The DESIM method can provide two types of distributed targeted information relevant to distributed teams during the pre-process coordination phase. First, DESIM allows teams to identify differences in mental models about how the world works and/or how the team works together and take the necessary remediation steps to align the team's diverse mental models or create a new model reflecting and preserving the diverse beliefs. By sharing the DESIM-generated products, conversations are enabled around the graphical influence diagrams describing the causal relationships among the environmental factors that are understood by each team member: in other words, beliefs about how the environment came to its current state or will progress to other states. This approach contrasts with simply modeling and sharing the values of the environmental factors, which depict the state of the environment (the "what") but not the mechanisms that resulted in the environment's state (the "how"). Moreover, for each team member, the strengths of these causal relationships can also be represented quantitatively on the diagrams. Understanding distributed team members' beliefs about the environment and factors that affect its changing state can positively impact task performance (Lim & Klein, 2006).

The second type of targeted information that DESIM can provide during the pre-process phase consists of depictions of the ranges of likely outcomes for each possible decision option (that is, the decision space). These depictions are created using the iterative FCM process described earlier in the chapter, which produces computational models that create distributions of plausible outcomes. Doing so can develop a team's common understanding of the relative desirability of the possible courses of action and the factors that lead to better versus worse outcomes. Thus, team members can be better prepared to take actions during the next phase.

The next phase in the coordination space framework is in-process coordination, which consists of parsing information during task execution, for example via a knowledge manager (Fiore et al., 2003). This approach is similar to Perusich and McNeese's (2006) use of fuzzy cognitive maps to filter and direct information to experiment participants acting as crew members of a simulated command and control aircraft. DESIM products can use awareness of their team's causal mental model(s) as guidance for disseminating information that is relevant for each team member's roles and beliefs.

We have used DESIM to develop models of how processes work, but it would be possible to create models of how the team works together to perform in-process coordination during task execution, that is, capture episodic or transactive memory (Fiore et al., 2003). DESIM could depict interpersonal processes that make up team behavior, with different cases based on different contexts. For example, the team leader might need three different models for how the team works in three different situations. By eliciting models from team members under these three situations, we could identify any incongruities among them. The DESIM process allows us to quantify these different models and run them as executable models to determine whether their outcomes would be different given the same assumptions about environmental factors as input values. We could not only determine what conflicts exist among the team members, but could also determine which conflicts would need to be resolved to improve team performance because they have a large impact on outcomes.

Post-process coordination includes information disseminated after interaction and post-interaction assessment of performance (Fiore et al., 2003). The long-term memory models captured by the DESIM process can also assist in such performance assessment by providing a standard structure against which performance can be assessed. Specifically, actual performance can be compared to what the models predict would be the performance, because the models codify what each individual expected would happen.

General Implications

DESIM can improve joint understanding of each other's causal mental models by enabling conversations around accessible graphical representatives of members' diverse causal mental models. By sharing this information to make diversity more explicit, teams could be better able to align causal mental models, which Mathieu et al. (2000) have found leads to performance increases. Alternately, teams could harness the strength of their diversity so that they can consider the issues explicitly from multiple perspectives and develop more innovative or robust solutions as a consequence (De Dreu & West, 2001). The key to both approaches is to make causal mental models explicit and understandable to team members.

DESIM is an example of a process and toolset that can be used to rigorously examine causal mental models held by distributed team members and thereby enable them to better understand and/or align those models across the distributed team. DESIM's rigor comes from a mathematical process for crowdsourcing pieces of mental models and recombining the results to obtain unbiased estimates of edge weights, whose distributions can be examined for agreement and patterns. In addition to the example of the organizational effectiveness evaluation presented in this chapter, we have used DESIM to explore attitudes regarding information technology acceptance in healthcare, military decision making, geopolitical events, and electric vehicle purchase decisions (Pfaff et al., 2015; Pfaff, Drury, Klein, & Boston-Clay, 2016; Pfaff et al., 2017). DESIM is a context-agnostic process that therefore is appropriate for use in a wide variety of domains.

DESIM is not well suited for analyzing situations that are both infrequently encountered and have not been addressed in training, such that subject matter experts have not had a chance to form causal mental models around those situations. We have found DESIM to be most useful when there is a high likelihood of complex model(s) or segmentation of beliefs among subpopulations. At the other end of the spectrum, DESIM is not a good candidate for analyzing simple models with a high degree of agreement among subject matter experts, because the effort needed for a full DESIM analysis in these cases is usually unwarranted. For these situations, or when there is time only for an abbreviated investigation, participatory modeling sessions using a freely available tool such as MentalModeler (Gray et al., 2013) can be beneficial and quickly performed. Comparing the structures of the resulting models from several experts can result in identifying points of similarities and differences that can be explored further via focus groups, followed by broad dissemination of findings that encourage further conversation and convergence by team members on a shared causal mental model.

REFERENCES

Cañas, A. J., Hill, G., Carff, R., Suri, N., Lott, J., & Gómez, G. (2004). CmapTools: A knowledge modeling and sharing environment. In A. J. Cañas, J. D. Novak, & F. M. González (Eds.), *Concept maps: Theory, methodology, technology. Proceedings of the first international conference on concept mapping* (pp. 125–133). Pamplona, Spain: Universidad Pública de Navarra.

De Dreu, C. K. W., & West, M. A. (2001). Minority dissent and team innovation: The importance of participation in decision making. *Journal of Applied Psychology, 86*(6), 1191–1201. http://doi.org/10.1037//0021-9010.86.6.1191

Dourish, P., & Bellotti, V. (1992). Awareness and coordination in shared workspaces. In *Proceedings of the 1992 ACM conference on Computer-supported Cooperative Work (CSCW '92)* (pp. 107–114). New York, NY: ACM. http://doi.org/10.1145/143457.143468

Drury, J. L., Klein, G. L., Pfaff, M. S., & More, L. D. (2009). Dynamic decision support for emergency responders. In *Proceedings of the 2009 IEEE conference on Technologies for Homeland Security (HST '09)* (pp. 537–544). New York, NY: IEEE.

Drury, J. L., & Williams, M. G. (2002). A framework for role-based specification and evaluation of awareness support in synchronous collaborative applications. In *Proceedings of the eleventh IEEE international Workshops on Enabling Technologies: Infrastructure for Collaborative Enterprises (WET ICE) 2002* (pp. 12–17). New York, NY: IEEE.

Endsley, M. R. (1988). Situational awareness global assessment technique (SAGAT). In *Proceedings of the IEEE National Aerospace and Electronics conference* (pp. 789–795). New York, NY: IEEE.

Endsley, M. R. (1995). Toward a theory of situation awareness in dynamic systems. *Human Factors Journal, 37*(1), 32–64.

Endsley, T., Reep, J., McNeese, M. D., & Forster, P. (2015). Crisis management simulations: Lessons learned from a cross-cultural perspective. In *Proceedings of the 6th Int'l conference on Applied Human Factors and Ergonomics (AHFE 2015)*. Amsterdam: Elsevier. http://doi.org/10.1016/j.promfg.2015.07.918

Espinosa, J., & Clark, M. (2014). Team knowledge representation: A network perspective. *Human Factors, 56*, 333–348. http://doi.org/10.1177/0018720813494093

Fiore, S. M., Salas, E., Cuevas, H. M., & Bowers, C. A. (2003, July–December). Distributed coordination space: Toward a theory of distributed team process and performance. *Theoretical Issues in Ergonomic Science, 4*(3–4), 340–364.

Fiore, S. M., Smith-Jentsch, K. A., Salas, E., Warner, N., & Letsky, M. (2010). Towards an understanding of macrocognition in teams: Developing and defining complex collaborative processes and products. *Theoretical Issues in Ergonomics Science, 11*(4), 250–271. http://doi.org/10.1080/14639221003729128

Forbes Insights Team. (2011). *Global diversity and inclusion: Fostering innovation through a diverse workforce*. New York: Forbes. Retrieved from https://i.forbesimg.com/forbesinsights/StudyPDFs/Innovation_Through_Diversity.pdf

Gallo, A. (2018, January 3). Why we should be disagreeing more at work. *Harvard Business Review*. Cambridge, MA: Harvard University. Retrieved from https://hbr.org/2018/01/why-we-should-be-disagreeing-more-at-work

Gaver, W. W., Moran, T., MacLean, A., Lövstrand, L., Dourish, P., Carter, K. A., & Buxton, W. (1992). Realising a video environment: EUROPARC's RAVE system. In *Proceedings of the conference on human factors in computing systems (CHI '92)* (pp. 27–35). New York, NY: ACM.

Gentner, D., & Stevens, A. L. (1983). *Mental models*. Hillsdale, NJ: Lawrence Erlbaum Associates.

Gray, S. A., Gray, S., Cox, L. J., & Henly-Shepard, S. (2013). Mental modeler: A fuzzy-logic cognitive mapping modeling tool for adaptive environmental management. In *Proceedings of the 46th Hawaii international conference on system sciences* (pp. 965–973). New York, NY: IEEE.

Gross, T. (2013). Supporting effortless coordination: 25 years of awareness research. *Computer Supported Cooperative Work, 22*(4–6), 425–474.

Gross, T., Stary, C., & Totter, A. (2005, June). User-centered awareness in computer-supported cooperative work-systems: Structured embedding of findings from social sciences. *The International Journal of Human-Computer Interaction (IJHCI), 18*(3), 323–360.

Gutwin, C., & Greenberg, S. (2002). A descriptive framework of workspace awareness for real- time groupware. *Computer Supported Cooperative Work, 11*(3–4), 411–446.

Harker, P. T. (1987). Incomplete pairwise comparisons in the analytic hierarchy process. *Mathematical Modelling, 9*(11), 837–848. http://doi.org/10.1016/0270-0255(87)90503-3

Hinsz, V. B., & Ladbury, J. L. (2012). Combinations of contributions for sharing cognitions in teams. In E. Salas, S. M. Fiore, & M. P. Letsky (Eds.), *Theories of team cognition: Cross-disciplinary perspectives* (pp. 245–270). New York, NY: Taylor and Francis Group.

Howe, J. (2006, June). The rise of crowdsourcing. *Wired.* Retrieved from http://archive.wired.com/wired/archive/14.06/crowds_pr.html

Hutchins, E. (1995). *Cognition in the wild.* Cambridge, MA: MIT Press.

Jonker, C. M., van Riemsdijk, M. B., & Vermeulen, B. (2010). Shared mental models: A conceptual analysis. In *Proceedings of 9th Int. conference on Autonomous Agents and Multiagent Systems (AAMAS 2010)* (pp. 132–151). New York, NY: ACM.

Klein, G. L., Drury, J. L., Pfaff, M. S., & More, L. D. (2010). COAction: Enabling collaborative option awareness. In *Proceedings of the 15th International Command and Control Research and Technology Symposium (ICCRTS).* Washington, DC: Department of Defense.

Kosko, B. (1986). Fuzzy cognitive maps. *The International Journal of Man-Machine Studies, 24*(1), 65–75.

Letsky, M., Warner, N., Fiore, S. M., Rosen, M., & Salas, E. (2007). Macrocognition in complex team problem solving. In *Proceedings of the 12th International Command and Control Research and Technology Symposium (ICCRTS).* Washington, DC: Department of Defense.

Lim, B.-C., & Klein, K. J. (2006). Team mental models and team performance: A field study of the effects of team mental model similarity and accuracy. *The Journal of Organizational Behavior, 27,* 403–418. http://doi.org/10.1002/job.387

Lin, J., Sadeh, N., Amini, S., Lindqvist, J., Hong, J. I., & Zhang, J. (2012). Expectation and purpose. In *Proceedings of the 2012 ACM Conference on Ubiquitous Computing—UbiComp '12* (pp. 501–510). New York: ACM Press. http://doi.org/10.1145/2370216.2370290

Liu, Y., Moon, S. P., Pfaff, M. S., Drury, J. L., & Klein, G. L. (2011). Collaborative option awareness for emergency response decision making. In *Proceedings of the 8th Int'l conference on Information Systems for Crisis Response and Management (ISCRAM).* New York, NY: ACM.

Mathieu, J. E., Heffner, T. S., Goodwin, G. F., Salas, E., & Cannon-Bowers, J. A. (2000). The influence of shared mental models on team process and performance. *Journal of Applied Psychology, 85*(2), 273–283.

McNeese, M. D. (2019). *Personal communication regarding the changing nature of teaming,* 24 April (original communication freely available upon request from the authors).

McNeese, M. D., Rentsch, J. R., & Perusich, K. (2000). Modeling, measuring, and mediating teamwork: The use of fuzzy cognitive maps and team member schema similarity to enhance BMC^{3}I decision making. In *Proceedings of the IEEE Int'l conference on systems, man, and cybernetics* (1081–1086). New York, NY: IEEE.

McNeese, M. D., Zaff, B. S., Citera, M., Brown, C. E., & Whitaker, R. (1995). AKADAM: Eliciting user knowledge to support participatory ergonomics. *International Journal of Industrial Ergonomics*, *15*(1995), 345–363.

Mohammed, S., Hamilton, K., Tesler, R., Mancuso, V., & McNeese, M. (2015). Time for temporal team mental models: Expanding beyond "what" and "how" to incorporate "when". *European Journal of Work and Organizational Psychology*, *24*(5), 693–709. http://doi.org/10.1080/1359432X.2015.1024664

Osei-Bryson, K.-M. (2004). Generating consistent subjective estimates of the magnitudes of causal relationships in fuzzy cognitive maps. *Computers & Operations Research*, *31*(8), 1165–1175. http://doi.org/10.1016/S0305-0548(03)00070-4

Özesmi, U., & Özesmi, S. L. (2004). Ecological models based on people's knowledge: A multi-step fuzzy cognitive mapping approach. *Ecological Modelling*, *176*(1–2), 43–64.

Perry, M. (2017). Socially distributed cognition in loosely coupled systems. In S. J. Crowley & F. Vallee-Tourangeau (Eds.), *Cognition beyond the Brain: Computation, interactivity and human artifice* (2nd ed., pp. 147–169). London: Springer-Verlag.

Perusich, K. A., & McNeese, M. D. (2006). Using fuzzy cognitive maps for knowledge management in a conflict environment. *IEEE Systems, Man and Cybernetics*, *36*(6), 810–821.

Pfaff, M. S., Drury, J. L., & Klein, G. L. (2015). Crowdsourcing mental models using DESIM (Descriptive to Executable Simulation Modeling). In *Proceedings of the naturalistic decision making conference*. McLean, VA: The MITRE Corporation.

Pfaff, M. S., Drury, J. L., Klein, G. L., & Boston-Clay, C. (2016). Modeling knowledge using a crowd of experts. In *Proceedings of the Human Factors and Ergonomics Society Annual Meeting*, *60*(1), 183–187.

Pfaff, M. S., Klein, G. L., Drury, J. L., Moon, S. P., Liu, Y., & Entezari, S. O. (2013). Supporting complex decision making through option awareness. *Journal of Cognitive Engineering and Decision Making*, *7*(2), 155–178.

Pfaff, M. S., Klein, G. L., & Egeth, J. D. (2017). Characterizing crowdsourced data collected using DESIM (Descriptive to Executable Simulation Modeling). In *Proceedings of the Human Factors and Ergonomics Society Annual Meeting*, *61*(1), 178–182. Newbury Park, CA: Sage Publishing. https://doi.org/10.1177/1541931213601529

Prell, C., Hubacek, K., Reed, M., Quinn, C., Jin, N., Holden, J., . . . Sendzimir, J. (2007). If you have a hammer, everything looks like a nail: Traditional versus participatory model building. *Interdisciplinary Science Reviews*, *32*(3), 263–282.

Rouse, W. B., & Morris, N. M. (1986). On looking into the black box: Prospects and limits in the search for mental models. *Psychological Bulletin*, *100*, 349–363.

Saaty, T. L. (1990). How to make a decision: The analytic hierarchy process. *European Journal of Operational Research*, *48*(1), 9–26.

Salas, E., Dickinson, T. L., Converse, S. A., & Tannenbaum, S. I. (1992). Toward an understanding of team performance and training. In R. W. Swezey & E. Salas (Eds.), *Teams: Their training and performance* (pp. 3–29). Norwood, NJ: Ablex.

She, M., & Li, Z. (2017). Team situation awareness: A review of definitions and conceptual models. In D. Harris (Ed.), *Engineering psychology and cognitive ergonomics: Performance, emotion and situation awareness, EPCE 2017, lecture notes in computer science*, *10275* (pp. 406–415). London: Springer-Verlag.

Sieck, W., Rasmussen, L., & Smart, P. R. (2010). Cultural network analysis: A cognitive approach to cultural modeling. In D. Verma (Ed.), *Network science for military coalition operations: Information extraction and interaction* (pp. 237–255). Hershey, PA: IGI Global.

Vennix, J. (1999). Group model-building: Tackling messy problems, *System Dynamics Review, 15*(4), 379–401.

7 Quantitative Modeling of Dynamic Human-Agent Cognition

James Schaffer, James Humann, John O'Donovan, and Tobias Höllerer

CONTENTS

INTRODUCTION

Information systems have evolved to the point of being potential collaborators rather than manipulable tools. This has the potential to decrease human mental effort and increase the amount of data that can be incorporated into the human decision-making process. Intelligent agents can allow easy access to stored procedural knowledge and may alleviate the need to become an expert before taking action in a particular domain. Despite this, intelligent agents also face the danger of pushing the user out of the loop, such as pathfinding algorithms for automobile navigation (Ahuja, Mehlhorn, Orlin, & Tarjan, 1990) and collaborative filtering for movie recommendations (Breese, Heckerman, & Kadie, 1998), requiring only a yes-no confirmation but not revealing their underlying operations. Ideally, the complexity of these algorithms could be reduced to the level of common information tools such as winnowing interfaces, but this is not always possible. The conundrum of usefulness vs. simplicity was identified by Norman as early as 1986—he writes, "simple tools have problems because they can require too much skill from the user, intelligent tools can have problems if they fail to give any indication of how they operate and of what they are doing" (1986).

Designing interaction paradigms for intelligent agents remains an open problem (Gunning, 2017). The primary challenge is that most accurate algorithmic solutions for complex problems would require significant investment from a user to gain complete understanding. Even then, nonlinear decision boundaries utilized by an algorithm are difficult to visualize and explain, although there is progress on this front (Lakkaraju, Kamar, Caruana, & Leskovec, 2017; Ribeiro, Singh, & Guestrin, 2016a, 2016b). Another contributing factor is that algorithm technology continues to rapidly improve while models of human interaction and cognition during use of these systems lags behind. This might be because human-agent interaction (HAI) is a chaotic system (Gregersen & Sailer, 1993), making predictions of the convergence unlikely or impossible, even if ideal quantitative measurements could be taken. This problem is further complicated by the potential of multi-agent systems, which stand to be even more difficult to model and harder to understand than single, "monolithic" systems. Despite these challenges, these problems can still be addressed, even if only uncertain or approximate solutions can be given (for example, predicting rain this afternoon with 51% or higher accuracy is much better than no prediction at all). This chapter defines and assesses the value of different cognitive and behavioral measurements in an attempt to explain variability in the human-agent system.

We propose profiling complex, automated algorithms using what we refer to as the explanation, control, and error (ECR) profile. We profile human users based on trust propensity, cognitive reflection, domain knowledge, and self-reported knowledge. We use the human and machine profile to investigate the human cognitive (trust, situation awareness, beliefs about an agent, perceptions of the agent, and cognitive load) and behavioral reactions to variations in these profiles. The factors investigated are then used in a statistical model to explain two types of human decision-making behaviors: adherence and decision outcomes. Specifically, we study how users interact with non-embodied, monolithic systems under two different task paradigms. We follow this with a discussion of how to extend the analysis to systems with multiple agents. Formally, we ask the following three questions about dynamic human-agent cognition:

(1) How do a person's cognitive traits affect usage of an intelligent agent and resulting decision outcomes?
(2) Which cognitive or system factors explain variability in decision making (interaction, adherence, success) in the HAI system?
(3) What is the relationship between correct beliefs about agents, their use, and trust?

In order to answer these questions, we generate a statistical map of all the factors mentioned through an exploratory factor analysis. We model the human-agent system by considering the agent's profile (explanation, control, error) as predictors of adherence and decisions while controlling for cognitive traits (domain knowledge, cognitive reflection, reported knowledge, and trust propensity). This results in the ability to predict the outcomes (decision making, adherence) in terms of the *starting point* of the HAI system. Moreover, we consider inter-task states and behaviors (perceptions, cognitive load, trust, situation awareness, interaction) as partial or full mediators of the starting point variables. The final measurement model is shown in Figure 7.1.

Two exploratory structural equation models (SEM) (Ullman & Bentler, 2003)—one from each study—were fit (by testing around 85 hypotheses). Controlling for multiplicity was done using the Benjamini-Hochberg procedure (Benjamini & Hochberg, 1995) using the exploratory value of $Q = 0.10$, which penalizes more for false positives than false negatives (in other words, we do not want to miss any potentially interesting effects that could be the basis for future studies). This quantitative map of these two studies will not only lead to better understanding of how humans react to intelligent agents, but also inform the design of future research in this area.

In this section we attempt to clearly define the semantic meaning of each factor studied. Then, we give a brief overview of the terminology used. Finally, we follow with a discussion of related work in each area.

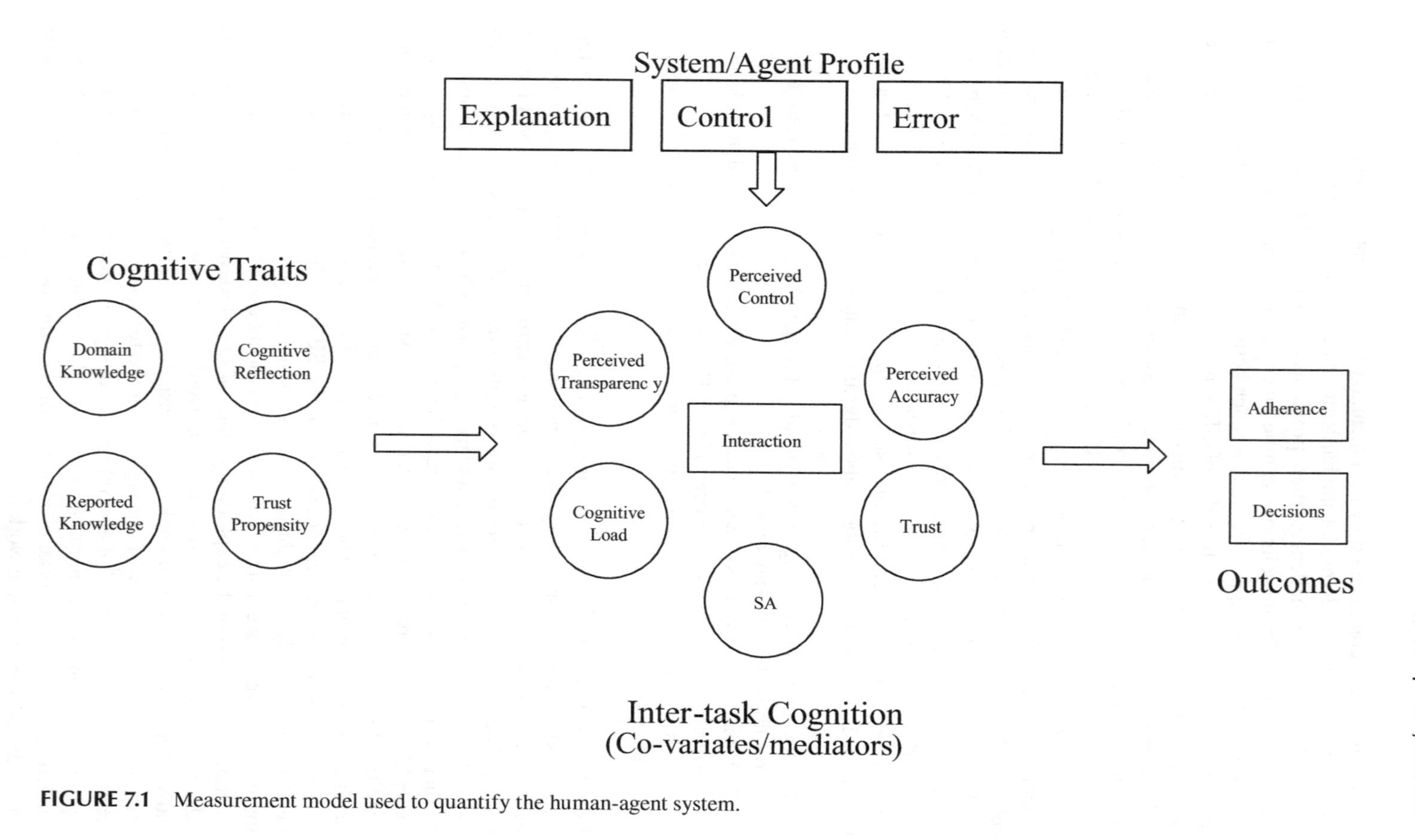

FIGURE 7.1 Measurement model used to quantify the human-agent system.

Example Application Domains

The context of this research is human interaction with intelligent agents when performing a task or making a decision. This interaction occurs in a wide and increasing variety of contexts as artificial agents' capabilities are bolstered by increasing access to data and computing power. We have alluded to familiar example use cases in route planning by a smartphone or GPS or movie recommendations based on inferred user preferences from online streaming providers such as Netflix. A more complex example is automated suggestions for disease diagnosis to supplement a doctor's expert analysis. Later in this chapter, we describe in detail a movie recommendation system and a restaurant ordering recommendation system to aid a user in maximizing the value he gets from a social meal. The common thread among all these examples is goal-oriented human interaction with an agent whose algorithms, reliability, and data store may not be fully transparent.

Terminology and Cognitive Factors

- Embodiment vs. artifact: We distinguish between two types of agents, those that are embodied visually or physically and those that are not (artifacts).
- Multi-agent systems: Multi-agent systems distinguish themselves from single-agent systems by maintaining separate internal states and knowledge of the environment.
- Subjective and objective task domains: An objective task domain has a criterion for success that can be measured and verified by a third party, such as the goal of removing body fat in the fitness domain. A subjective task domain attempts to model and satisfy the preferences of an individual person, such as the goal of providing an appropriate item to a customer.
- Explanation, control, and error: This is a simple method for profiling an agent. Explanation refers to the degree that operations are communicated to a user, control refers to the degree that agent behavior can be re-directed, and error refers to the probability that the agent's output does not solve the task domain.
- Situation awareness: Situation awareness is defined as the match between a person's mental model and the state of the environment. Situation awareness is defined either globally or with respect to a particular object in the environment.
- Trust, user experience, system perceptions: Trust is defined as a person's willingness to accept an agent's recommendations. User experience is the feeling (positive or negative) that a person has when interacting with an agent. System perceptions are defined as more nuanced forms of user experience, e.g., a person may have a good experience overall but may identify that an agent is bad at explaining itself.
- Cognitive load: A person's state of frustration while attempting to process task factors.

- Reported and true domain knowledge: Domain knowledge is defined as the total number of insights about a domain that a person has. Reported knowledge is their self-assessed knowledge level, which may not accurately reflect their true domain knowledge.

Embodiment vs. Artifact

Embodied agents are agents that are embedded within and control a particular physical system or visual entity, whether networked or un-networked (e.g., automated drones, Amazon's Alexa). Agents can also be a technological artifact of an algorithm (or collection thereof) that exhibits complex behavior but does not necessarily manifest in a physical or visual form (e.g., recommender systems on Amazon, Netflix, etc.). Virtual embodied agents (typically referred to as just virtual agents) are known to strongly influence the behavior of users (Hertzum, Andersen, Andersen, & Hansen, 2002) when compared to their nonembodied "artifact" counterparts (Komiak & Benbasat, 2006). This is because people react to virtual agents similarly to the way they react to real people. They form an opinion of the agent within the first 13 seconds of interaction and become more conscientious about behaviors (Cafaro et al., 2012). Despite this, trust relationships with non-embodied agents, especially recommender systems, continue to be studied (Benbasat & Wang, 2005; Knijnenburg, Bostandjiev, O'Donovan, & Kobsa, 2012a; O'Donovan & Smyth, 2005; Pu, Chen, & Hu, 2011), perhaps because virtual agents remain expensive and their performance is considered an open question (Choi & Clark, 2006; Veletsianos, 2007).

A physical embodied agent is tangible (e.g. drones, robots, and to some extent Amazon's Alexa), while intangible agents, such as the recommender systems that are embedded on modern e-commerce websites, reside in digital space. Human interaction with tangible agents can differ dramatically from intangible agents, even if the recommendations and overall system goal are the same. Interactions with physical agents can be colored by social cues, cultural norms, differing expectations relative to computers, and levels of acceptance of anthropomorphic form factors (Breazeal, 2004). It was shown in Podevijn et al., (2016) that humans have differing physiological effects when dealing with physical robot swarms than with virtual robot simulations. There is also evidence that users have a better subjective experience when performing a task aided by physical robots than by simulated robots (Wainer, Feil-Seifer, Shell, & Mataric, 2006). Humans have been shown to be more "polite" to physical robots than to virtual agents, being more likely to respond to greetings and afford the robot personal space while performing tasks (Bainbridge, Hart, Kim, & Scassellati, 2008). Perhaps most significantly, users may be more trusting of physical agents, as they were more likely to perform an unintuitive task when instructed by a robot than when instructed by a virtual agent (Bainbridge et al., 2008).

Monolothic vs. Multi-Agent Systems

Popular examples of monolithic systems include expert systems or recommender systems, which we expect many people to have come across on e-commerce sites such as Amazon or Netflix. There is a relative wealth of research on user behavior and cognition with monolithic systems, which will be discussed in the following sections. Here, we focus on multi-agent systems, which is an emerging research area.

Multi-agent systems consist of multiple agents with separate internal states and knowledge of the environment. This knowledge shielding is critical to the definition of software multi-agent systems, as a group of software agents with perfect group knowledge and communication would be indistinguishable from a monolithic system (Panait & Luke, 2005). Systems can include multiple intangible agents, such as multi-disciplinary design aids (Ren, Yang, Bouchlaghem, & Anumba, 2011). Tangible multi-agent systems, such as groups of robots, are more easily recognizable as their practical restrictions on communication and distribution in space ensure that their states and knowledge cannot be uniform.

As autonomy becomes embedded in more objects of diverse form factors, humans are more likely to interact with multiple agents simultaneously. Increasing the number of agents in the system can place control and attention burdens on the human user, but there are also many benefits to distributed systems, such as adaptability and scalability (Humann, Khani, & Jin, 2016; Prokopenko, 2013; Requicha, 2013). In complex distributed systems, agents can reduce the cognitive load on humans by making decisions locally and at a lower hierarchical level. This is useful in many practical applications, such as power grid management or control of customizable manufacturing processes (Marik & McFarlane, 2005). This delegation to low-level agents allows the human to focus on decision making at a higher hierarchical level while tracking fewer details of the inner workings of the system. Other benefits include adaptability and re-configurability, as agents continuously adapt to one another and new agents that are introduced into the system. Agents can also support multi-disciplinary decision making, as each agent can represent specialized knowledge in a domain, relieving the user of the responsibility to be an expert in every relevant domain.

Multi-agent systems can be used as simulations, surrogates for the behavior of humans. This is especially helpful in design of systems that are meant to have many users simultaneously. This approach has been used to model organizational processes (Jin & Levitt, 1996), evacuation procedures (Mikhailov, 2011; Stuart et al., 2013), seating layout design (Humann & Madni, 2014), and many others. The results of simulation can predict how the systems will be used in practice, and allow for design changes to be made up front, rather than waiting for problems to arise in use.

In multi-agent intelligent systems, the intelligence can be distributed and devolved, so that the recommendations to the user are not coming from a single source. Recommendations may not always be meant for the user either; in a multi-agent system, agents could be making intelligent recommendations for use by other agents. For example, in Humann and Spero (2018), a surveillance task of classifying threats within a field was carried out by humans interacting with two different classes of unmanned aerial vehicles (UAVs). As a first pass, fast high-altitude vehicles would tag points of interest within the field. Their sensitivity to risk variables (e.g. heat, metal content, movement) could be increased to eliminate false negatives, but at the cost of increasing false positives. Thus the user must set the sensitivity at such a level that they can work efficiently, be confident that they are avoiding false negatives, and use further analysis to root out the false positives. At every step in this process, there is an opportunity for agents to explain their recommendations to the user, but in large systems, this could quickly result in information overload.

When human control of each agent is infeasible, the agents must be made more autonomous or hierarchical control may be introduced. In one example of hierarchical control, "RoboLeader" is an autonomous virtual intermediary between a human operator and tangible robotic agents (Rosenfeld, Agmon, Maksimov, & Kraus, 2017; Snyder, Qu, Chen, & Barnes, 2010). The human interacts with RoboLeader by issuing high-level commands, and RoboLeader is responsible for controlling a team of robots that is searching for survivors in a simulated rescue mission. This hierarchy shields the complexity of multi-agent control from the human in most cases, while still allowing the human to take direct control of individual robots in special cases.

Explanation, Control, and Error

Research on virtual monolithic agents has led us to the theory that, at a fundamental level, all agents can be profiled by their levels of explanation, control, and error (ECR). Explanation level is the amount of output (and thus visual) bandwidth that is allocated for indications of operation. For instance, showing intermediate sorting steps would be an explanation of a sorting algorithm. Control level is the degree to which the system requires or allows input from the user. The ideas of control and automation are intrinsically linked. Increased automation necessarily leads to systems that more specifically target a particular task, reducing flexibility and reusability, but requiring less control. For instance, requiring the user to select the kernel of a support vector machine (as can be done in Weka; Holmes, Donkin, & Witten, 1994) decreases the level of automation and increases the cognitive demands on the user, but also increases the overall flexibility of the system. Explanation features are sometimes intentionally designed to accommodate control features, such as the selection of an alternate route in a GPS system. Automation can also be dynamic, turning on or off when the system detects it is in a critical state. Finally, all computational functions and algorithms solve a well-defined problem, but due to limitations in information or processing power, errors can occur. For instance, recommender algorithms attempt to predict user preferences in sets of items, but complete knowledge of a user's preferences can only be estimated from the user's item profile, which only partially defines their tastes. In other applications, processing may be under a time limit, which means systems must sometimes settle for approximate solutions.

Explanation and control from automated algorithms has been studied since at least 1975 (Shortliffe et al., 1975). This section presents work on explanation and control features in three research areas: recommender systems, expert systems, and scientific computing. We will also survey research where the accuracy of decision support systems was experimentally manipulated.

Explanation in Recommender Systems

Over the last 15 years, research has shown that explanation of a recommender system's reasoning can have a positive impact on trust and acceptance of recommendations. Recent keynote talks (Chi, 2015) and workshops (O'Donovan et al., 2015) have helped to highlight the importance of usability. Many recommender systems function as *black boxes*, providing no transparency into the working of the recommendation process, nor offering any additional information beyond the recommendations themselves (Herlocker, Konstan, & Riedl, 2000). This may negatively affect user perceptions of recommendation systems and the trust that users place in predictions. To address this issue, static or interactive/conversational explanations can be given to improve the transparency and control of recommender systems (Tintarev, Kang, Höllerer, & O'Donovan, 2015).

Bilgic and Mooney (2005) furthered this work and explored explanation from the promotion vs. satisfaction perspective, finding that explanations can actually improve the user's impression of recommendation quality. Later work by Tintarev and Masthoff (2007) surveyed literature on recommender explanations and noted several pitfalls to the explanation process, notably including the problem of confounding variables. This remains a difficult challenge for most interactive recommender systems (Tintarev et al., 2014), where factors such as user cognitive ability, mood and other propensities, experience with the interface, specific interaction patterns, and generated recommendations can all impact on the user experience with the system. Sinha and Swearingen (2002) noted that users liked and felt more confident about recommendations they perceived as transparent. The importance of system transparency and explanation of recommendation algorithms has also been shown to increase the effectiveness of user adoption of recommendations by Knijnenburg et al. (2012a).

Explanation in Expert Systems

Work in knowledge-based or "expert systems" has illuminated the effects of exposing explanations from complex agents. Gregor and Benbasat (1999) provide an excellent summary of the theory of crafting explanations for intelligent systems. User studies which test the effects of explanation typically vary explanation level and quantify concepts such as adherence or knowledge transfer. Key findings show that explanations will be more useful when the user has a goal of learning or when the user lacks knowledge to contribute to problem solving. The impact of explanation on both novices and experts has also been extensively studied (Arnold, Clark, Collier, Leech, & Sutton, 2006): novices are much more likely to adhere to the recommender/expert system due to a lack of domain knowledge, and expert users require a strong "domain-oriented" argument before adhering to advice. Experts are also much more likely to request an explanation if an anomaly or contradiction is perceived. Most of these studies focus on decision-making domains (financial analysis, auditing problems) and were conducted before the explosion of data that is available to modern tools. When browsing or analyzing data that is too large to be analyzed by hand, decision makers have no choice but to utilize automated filtering techniques as part of their search strategy. This creates new questions about what might change in the dynamics between humans and automated algorithms.

Automation and Error

Intelligent assistants often vary in their degree of automation and effectiveness, but these have not garnered as much attention as the explanation issue. The pros and cons of varying levels of automation have been studied in human-agent teaming (Chen & Barnes, 2014). Less prompting of the user for intervention may reduce cognitive load, but might also reduce awareness of system operations. Although system effectiveness continues to improve (e.g. Koren & Bell, 2015), it is not conclusive that improved algorithms result in improved adherence to system recommendations. For instance, no effect of system error was found in Salem et al. (2015). In contrast, Yu et al. (2017) found that system errors do have a significant effect on trust and Harman et al. (2014) found that users trust inaccurate recommendations more than they should. These research studies also call the relationship between trust and adherence into question. In this study, we attempt to clarify this relationship through simultaneous measurement of trust and adherence while controlling for system error and automation.

Situation Awareness

The theory of situation awareness (SA) can answer some questions about human decision making in contexts where intelligent agents are present (Endsley, 1995b; Parasuraman, Sheridan, & Wickens, 2008). Maximal SA is a requirement for optimal decision making. If an analyst cannot understand what an intelligent agent is doing and an error is made, it could potentially result in catastrophic errors. For example, the Air France 447 crash[1] was caused by a combination of system error and lack of transparency. Measurement methodologies for SA have been established (Endsley, 1995a), although new SA question items must be devised for each new domain.

SA-Based Agent Transparency

The theory of SA has also been applied to the problem of agent transparency (Chen et al., 2014). Chen's theory is called SA-based agent transparency (SAT), which is based on Endsley's three levels of SA and other theories. Chen refers to Endsley's SA as "global" SA, while SAT is relevant only to transparency requirements relevant to understanding the intelligent agent's task parameters, logic, and predicted outcomes.

Incorporating all three levels of SA into SAT should help a user gain understanding of an agent's reasoning and operation and help the user make informed decisions about "intervention," or what we call here as the manipulation of a "control" parameter. Chen notes that automation reliability strongly influences a user's attitude toward automation which can have significant impacts on trust, and thus has an impact on the degree to which that automation is leveraged. Over-trusting automation leads to automation bias (Cummings, 2004) and under-trusting results in disuse of the automation. Chen notes that information visualization and the display of uncertainty are key factors in understanding automation and discussed this in more detail in Chen, Barnes, and Harper-Sciarini (2011a).

Trust, User Experience, and System Perceptions

The word "trust" has been used to describe a number of phenomena in many different domains and therefore it is carefully defined in this section. In this work, the word trust refers to the user's *perception* that he or she can blindly rely on the system. This view was strongly influenced by the research of McKnight, which distinguishes the concepts of trust in technology and trust in other people (McKnight, Carter, Thatcher, & Clay, 2011; McKnight, Carter, & Clay, 2009), showing that people can discriminate between the two.

Trust propensity and its relationship to trust has been studied extensively in psychology, notably Colquitt et al. (2007) and Gill et al. (2005). Behavioral outcomes are affected by trust propensity when partially mediated by trust and *trustworthiness*, which is information about a trustee. The effects of trust propensity on behavioral outcomes disappears when information about the trustee becomes more reliable. Other studies in e-commerce have also found similar mediating effects between trust and trust propensity (Lee & Turban, 2001). Both trust and trust propensity need to be measured simultaneously to isolate system properties that instill trust from effects caused by highly trusting users.

Cognitive Load

The term "cognitive load" originates from education and learner theory (Sweller, 1994) and problem solving (Sweller, 1988) and is loosely defined as a "multidimensional construct representing the load that performing a particular task imposes on the learner's cognitive system." Information overload (Eppler & Mengis, 2004), a related concept, shares many of the same properties. Greater cognitive effort by users of systems leads to increased error when performing tasks. Paas et al. (2003) surveys numerous methods of measuring cognitive load during participant tasks, noting that cognitive load can be assessed by measuring mental load (portion of cognitive load that originates from task to subject relationship characteristics), mental effort (the actual effort exerted as demanded by task requirements), and performance. Participant self-reported rating scale techniques have been successful, as participants seem capable of accurately reporting their mental burden. Physiological techniques, such as the measurement of heart rate, brain activity, and pupil dilation, have also been successful. Finally, other kinds of performance measures can be applied, such as measuring the participant's effectiveness at managing a secondary task periodically while performing the primary task.

Cognitive Reflection

Work on attention and cognitive reflection (CRT) by Daniel Kahneman (1973) has been successful in discriminating between "fast" and "slow" thinking using a variety of questions that effectively trick the human processing system. Since then, CRT tests have been frequently used due to a correlation with human intelligence and decision making (Toplak, West, & Stanovich, 2011; Welsh, Burns, & Delfabbro, 2013). This work hypothesizes that CRT would be a strong predictor of a person's decision behavior when interacting with an agent.

Reported and True Domain Knowledge

Consequences of self-reported ability have been recently discovered in studies of cognitive psychology (Hoorens, 1993; Kruger, 1999). The Dunning-Kruger effect predicts that low-ability individuals maintain an over-estimated belief in their own ability. This work also illustrates how quantitative metrics collected through questionnaires do not always measure their face value. For instance, the Dunning-Kruger effect shows us that asking a user how much he knows about a particular topic will only quantify the number of "unknown unknowns" relative to the user, rather than the user's actual ability in that area (Merritt, Smith, & Renzo, 2005). Deficits in knowledge are a double burden for these users, not only causing them to make mistakes but also preventing them from realizing they are making mistakes (Kruger & Dunning, 1999).

The Dunning-Kruger effect is part of a larger group of cognitive effects sometimes referred to as "illusory superiority." Other effects in this category create an additional concern for the success of intelligent assistants. For instance, it is known that estimating the intelligence of others is a difficult task (the Downing effect) (Davidson & Downing, 2000) that requires high intelligence. This explains the tendency of people to be very likely to rate themselves as "above average," even though not everyone can be so. We might expect that lower-intelligence users would fail to accurately gauge the intelligence of information systems, leading to disuse.

The research on self-reported ability leads us to hypothesize that overconfident individuals are less likely to interact with or adhere to intelligent assistants, due to the over-estimation of their own ability and their inability to assess the accuracy of the system.

Introduction to Task Paradigms

We considered two tasks to fit and validate the cognitive measurement model presented here. The first was a subjectively validated task—Movie Recommendation (MR)—and the second was an objectively validated game theoretic task—the Diner's Dilemma (DD).

In the MR task participants interacted with an interface dubbed "Movie Miner" to find a set of movies to watch in the future. This is a common setup in studies of recommendation, however, we improve upon these studies by including better modeling of decision satisfaction (Schaffer, O'Donovan, & Höllerer, 2018) and behavioral (rather than reported; Pu et al., 2011) adherence. Behavioral adherence modeling is only possible if the task is unrestricted and participants have the freedom to choose between alternative tools. Thus, participants were given two tools to work with and their behavior was not restricted. The methodology was thus very similar to typical online browsing sections, such as on Amazon, where a customer is browsing a product catalog and adding items to their "shopping cart." In summary, we kept the following three goals in mind: (1) to make the system as familiar to modern web users as possible, (2) to make the system as similar to currently deployed recommender systems as possible, and (3) to ensure that the study can be completed without forcing the users to accept recommendations from the system, so adherence can be measured. The use of novelty in any design aspect was minimized so that results would have more impact on current practice.

The second study, DD, was chosen due to its wide applicability, limited complexity, and research base. The Diner's Dilemma is an n-player, iterated version of the basic prisoner's dilemma. In the basic prisoner's dilemma, two players decide to take an action (cooperate or defect) without communication beforehand, where defection leads to a higher outcome for an individual regardless of the other players' actions, but with mutual cooperation leading to a higher outcome than mutual defection. The iterated form of this game can show evolution in player strategies as they learn the other player's tendencies to defect or cooperate.

Multi-player versions of this game, such as the Diner's Dilemma, are more complex, which has made them suitable for studying the effects of increased information available to players through a user interface (Gonzalez, Ben-Asher, Martin, & Dutt, 2013; Martin, Juvina, Lebiere, & Gonzalez, 2011). In this experiment, the iterated three-player version was used, which limits the complexity of the game such that it is within the comprehension of human players, but is still sufficiently complex to warrant a computational aid. In this chapter, our DD gives the user a choice between ordering a hot dog or lobster when dining out with friends, under the agreement that the table's total bill will be split evenly among the diners. The *defect* strategy is to order the expensive and satisfying lobster, hoping that others will order hot dogs and subsidize the user's bill. The *cooperate* strategy is to order the inexpensive hot dog.

Participants in both studies were recruited on Amazon Mechanical Turk (AMT). AMT is a web service that gives tools to researchers who require large numbers of participants and are capable of collecting data for their experiment in an online setting. AMT has been studied extensively for validity; notably Buhrmester, Kwang, and Gosling (2011) found that the quality of data collected from AMT is comparable to what would be collected from laboratory experiments (Hauser & Schwarz, 2015). Furthermore, since clickstream data can be collected, satisficing—the act of rapidly "tab-clicking" through study questionnaires—is easy to detect.

MOVIE RECOMMENDATION METHODOLOGY

This section details the methodology used in the Movie Recommendation (MR) study.

Task and User Interface Design

The MR interface was closely modeled after modern movie "browsers" (such as IMDb or MovieLens) that typically have recommender functionality. On the left side of the interface, the system featured basic search, sort, and filter for the entire movie dataset. The right side of the interface provided a ranked list of recommendations derived from collaborative filtering, which interactively updated as rating data was provided.

The user interface provided the following functionality: mousing over a movie would pop up a panel that contained the movie poster, metadata information, and a plot synopsis of the movie (taken from IMDb); for any movie, users could click anywhere on the star bar to provide a rating for that movie, and they could click the green "Add to Watchlist" button to save the movie in their watchlist (CS was measured on their chosen movies at the end of the task). Clicking the title of any movie would take a user to the IMDb page where a trailer could be watched (this was also available during the CS feedback stage).

On the left (browser) side of this interface, users had three primary modes of interaction which were modeled after the most typical features found on movie browsing websites:

(1) Search: Typing a keyword or phrase into the keyword matching box at the top of the list returned all movies that matched the keyword. Matches were not personalized in any way (a simple text matching algorithm was used).
(2) Sort: Clicking a metadata parameter (e.g. Title, IMDb Rating, Release Date) at the top of the list re-sorted the movies according to that parameter. Users could also change the sort direction.
(3) Filter: Clicking "Add New Filter" at the top of the list brought up a small popup dialog that prompted the user for a min, max, or set coverage value of a metadata parameter. Users could add as many filters as they wanted and re-edit or delete them at any time.

The recommendation side operated identically to the browser side, except that the list was always sorted by the collaborative filtering prediction and the user could not override this behavior.

When explanations were provided, they appeared on mouse over and could not be hidden. First, each explanation stated: "Movie Miner matches you with other people who share your tastes to predict your rating." The rest of the explanation was generated by examining the three users in the database that were most similar to the user at the current point in time and taking the intersection of their rated movies with the user's profile. This identifies the movies that are *most responsible* for an item appearing at its respective location in the recommendation list.

The MovieLens 20M dataset was used for this experimental task. The MovieLens dataset has been widely studied in recommender systems research (Miller, Albert, Lam, Konstan, & Riedl, 2003; Jung, 2012; Harper & Konstan, 2016). Due to update speed limitations of collaborative filtering, the dataset was randomly sampled for 4 million ratings, rather than the full 20 million.

Agent Design

A traditional user-user collaborative filtering approach was chosen for the agent. Details for this can be found in Resnick et al. (1994). Collaborative filtering was chosen due to the fact that it is well understood in the recommender systems community and it achieves extremely high performance on dense datasets such as MovieLens (Koren & Bell, 2015). The results from this study should generalize reasonably well to other collaborative-filtering-based techniques, such as matrix factorization and neighborhood models. We made two minor modifications to the default algorithm based on test results from our benchmark dataset: Herlocker damping and rating normalization.[2]

ECR Manipulation

Two levels of control, two levels of explanation, and two levels of recommendation error were manipulated. All manipulations (three parameters, two values taken,

$2^3 = 8$ manipulations) were used as between-subjects treatments in this experiment. Text-based explanations were chosen due to their similarity to real-world systems such as Netflix and Amazon. To add user control, we chose to allow users to define filters on the list of recommendations. This approach is similar to real-world systems that are currently deployed on MovieLens, IMDb, and so on. To vary recommendation error, noise was added to the algorithm. This approach was validated by verifying that the random noise was reducing accuracy by performing a five-fold cross validation on our ratings data set. The error-free recommender achieved a mean absolute error of 0.144, while the noisy version did considerably worse at 0.181 (nearly a 26% difference).

Explanation Manipulation

- Opaque (*Explanation* = 0): The opaque recommender simply provided the recommendations without any explanation.
- Justification (*Explanation* = 1): The justification explained how ratings were calculated with the following blurb: "Movie Miner matches you with other people who share your tastes to predict your rating." This was followed by a list of the items in the user's profile that most affected the recommendation.

Control Manipulation

- Automatic (*Control* = 0): The recommender would update and re-sort its recommendations automatically. Participants could only affect the recommender's behavior by changing their user profile.
- Customizable (*Control* = 1): On top of the partial control features, users were allowed to define custom filters on recommender results to narrow the recommendations. Additionally, users could remove individual movies (indicating they were "not interested") from the recommendation list.

Error Manipulation

- Collaborative Filtering (*Error* = 0): Collaborative filtering: user-user similarity, Herlocker damping, and normalized across the 0.5–5 star rating scale.
- Collaborative Filtering with Noise (*Error* = 1): A vector of noise (of up to two stars difference) was calculated at session start and the vector was added in to the recommendation vector before normalization. From the participant's perspective, the list of recommendations thus appeared to be reordered as affected by this noise.

Procedure

Participants made their way through four phases: the pre-study, the ratings phase, the watchlist phase, and the post-study (Figure 7.4, top). The pre-study and post-study were designed using Qualtrics.[3] In the "ratings" phase, participants accessed Movie Miner and were shown only the blue "Movie Database" list and the ratings box. We asked participants to find and rate *at least* ten movies that they believed would best represent their tastes, but many participants rated more than the minimum. In the "watchlist phase," participants were shown the brown "Recommended for You" list and the watchlist box. Instructions appeared in a pop-up window and were also shown at the top of the screen when the pop-up was closed. Participants were told to freely use whichever tool they preferred to find some new movies to watch. They could add movies to their watchlist with the green button that appeared on each individual movie (regardless of the list that it appeared in). We asked them not to add any movies that they had already seen, required them to add at least five movies (limited to seven maximum), and required them to spend at least 12 minutes interacting with the interface. A 12-minute session in which five to seven items are selected was deemed sufficient time to select quality items, given that people only browse Netflix for 60 to 90 seconds to find a single item before giving up (Gomez-Uribe & Hunt, 2016).

Accommodating Subjective Decision Making

Quantifying decision satisfaction is problematic because it can be influenced by user experience, mood, health, and so on. To improve modeling of subjective decision satisfaction, we used the satisfaction baseline approach (Schaffer et al., 2018). Baseline satisfaction was measured shortly after the pre-study by getting participant feedback on movies that were chosen from the database at random.

Ten random movies were shown, one at a time, and the responses (question items, bs1–4, given in Table 7.4) were averaged together. Satisfaction with selected items was measured after the "watchlist" phase. For this, the recommender interface was removed and the questions items were shown for each item chosen by the participant. Note that the question items are phrased in terms of the recommendations (Table 7.4, ds1–4), not the interface. This is to help the participant distinguish between the browsing tools and the features of the recommender system. By modeling changes between baseline satisfaction and satisfaction with selected items, it is possible to quantify the *change* in satisfaction, which is what the user interface would actually influence.

DINER'S DILEMMA METHODOLOGY

This section details the methodology used in the Diner's Dilemma (DD) study. A screenshot of the interface used is shown in Figure 7.2.

TABLE 7.1
Diner's Dilemma Choice Payoff Matrix

Player Chooses:			
Hot Dog	**Lobster**		
Two co-diners	20.00	24.00	Cooperate
One co-diner	12.00	17.14	Cooperates
	8.57	13.33	Neither cooperates

Task and User Interface Design

In the Diner's Dilemma, several diners eat out at a restaurant over an unspecified number of days with the agreement to split the bill equally each time. Each diner has the choice to order the inexpensive dish (hot dog) or the expensive dish (lobster). Diners receive a better dining experience (here, quantified as *dining points*) when everyone chooses the inexpensive dish compared to when everyone chooses the expensive dish. To be a valid dilemma, the quality-cost ratio of the two items available in a valid Diner's Dilemma game must meet a few conditions. First, if the player were dining alone, ordering hot dog should maximize dining points. Second, players must earn more points when they are the sole defector than when all players cooperate. Finally, the player should earn more points when the player and the two co-diners all defect than when the player is the only one to cooperate. This "game payoff matrix" means that in one round of the game, individual diners are better off choosing the expensive dish regardless of what the others choose to do. However, over repeated rounds, a diner's choice can affect the perceptions of other co-diners and cooperation may develop, which affects long-term prosperity of the group. Hot dog/lobster cost and values for the game are shown in Figure 7.2, under each respective item, resulting in the payoff matrix that is shown in Table 7.1.

Participants played the Diner's Dilemma with two simulated co-diners. The co-diners were not visually manifested as to avoid any confounding emotional responses from participants (see Choi, de Melo, Khooshabeh, Woo, & Gratch, 2015). The co-diners played variants of tit-for-tat (TFT), a proven strategy for success in the Diner's Dilemma wherein the co-diner makes the same choice that the participant did in the previous round. To make the game more comprehensible for participants, simulated co-diners reacted only to the human decision and not to each other. In order to increase the information requirements of the game, some noise was added to the TFT strategy in the form of increased propensity to betray (respond to a hot dog order with a lobster order) or forgive (respond to a lobster order with a hot dog order). Participants played three games with an undisclosed number of rounds (approximately 50 per game) and co-diner strategies switched between games. This means that the primary task for the user was to figure out what strategies the co-diners were employing and adjust accordingly. In the first game, co-diners betrayed often and the best strategy was to order lobster. In the second game, co-diners betrayed at a reduced rate and also forgave to some degree, which made hot dog the best choice. In the final game, co-diners were very forgiving and rarely ordered lobster even when betrayed, which again made lobster the best choice. The mean performance of participants in each game is shown in Table 7.2.

Participants played the game through the interface shown in Figure 7.2. This interface contains four components: the last round panel (left), the control panel (center), the history panel (bottom), and the virtual agent, the Dining Guru (right). Across treatments, all panels remained the same except for the Dining Guru, which varied (see Figure 7.3).

Participants were provided with a basic interface containing all of the information that the Dining Guru used to generate its advice. The last round panel was shown on the left side of the interface and the control panel was shown in the middle. Together, these panels displayed the current dining points, the food quality and cost of each

TABLE 7.2
Performance of the Dining Guru (DG) across All Games Compared to Participants

Game #	Optimal Choice	# Rounds	DG/No Error	DG/Weak Error	DG/Error	Participants
1	LOBSTER	55	0.92	0.75	0.63	0.70
2	HOT DOG	60	0.65	0.62	0.56	0.46
3	LOBSTER	58	0.79	0.69	0.60	0.62
All		17	0.78	0.68	0.60	0.59

Note: The ratio of optimal moves made by the error-free Dining Guru (DG/No Error), weak-error Dining Guru (DG/Weak Error), full-error Dining Guru (DG/Error), and participants are given. DG/Error performed as well as the participants on average.

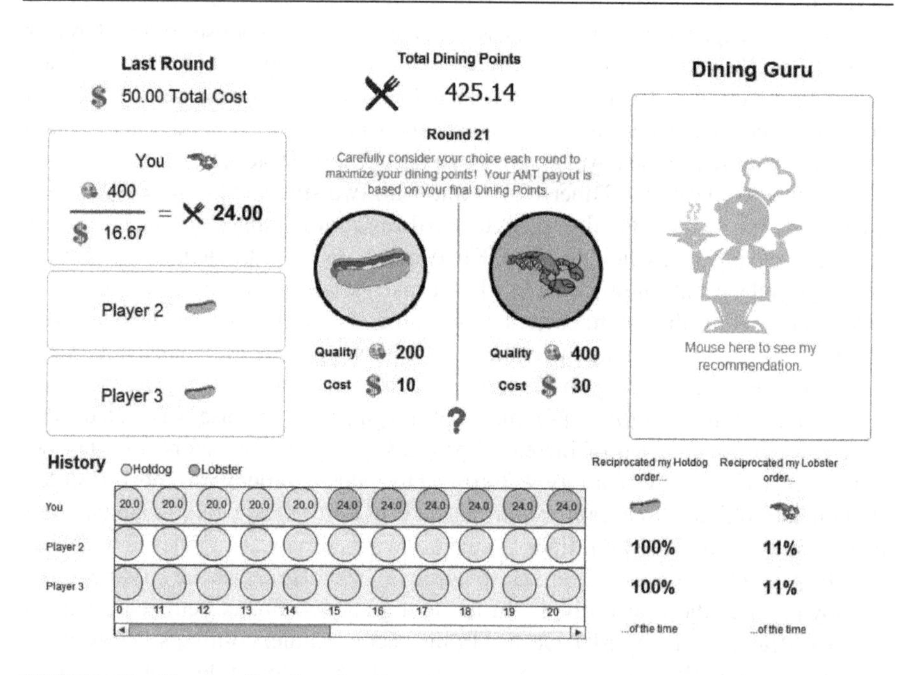

FIGURE 7.2 The user interface for the game.

menu item, the current round, and the results from the previous round in terms of dining points. These panels allowed the participant to make a choice in each round. On the lower portion of the screen a history panel was provided. This panel contained information about who chose what in previous rounds and reciprocity rates.

Agent Design

The Dining Guru was shown on the right side of the screen. In each round, the Dining Guru could be examined by a participant to receive a recommendation

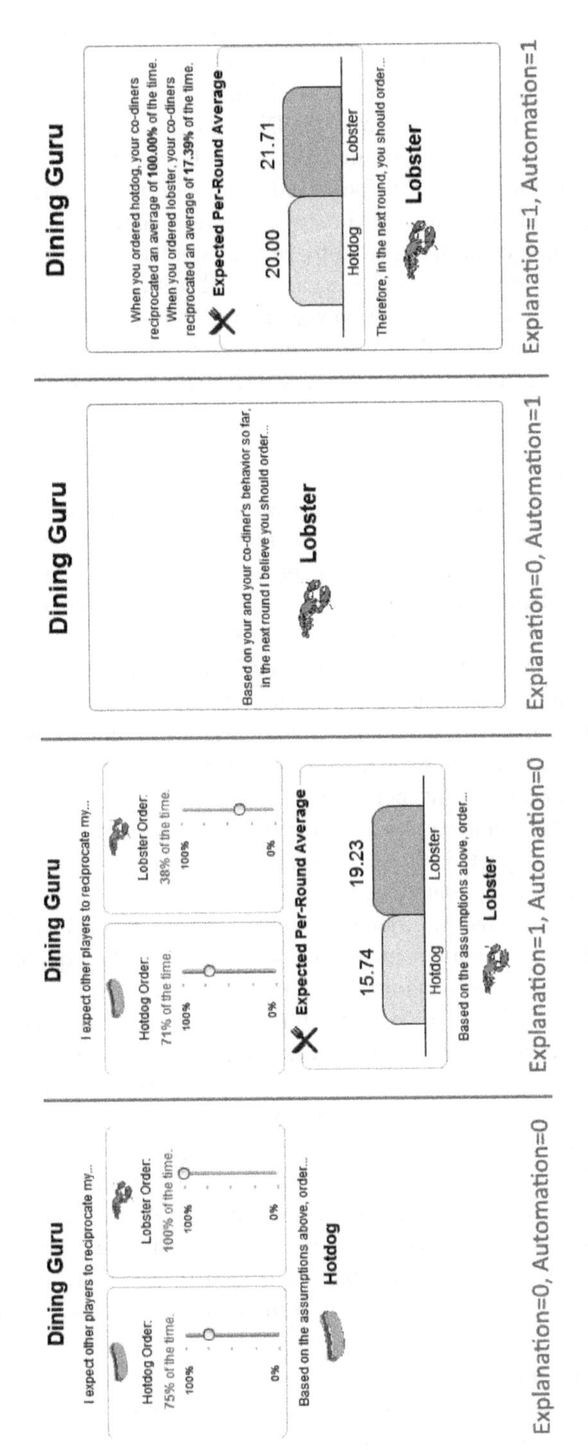

FIGURE 7.3 Variations of the Dining Guru, based on automation and explanation treatments.

about which item (hot dog or lobster) would maximize their dining points. As with the simulated co-diners, the Dining Guru was not given any dynamic personality beyond being presented as an agent—a static drawing was used to communicate this. Users were required to mouse over the Dining Guru to invoke a recommendation, which made it possible to measure adherence. Recommendations were generated by calculating the expected value of ordering hot dog or lobster in the future, based on the maximum likelihood estimates of the rates of forgiveness and betrayal from co-diners. Due to the fixed strategy of the simulated co-diners, the Dining Guru made the "best possible" choice in each round, with most of the errors occurring in earlier rounds when information was incomplete. A "manual" version of the Dining Guru was given some treatments, which required participants to supply the Dining Guru with estimates of hot dog and lobster reciprocity rates (see Figure 7.3).

ECR Manipulation

Two levels of control (*Automation* = 0, *Automation* = 1), two levels of explanation (*Explanation* = 0, *Explanation* = 1) and three levels of recommendation error (*Error* = 0, *Error* = 0.5, *Error* = 1) were manipulated between subjects (see Figure 7.3 for a visual). All manipulations (three parameters, $3 \times 2^2 = 12$ manipulations) were used as between-subjects treatments in this experiment.

The explanation for the Dining Guru was designed to accurately reflect the way that it was calculating recommendations. Since the Dining Guru calculates maximum-likelihood estimates of co-diner behavior and cross references this with the payoff matrix to produce recommendations, the explanation thus needed to contain estimates for the expected points per round of each choice. Additionally, in the manual version, a text blurb appeared explaining the connection between co-diner reciprocity rates and the expected per-round average. Reciprocity rates were provided by participants in the non-automated version, so the explanatory version only required the addition of the expected per-round averages. A bar graph was used to represent the averages so that participant attention would be drawn to the explanation.

Three levels of error were manipulated: no error, weak error, and full error. In the no-error treatment (*Error* = 0.0), the Dining Guru produced recommendations that could be considered flawless, which if followed would result in mostly optimal moves. The weak error (*Error* = 0.5) version would randomly adjust the reciprocity estimates up or down by up to 25%. For instance, if the true hot dog reciprocity rate was 65%, the Dining Guru would use a value anywhere between 40% and 90%. Finally, the "full" error (*Error* = 1.0) condition adjusted reciprocity estimates by up to 50% in either direction. A practical consequence of this was that the Dining Guru would flip its recommendation almost every round. The error in the recommendations was reasonably hidden from participants and indeed was only noticeable when either explanation was present or the Dining Guru was not automated.

Explanation Manipulation

- Opaque (*Explanation* = 0): The opaque recommender simply provided the recommendations without any explanation.

- Justification (*Explanation* = 1): The justification explained the relationship between co-diner reciprocity rates and optimal choices. This was communicated visually with a bar graph, showing the expected points per round for each choice based on historical co-diner data.

Control Manipulation

- Automatic (*Control* = 0): The recommender would update its recommendations automatically. Participants would have to mouse over the Dining Guru each round to check the most up-to-date recommendation.
- Customizable (*Control* = 1): The recommender required the user to provide estimated reciprocity rates for each co-diner. The estimates were provided by moving two sliders, which took no value until users first interacted with them. Users could freely experiment with the sliders, which means that they could be used to understand the relationship between the payoff matrix and co-diner reciprocity rates.

Error Manipulation

- Maximum Likelihood Estimation (*Error* = 0): Dining Guru produced recommendations that would be unbeatable, which if followed would result in the maximum number of optimal moves.
- Maximum Likelihood Estimation with Weak Noise (*Error* = 0.5): Randomly adjusts estimates for reciprocity rates up and down 25%, resulting in occasionally inaccurate recommendations.
- Maximum Likelihood Estimation with Noise (*Error* = 1.0): Randomly adjusts estimates for reciprocity rates up and down 50%, resulting in frequently inaccurate recommendations.

Procedure

An overview of the procedure for the DD study is given in Figure 7.4 (bottom). Before playing the game, participants were introduced to game concepts and the Dining Guru by playing practice rounds (training phase). Several training questionnaires, which could be resubmitted as many times as needed, were used to help participants learn the game. The Dining Guru was introduced as an "AI adviser" and participants learned how to access it and what its intentions were. Participants were told that the Dining Guru was not guaranteed to make optimal decisions and that taking its advice was their choice. Participants played three games of Diner's Dilemma against three configurations of simulated co-diners with varying behavior characteristics.

TASK COMPARISON

This section details the differences in measurements and tasks between the DD and MR studies.

Item Operationalization

A comparison of the differing procedures is shown in Figure 7.4. This figure indicates where measurements were taken for each study, which was largely the same. The most critical difference is that the DD study contained a training phase, so the domain knowledge test measured retention rather than the participant's stored knowledge.

In the MR study, quantity (and type) of interaction with each tool was measured, constituting recommender and browser interaction. Adherence was measured as the percentage of items in each participant's watchlist that originated from the recommender side of the interface. Decision satisfaction was modeled as a two-wave, multi-item factor, so was modeled via confirmatory factor analysis (CFA). In the DD study, the quantity of interactions with the Dining Guru constituted recommender interaction. Absence of interaction with the Dining Guru was treated equivalent to browser interaction in the MR study. Adherence occurred for each round where the user choice matched the last recommendation given by the Dining Guru. The final adherence measurement was scaled between 0 and 1, where 0 indicates no adherence and 1 indicates complete adherence. Some users never accessed the Dining Guru, which caused their adherence score to become 0. Decision optimality was quantified as the total percentage of rounds where optimal decisions were made. An optimality score of 1 indicates the player ordered 100% lobster in games 1 and 3, and 100% hot dog in game 2.

We used a SAGAT-style freeze (Endsley, 2000) to assess situation-awareness-based agent transparency (SAT). For MR, this was done eight minutes into the watchlist phase of the task, whereas for DD it was done partway through game 2. Awareness of game factors (SAG) was also taken in the DD study. The SAG questionnaire contained five questions related to the current game state. The situation awareness question items each contained a slider (min: 0%, max: 100%) and asked participants to estimate their current cooperation rate (1) and the hot dog (2, 3) and lobster (4, 5) reciprocity rates for each co-diner. The game interface was not available at this time. The SAG score was calculated by first summing up the errors from each of the five estimation questions and then inverting the scores based on the participant with the highest error, such that higher SAG scores are better.

CFA was used to eliminate measurement error when possible. Factor fit was improved iteratively by removing items until Cronbach's alpha was maximized, resulting in the list of items shown in Table 7.4. Internal reliability fit metrics for each factor are shown in Table 7.6. Domain knowledge, SAT, and SAG were expected to be multidimensional at the outset, and thus were parceled instead of factored—the question items used for the parcels is shown in Table 7.5. For the parceling, question items were summed and the loading of the factor on the parcel was set to 1. The variance of the parcel was freed to maximize fit.

In this study, we originally intended to model perceived transparency, control, effectiveness, and trust as separate factors. However, during the confirmatory factor analysis, inter-item correlations indicated we only had a single factor for MR and two factors for DD: perceived control and trust. Moreover, the perceived control factor complicated pathways in the model and was not strongly predictive of each outcome.

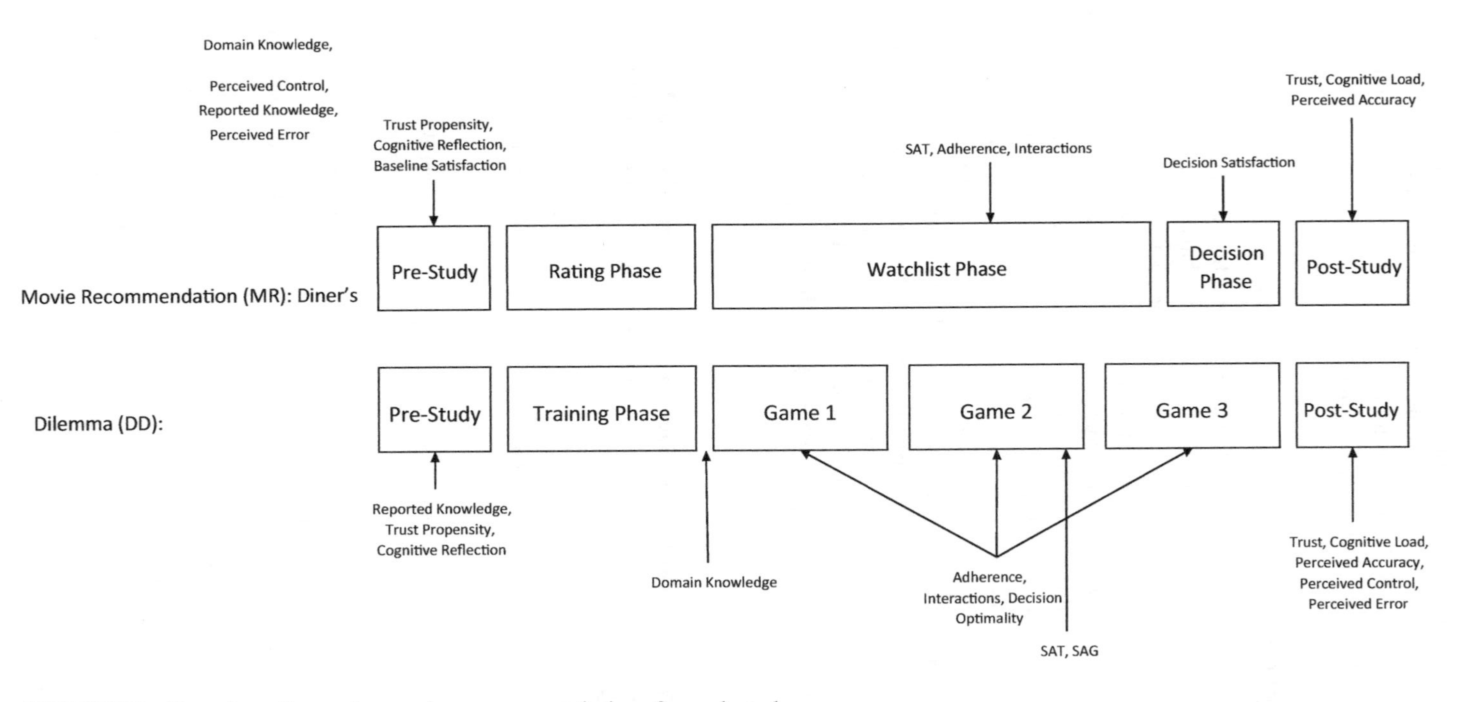

FIGURE 7.4 Overview of procedure and measurement timings for each study.

Thus, we collapsed these factors into a single factor, which we referred to as user experience with the agent (UXA). Doing so not only increased factor fit, but model fit as well. Once this was done, all factors in Table 7.6 achieved discriminant validity using the Campbell and Fiske test (Campbell & Fiske, 1959).

Task Differences

The main difference between the MR and DD studies lies in the nature of the task and the criterion for decision success. Both task spaces have been studied extensively. The DD task is a variant of the iterated Prisoner's Dilemma, whose applicability to real-world situations has been well established[4] (Stephens, McLinn, & Stevens, 2002; Ainslie, 2001; Varian, Bergstrom, & West, 1996). Decision success for each study was based on different parameters, with success in the MR study being subjective and success in the DD study being objective. It should also be noted that the treatment manipulations for explanation and control were minimal. This was done due to an understanding that decision makers are sensitive to the environment in which decisions are made (Payne, Bettman, & Johnson, 1993) and also to increase the relevance of the results (it is easier to implement a text-based explanation that a visual one). Differences between effects in the studies can thus be attributed to differences in the task parameters and decision criterion (Table 7.3), while similarities in effects thus have strong support for their generalization.

TABLE 7.3
Comparison of Task Parameters, Decision Success Criteria, and Treatment Differences between the Movie Recommendation (MR) and Diner's Dilemma (DD) Studies

Study	MR	DD
Decision task	Catalog browsing	Binary choice
Number of decision iterations	5–7	173
Agent support	Collaborative filtering	Maximum likelihood estimation
Alternative	Winnowing interface	History visualization
Embodiment	Artifact	Picture
Decision Criteria	Subjective	Objective
Domain	Movie metadata	Game rules
Explanation manipulation	Text-based explanation of agent's calculation	Text-based explanation of agent's calculation
Control manipulation	Optional metadata filters to be applied on ranked recommendation list	Requires specification of input parameters but allows exploration of metadata space
Error manipulation	Noise added to recommendation score, changing top recommendations	Noise added to expected values of binary choice, changing per-round recommendations

RESULTS

We collected more than 1,055 samples of participant data using Amazon Mechanical Turk: 526 samples for MR and 529 samples for DD. Participant data was checked carefully for satisficing and these records were removed (approximately ten per study), resulting in the 1,055 complete records. Participants were paid $3.00 for the DD and $1.50 for MR. In either case, participants spent between 25 and 60 minutes completing the study. Participants were between 18 and 71 years of age and were 50% male, however, DD attracted more male participants (54%) while MR attracted more female (55%).

Means and variances of non-factor measurements are given in Table 7.7 (factors are not listed here—all factors are modeled to have a mean of 0 and standard deviation of 1). Scores on tests were normalized between 0 and 1. Note that in the DD study, absence of interaction with the agent was considered interaction with the alternative. Decision optimality was modeled as two-wave decision satisfaction in the MR study, meaning that it was modeled as a factor and thus does not appear in the table.

FIT SEM MODELS

Data from each study was fit using an exploratory SEM, with the exception that decision satisfaction from the MR study was analyzed separately in a Raykov change model (Raykov, 1992) (Table 7.8). This is because baseline satisfaction needs to be taken into account when evaluating subjective satisfaction (Schaffer et al., 2018). A visual comparison of the results from both studies is shown in Figure 7.5, along with fit statistics and regression statistics. Due to being non-normal, treatment variables take the value of 0 or 1 and coefficients reported in the figure are B values (effect sizes in the units of the original measurement), which predict a change in standard deviation of the regressand when the treatment is switched on. All dependent and latent variables were standardized to have a mean of 0 and variance 1 and coefficients reported are β values, which predict a change in standard deviation of the regressand with a standard deviation change in the regressor. Both models were built using R 3.0.3, lavaan 0.5–17.

Multiplicity control was enforced in our chosen SEM using the Benjamini-Hochberg procedure with $Q = 0.10$ (Benjamini & Hochberg, 1995), which is recommended for exploratory SEM analysis (Cribbie, 2007). This procedure indicates how many of the tested relationships in the all-factor SEM are expected to be false positives. The MR and DD tasks had 86 and 87 hypotheses, respectively. These hypothesis numbers are derived from the exploratory way in which the SEMs were built, that is, specifying some factors/variables as downstream from others and testing the presence of significant predictive or causal relationships. Effects that failed the false discovery rate test were trimmed from the models—these were typically regressions on target variables whose regressor was already significantly correlated with another variable that predicted on the regressand. For example, reported and domain knowledge had a significant negative correlation in both studies. When controlling for one

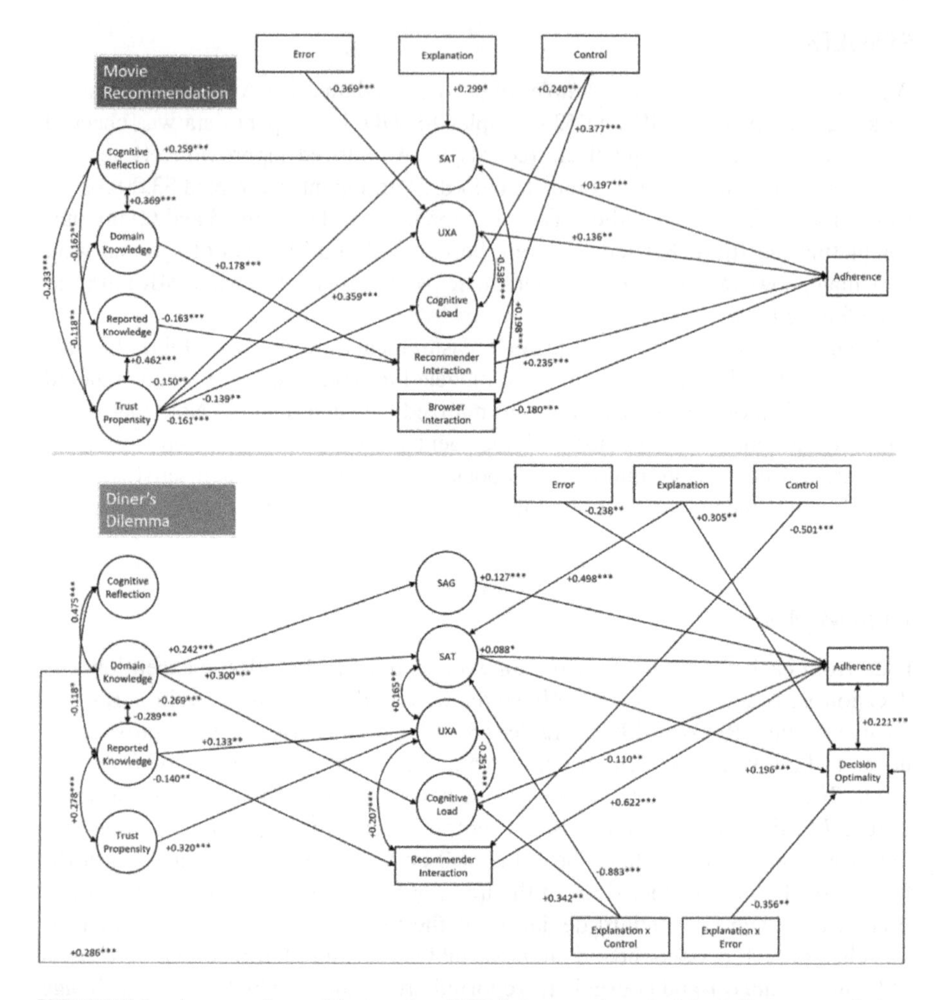

FIGURE 7.5 A comparison of the two fitted SEMs from each study. Movie Recommendation model fit: $N = 526$ with 77 free parameters = approximately 6.5 participants per free parameter, $RMSEA = 0.054$ (CI: [0.050, 0.057]), $TLI = 0.919$, $CFI = 0.926$ over null baseline model, $\chi^2(512) = 1285.408$. Diner's Dilemma model fit: $N = 529$ with 72 free parameters = approximately 7 participants per free parameter, $RMSEA = 0.030$ (CI: [0.025, 0.035]), $TLI = 0.969$, $CFI = 0.973$ over null baseline model, $\chi^2(378) = 557.889$.

or the other, effects on regressands (e.g., SAT) become less significant and fail the false discovery rate threshold.

DISCUSSION

This section first discusses the statistical effects observed in each study in detail. Then, we compare the results from two studies. Finally, we discuss the implications of these effects in the context of multi-agent and physical systems and highlight future research challenges.

Movie Recommendation: Statistical Effects

In the MR data, we found that splitting user experience into different subjective user-experience factors (similar to the ResQue framework (Pu et al., 2011) and the model in Knijnenburg et al. (2012a)) decreased model fit, despite each sub-factor (perceived transparency, perceived control, perceived quality, and trust) having items with acceptable fit but high inter-correlation (about 0.95), implying poor discriminant validity. Generally, high correlations among factors are undesirable due to the decreased questionnaire item-to-information ratio. For instance, in this study, a three-item scale for "trust" would have captured nearly the same signal as the 12-item SSA model that was used. This may have occurred because participants had a unidimensional perception of the recommender (i.e. "I like this" or "I don't like this"), which was a surprising finding. We considered it important to compare our results with Knijnenburg et al. and the ResQue framework. The Knijnenburg data was available[5] and we examined the covariances of perceived quality, satisfaction, control, and understandability. The scales in Knijnenburg's study were slightly better in terms of discriminatory power: about a 0.7 Pearson correlation between perceived overall system satisfaction, quality, and control, but this correlation level is still quite high. The transparency sub-construct, "understandability," is much more discriminative (0.34), perhaps due to the user-centric phrasings used. Unfortunately, discriminant validity between factors in the ResQue framework were not reported. For interested readers, a detailed discussion of user experience modeling along this vein is discussed in Schaffer et al. (2018). In light of this analysis, we encourage other researchers to consider the inter-factor correlations and discriminant validity of their chosen factors.

As evidenced by the profiling traits of CRT, domain knowledge, reported knowledge, and trust propensity, users of intelligent agents can be broken into two groups of high and low task ability. Higher-ability users are more likely to understand the recommender but less likely to form positive perceptions of it, while lower-ability users reported being more trusting and over-estimating their task knowledge, subsequently interacting and adhering to the system to a lesser degree and having worse decision outcomes overall.

Despite the recommender system community's emphasis on user experience, we found that user experience with the agent (UXA) was perhaps the most trivial factor in predicting adherence. This is evidenced in Figure 7.5, where it can be observed that SAT, recommender interaction, and browser interaction all are better predictors of adherence than UXA. The exception to this was cognitive load, which does not correlate with adherence in the final fitted model. However, cognitive load and user experience were strongly negatively correlated, so any alternative model using cognitive load as a predictor of adherence instead of UXA is valid. This result reinforces the idea that cognitive load and user experience have an inverse relationship (see Jung (2012)).

Curiously, UXA was a negative predictor of decision satisfaction, as was adherence to recommendations. This again contradicts results from other studies of recommendation (Knijnenburg, Willemsen, Gantner, Soncu, & Newell, 2012; Pu et al., 2011), however, this study has the advantage of modeling change in satisfaction over the baseline (Schaffer et al., 2018)—Δ Decision Satisfaction, rather than just satisfaction at one point in the task. This modeling is more accurate because it accounts for

TABLE 7.4
Factors Fit from Participant Responses to Subjective Questions

Code	MR Item	DD Item
re1	I am an expert on movies.	I am familiar with abstract trust games.
re2	I am a film enthusiast.	I am familiar with the Diner's Dilemma.
re3	I closely follow the directors that I like.	I am familiar with the public goods game.
crt1	If it takes five machines five minutes to make five widgets . . .	(same as MR)
crt2	A bat and ball together cost $1.10, and the bat costs $1.00 more than the ball . . .	(same as MR)
crt3	In a pond there is a patch of lily pads that doubles in size every day . . .	(same as MR)
tp1	I think I will trust the movie recommendations given in this task.	I think I would trust an AI adviser if one were available.
tp2	I think I will be satisfied with the movie recommendations given in this task.	I think I would be satisfied if I adhered to advice from an AI adviser.
tp3	I think the movie recommendations in this task will be accurate.	I think AI advisers give accurate information.
pt1	How understandable were the recommendations?	The Dining Guru's recommendations were understandable.
pt2	Movie Miner succeeded at justifying its recommendations.	I did not understand the Dining Guru.
pt3	The recommendations seemed to be completely random.	The Dining Guru's recommendations were groundless.
pa1	I preferred these recommendations over past recommendations.	
pa2	How accurate do you think the recommendations were?	The Dining Guru was accurate.
pa3	How satisfied were you with the recommendations?	The Dining Guru's recommendations were satisfactory.
pa4	To what degree did the recommendations help you find movies for your watchlist?	The Dining Guru's recommendations helped me to maximize points.
pc1	How much control do you feel you had over which movies were recommended?	I had control over the Dining Guru.
pc2	To what degree do you think you positively improved recommendations?	I could affect what the Dining Guru recommended.
pc3	I could get Movie Miner to show the recommendations I wanted.	I had no control over the Dining Guru.
t1	I trust the recommendations.	I trusted the Dining Guru.

TABLE 7.4 (Continued)
Factors Fit from Participant Responses to Subjective Questions

Code	MR Item	DD Item
t2	I feel like I could rely on Movie Miner's recommendations in the future.	I could rely on the Dining Guru.
t3	I would advise a friend to use the recommender.	I would advise a friend to take advice from the Dining Guru if they played the game.
cl1	There was too much information on the screen.	It was hard to keep track of all of the information needed to play the game.
cl2	I got lost when performing the task.	I got lost while playing the game.
cl3	Interacting with Movie Miner was frustrating.	I got frustrated during the game.
cl4	I felt overwhelmed when using Movie Miner.	
ds1	How excited are you to watch <movie>?	
ds2	How satisfied were you with your choice in <movie>?	
ds3	How much do you think you will enjoy <movie>?	
ds4	What rating do you think you will end up giving to <movie>?	
bs1	How excited would you be to watch <movie>?	
bs2	Would you be satisfied with choosing <movie>?	
bs3	How much do you think you would enjoy <movie>?	
bs4	What rating do you think you would end up giving to <movie>?	

Note: Items that were removed due to poor fit are not shown. All items achieved good fit, except for perceived transparency in the DD task, which was borderline.

TABLE 7.5
Question Items Used for Parceled Factors in Both Studies. Sum of Correct Responses Were Used to Calculate the Parcel

Code	MR Item	DD Item
dom1	Online, which genre has the highest current average audience rating?	In a one-round Diner's Dilemma game (only one restaurant visit), you get the least amount of dining points when . . . (four options)
dom2	Online, which of these genres tends to be the most common among the movies with the highest average audience rating?	In a one-round Diner's Dilemma game (only one restaurant visit), you get the most amount of dining points when . . . (four options)

(*Continued*)

TABLE 7.5 (Continued)
Question Items Used for Parceled Factors in Both Studies. Sum of Correct Responses Were Used to Calculate the Parcel

Code	MR Item	DD Item
dom3	Online, which of these genres has the highest current popularity?	Suppose you know for sure that your co-diners reciprocate your hot dog order 100% of the time and reciprocate your lobster order 100% of the time. Which should you order for the rest of the game? (H/L)
dom4	Generally, which of these genres has the most titles released, for all time periods?	Suppose you know for sure that your co-diners reciprocate your hot dog order 0% of the time and reciprocate your lobster order 100% of the time. Which should you order for the rest of the game? (H/L)
dom5	Online, which of these decades has the highest current average audience rating?	Suppose you know for sure that your co-diners reciprocate your hot dog order 50% of the time and reciprocate your lobster order 50% of the time. Which should you order for the rest of the game? (H/L)
dom6	How many movies have an average audience rating great than 9/10?	How much does a hot dog cost? (slider response)
dom7	Popular movies tend to have an average rating that is lower/ average/higher.	How much does a lobster cost? (slider response)
dom8	Movies with an average rating of 9/10 or higher tend to have fewer/ average/more votes.	What is the quality of a hot dog? (slider response)
dom9		What is the quality of a lobster? (slider response)
dom10		Which situation gets you more points? (two options)
dom11		Which situation gets you more points? (two options)
sat1	What is the recommender trying to predict?	The Dining Guru updates automatically every round. (T/F)
sat2	Are the recommendations I see just for me?	When the Dining Guru is updated, it predicts the choice I should make in the next round. (T/F)
sat3	What are the recommendations affected by?	When the Dining Guru is updated, it predicts the choice I should make in all remaining rounds. (T/F)
sat4	What are the recommendations based on?	When does the Dining Guru recommend hot dog?
sat5	When does the recommender update?	How does the accuracy of the Dining Guru change as the game progresses?
sat6	What happens if I delete all drama movies from my ratings?	Generally, I can maximize the dining points I get per round by ordering a mix of hot dog and lobster, regardless of what the Dining Guru recommends. (T/F)

TABLE 7.5 (Continued)
Question Items Used for Parceled Factors in Both Studies. Sum of Correct Responses Were Used to Calculate the Parcel

Code	MR Item	DD Item
sat7	What if I were to highly rate movies in the sci-fi genre?	Generally, I can maximize the dining points I get per round by only ordering what the Dining Guru recommends. (T/F)
sat8	What happens if I rate more movies according to my tastes?	
sat9	What happens if I remove accurate ratings?	
sag1		What is your current cooperation rate? (slider 0–100%)
sag2		What is Player 2's hot dog reciprocity rate? (slider 0–100%)
sag3		What is Player 2's lobster reciprocity rate? (slider 0–100%)
sag4		What is Player 3's hot dog reciprocity rate? (slider 0–100%)
sag5		What is Player 3's lobster reciprocity rate? (slider 0–100%)

TABLE 7.6
Factors Corresponding to User Metrics

Factor	Description	MR α	DD α
Trust propensity (tp)	The participant's propensity to trust the agent's recommendations.	0.92	0.91
Cognitive reflection (crt)	A measurement of decision-making ability.	0.79	0.73
Reported expertise (re)	The participant's self-assessed domain knowledge.	0.82	0.80
Perceived control (pc)	The participant's subjective assessment of their degree of control over the agent.	0.86	0.96
Perceived transparency (pt)	The participant's subjective assessment of the agent's ability to explain itself.	0.61	0.44
Perceived accuracy (pa)	The participant's subjective assessment of the agent's accuracy.	0.91	0.90
Trust (t)	The participant's reported overall trust in the agent.	0.93	0.90
User experience with the agent (UXA) (ux)	A combination of items from pc, pt, pa, and t.	0.89	0.95

(Continued)

TABLE 7.6 (Continued)
Factors Corresponding to User Metrics

Factor	Description	MR α	DD α
Cognitive load (cl)	The participant's subjective assessment of frustration that occurred during the task.	0.82	0.75
Baseline satisfaction (bs)	The participant's self-reported satisfaction with random items (MR only).	0.93	
Decision satisfaction (ds)	The participant's self-reported decision satisfaction (MR only).	0.93	

Note: α is Cronbach's alpha—a measure of internal reliability (this would mean the items that make up the factor are highly correlated). Items that were removed due to poor fit are not shown.

TABLE 7.7
Observed Dependent Variables in the Movie Recommendation Study

Variable Name	Description	MR μ	MR σ	DD μ	DD σ
Recommender int.	Number of interactions with the agent	14	29	25	34
Browser int.	Number of interactions with the simple, alternative tool	37	23		
Adherence	Proportion of the recommendations used in decision making	0.67	0.36	0.33	0.25
Domain knowledge	Score on initial insight questionnaire	0.45	0.16	0.73	0.15
SAT	Score on recommender beliefs questionnaire	0.67	0.2	0.55	0.18
SAG	Score on game state estimation questionnaire			0.99	0.556
Decision optimality	Percentage of moves made that were optimal (0.0 to 1.0)			0.59	0.12

Note: Scores on tests were normalized.

each participant's inherent ease of satisfaction and represents the quantitative change from that level of satisfaction and satisfaction arising from different task factors. The data here indicates that users who would be most likely to take recommendations at face value (without further interaction or investigation) would also be more easily satisfied by a random selection of items. Moreover, it is the knowledgeable users who engage with the system (as evidenced by increased recommender interaction) and understand it (as indicated by increased SAT) that are able to do better, especially when the system allows the user to override its behavior (as evidenced by the effect of Control on Δ Decision Satisfaction).

We believe the results in this work help to demonstrate the value of domain knowledge measurement, SAT, and CRT tests for recommender systems research. These constructs significantly increased the amount of explainable variance in decision satisfaction and adherence without affecting the order of complexity of the regression model. Moreover, their correlations with the user experience construct were quite low. Given that there were high correlations between the different system perception constructs (control, transparency, accuracy) in this experiment, it might be advisable to reduce the number of subjective user experience questionnaire items and instead use participant time to assess cognitive and knowledge variables.

Many findings in this experiment would have been missed if these measures had been omitted. Users with correct beliefs about the recommender were more likely to adopt recommendations (Figure 7.6). SAT had the highest direct positive impact on adherence with a β coefficient of 0.23, followed closely by the presence of control. User experience did not predict adherence nearly as well as the SAT factor and the control treatment. Furthermore, the "perceived transparency" subconstruct was not nearly as effective at explaining adherence (the tested relationship was non-significant in all models). This highlights the need for the use of the objective SAT measure, instead of perceived transparency, within recommender systems research. When combined with its impact on Δ Decision Satisfaction, it highlights the need for recommender system designers to instill deep understanding of recommender operations to maximize engagement, usage, and outcomes.

Increased interaction with the browser side of the interface was linked to increased SAT but also to decreased adherence. To explain this, we examined browser interaction in more detail. We found that, similar to the recommendation side, 50% of browser interactions were filter/sort/search actions and the other 50% were rating actions. What this might suggest is that participants were using the browser tool to find representative movies for their profile. As the participant found more representative items, there was more opportunity to get dynamic feedback from the recommender. Over time, this improved SAT but also increased the chance that the participant found satisfactory items from the browser tool (interesting titles were likely adjacent in metadata space to the targeted titles).

Explanation, control, and recommendation error steered the decision system towards different outcomes. Explanation improved SAT to a degree, which in turn correlated with increased adoption of recommendations and better decision outcomes. However, explanation also nullified the positive effects of control if both were switched on (see Table 7.8); this effect is difficult to explain, because there is no corresponding increase in cognitive load, decrease in user experience, or decrease in

TABLE 7.8
Regressions in the Raykov Change Model That Identifies Factors That Contributed to Improved Decision Making in the Movie Recommendation Task

Regressand	Regression (←)	Coeff.	P(> |z|)
	← UXA	−0.124	*
Δ Decision Satisfaction	← Cognitive Load	−0.237	***
$R^2 = 0.10$	← SAT	0.127	**
	← Control	0.295	***
	← Explanation × Control	−0.268	*
	← Adherence	−0.110	*

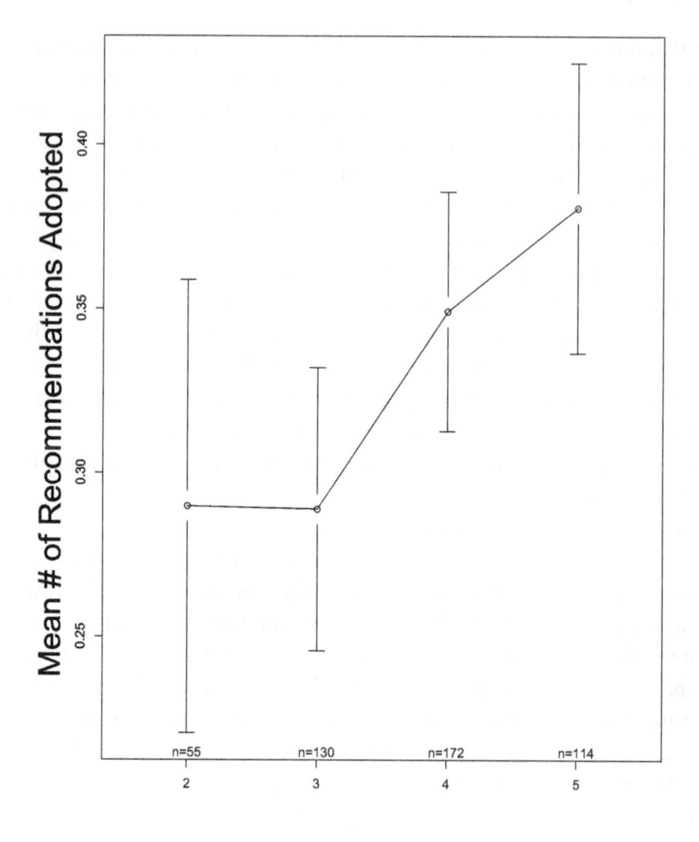

FIGURE 7.6 Relationship between adherence and understanding of the recommender in the Movie Recommendation study.

Note: As understanding increases, users adopt more recommendations.

recommender interaction caused by this particular configuration. A possible reason for this is that explanations boosted confidence in the recommendations, increasing adherence through improved SAT, and thus disengaging the participant. Next, we found that control played two roles. First, control (predictably) increased recommender interaction, which in turn correlated with increased cognitive load and adherence. Second, the presence of control features increased satisfaction with selected items regardless of interaction quantity. This leads us to believe that the ability to have control over a recommender system, whether or not that control is exercised, is a desirable feature of a recommender system because it leads users to be more satisfied with choices. Finally, reductions in recommendation error had the largest impact on user experience but had no direct effect on decision satisfaction. Since an alternative to the recommender was available in this task, it is likely that users switched to the browsing tool when the recommender failed to produce satisfactory results. Our data also indicates that control has a bigger impact on the user's satisfaction with his/her final watchlist rather than the accuracy of the recommender.

DINER'S DILEMMA: STATISTICAL EFFECTS

As with the MR study, we found that modeling the individual system perceptions was of little value, but for different reasons. While the perceived control construct was found to be externally discriminant from trust and perceived accuracy, the resulting construct was not predictive in the context of the rest of the model. This is because the alternative model containing a fit perceived control factor not only fit worse than our chosen model, but also complicated the story of reduced interaction caused by the control feature of the Dining Guru.

Perceived transparency had similar problems, but the factor also fit worse and was not predictive. Thus, we chose to model UXA instead, combining the items from trust and perceived accuracy. Also similar to the MR study, participants that considered themselves experts were much less likely to interact with the Dining Guru and adhere to recommendations. Unfortunately, these participants, who reported being trusting and yet performed worse on the domain knowledge test, subsequently ended up scoring fewer points in the game. Simultaneously, these users reported more trust than average with the Dining Guru but less interaction and adherence. Meanwhile, less trusting users demonstrated higher domain knowledge, which predicted more correct beliefs about the recommender and thus more adherence. This situation is strikingly similar to the MR task, but it is not yet clear how these users might be accommodated.

When both explanations and error were present, the model predicts that decision optimality drops below the mean. This is demonstrated in Figure 7.7. This indicates that explanations allowed users to better detect the errors in the Dining Guru, which may have steered them away from adherence in the error-prone treatments. Despite this, adhering to the Dining Guru in even full error condition would have put the user's performance at the mean, and adhering in the weak noise conditions would have put the users well above the mean (recall Table 7.2). This result implies that even relatively accurate decision support systems can be ignored if users are able to detect errors, regardless of the severity, in the agent.

The results from the DD data indicate many positive benefits of incorporating explanations into decision support systems. Explanation indirectly caused increased adherence through SAT and recommender interaction. It also had a direct effect on decision optimality, suggesting that the explanations were useful for helping the participant understand the game. Previously, explanations have been noted to increase trust (Tintarev & Masthoff, 2011), adherence (Arnold et al., 2006), and perceived control (Knijnenburg, Bostandjiev, O'Donovan, & Kobsa, 2012b). Now, this study demonstrates that explanations increase adherence through a mediating effect of the participants' beliefs about the recommender. Additionally, we have observed many important interaction effects between explanation and control or explanation and error. The results suggest that in some situations, explanation and control given together may overload users of the system. Furthermore, explanation features may draw attention to flaws in the system predictions, mitigating automation bias.

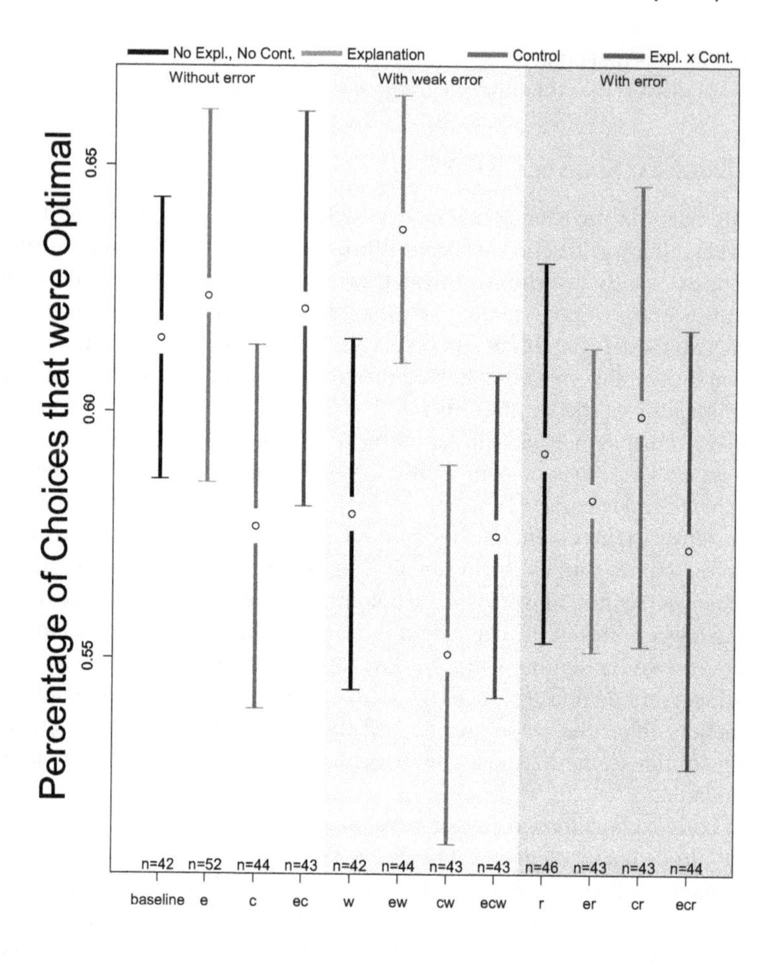

FIGURE 7.7 Mean percent of optimal choices made in each treatment in the Diner's Dilemma study.

Note: Explanations improve performance between "c" and "ec" as well as between "w" and "ew." Error bars are 95% confidence interval.

In some situations, this may lead to better decision making due to the rejection of incorrect predictions.

Comparative Analysis

In this section, the results from both the Movie Recommendation (MR) and Diner's Dilemma (DD) studies are compared. We highlight which results were replicated across both studies. A summary of effects linked to personal user characteristics is shown in Table 7.9. Across both studies, trust propensity predicted

TABLE 7.9
Support for Effects Related to User Profiling Factors

Effect	MR	DD
Trust propensity predicts higher perceptions of an agent	**Yes (***)**	**Yes (**)**
Trust propensity predicts more incorrect beliefs about agent	Yes (**)	No
Cognitive reflection predicts more correct beliefs about agent	Yes (**)	No
Cognitive reflection significantly correlates with domain knowledge	**Yes (***)**	**Yes (***)**
Self-reported expertise predicts less agent interaction	**Yes (***)**	**Yes (***)**
Domain knowledge predicts more agent interaction	Yes (***)	No
Correct beliefs about an agent significantly correlates with trust	No	Yes (**)

Note: Results supported by both studies are shown in bold.

greater perception of the decision support system. CRT also covaried significantly with initial insight tests regardless of domain (in fact, results from both studies suggest humans can be split into high CRT/high knowledge and high trust/high "reported expertise" groups). This suggests the Dunning-Kruger effect (Kruger & Dunning, 1999) is an important cognitive factor to consider when designing human-agent systems. In the MR study, there was a link between trust propensity, user experience, and low SAT, but this was not seen in the DD study, which indicated a covariance between SAT and trust. A link between trust propensity and recommender perceptions was also reported in Knijnenburg et al. (2012a). The agent present in the DD study was relatively simple when compared with the agent from the MR study, which may explain this discrepancy. Finally, users of higher domain knowledge in the MR study interacted more with the recommender, but domain knowledge was not a predictive factor in interaction with the Dining Guru. This may be explained by differences in each agent's facilities: the collaborative filtering algorithm provided information (the recommendation score) that was not present on the browser side of the interface, but the Dining Guru only aided in summarizing information that was already available, perhaps making it less useful to more capable players.

A summary of effects for the participant's cognitive states are shown in Table 7.10. Across both studies, SAT was an effective mediator of the effects of explanation on adherence and was also predictive of decision outcomes. This extends our understanding of the importance of explanations past the subjective realm of system perceptions and trust. Moreover, the positive effects on decision outcomes provide quantitative data to suggest that accurate but incomprehensible systems (e.g., deep learning) may require additional research in transparency. Where decision success was objective (DD), higher domain knowledge directly predicted better decision performance. Finally, we found that when controlling for domain knowledge in the DD study, CRT was an unnecessary predictor. This is likely due to limitations in our knowledge test for the MR study, which may have been less effective at capturing task concepts, or simply because Δ Decision Satisfaction in the MR study was independent of each participant's knowledge.

TABLE 7.10
Support for Effects Related to SAT and Domain Knowledge

Effect	MR	DD
Correct beliefs about an agent predicts increased adherence	**Yes (***)**	**Yes (*)**
Domain knowledge predicts better decision performance		Yes (***)
Correct beliefs about an agent predicts better decision performance	**Yes (**)**	**Yes (***)**
Cognitive load and user experience with an agent are negatively correlated	**Yes (***)**	**Yes (***)**
Higher user experience with an agent predicts worse decision performance	Yes (*)	No
Higher user experience with an agent predicts adherence	Yes (**)	No
More interaction with the agent predicts increased adherence	**Yes (***)**	**Yes (***)**

Note: Results supported by both studies are shown in bold.

TABLE 7.11
Support for Effects Caused by Altering the Agent's ECR Profile

Effect	MR	DD
Explanation causes correct beliefs about an agent	**Yes (*)**	**Yes (***)**
Explanation causes improved decision outcomes	**Yes (*, full mediation via SAT)**	**Yes (**)**
Control causes increased cognitive load	**Yes (**)**	**With explanation (**)**
Control causes increased adherence	Yes (***, full mediation via interaction)	No
Control increases decision performance	Yes (***)	No
Error decreases user experience with an agent	Yes (***)	No
Error decreases decision performance	No	With explanation (**)
Error decreases adherence	**Yes (**, full mediation)**	**Yes (**)**

Note: Results supported by both studies are shown in bold.

Across both studies, cognitive load was negatively correlated with user perceptions of the agent, indicating the potential for an agent to mentally relieve analysts. Higher user experience with the agent only led to increased adherence in the MR study. Moreover, higher UXA was linked to higher satisfaction with selected items in the MS study. As previously discussed, this satisfaction decrease was linked to lower engagement with the system, suggesting a need for maximizing interaction for the best outcome.

A summary of claims on explanation, control, and error of an agent is shown in Table 7.11. In both studies, the presence of explanations caused better decision success and better agent understanding. Recall that in both studies, explanation was given under varying levels of agent error. In the MR study, adherence dropped slightly when the agent made errors, but overall decision outcomes were not affected. In the DD study, error simultaneously predicts increased interaction with the agent, but also decreased adherence. When explanations were given alongside the erroneous recommendations, overall decision outcomes suffered. These results suggest that agent errors lead to complex situations. Explanations can potentially help users identify when an agent makes errors so that alternatives can be used instead. However, users may under-trust the system (DD) when, despite making errors, the average performance is still higher than the human operators. Moreover, if adequate alternative systems are accessible (MR), errors may not make their way into the final decision outcome, but will likely reflect on adherence to the agent, resulting in disuse.

Control features increased cognitive load across both the MR and DD studies. Control features in the MR study allowed users to customize the recommendation view to their tastes, getting the benefits of both traditional filtering and collaborative filtering. Control features in the DD study allowed users to explore the space of decision outcomes and also the "automation" from the Dining Guru. Users adhered less to the Dining Guru's recommendations when given control, due to decreased interaction, and thus decision optimality suffered (again, the Dining Guru performed significantly better than the mean in most treatments). Likewise, the control feature in the MR study gave the participant an increased ability to explore the movie catalog space, resulting in better Δ Decision Satisfaction. The negative outcomes associated with control in the DD study may have been due to a usability issue: the system required significantly more effort to use over the automated version, and unlike the explanation feature, there was no evidence that the control feature helped participants understand the game or the agent better. While the control features were designed to be analogous for their respective tasks, this aspect of the agent appears to be the most sensitive. Moreover, cognitive load appears to be a reliably unfortunate side effect of adding control features. We suggest that agent designers carefully consider and iteratively prototype control features to complement the task domain and agent.

Finally, agent errors predicted decreased adherence in both studies. In the MR study, this effect was fully mediated by user perception (with no direct effect found), indicating that users may have simply turned to the browser side of the interface when the recommender failed. In the DD study, the negative effects of error were partially mediated by recommender interaction, indicating only a minor decrease in adherence. As mentioned previously, this had a negative effect on participant performance, which was unfortunate. This analysis suggests that users may be overly sensitive to *perceived* errors on the part of an agent, which may be exacerbated when the user is overconfident (high reported expertise). In multi-trial tasks, perhaps agents can convince their users with a retrospective "if you had followed my advice . . ." argument. This is a promising area for future work.

Extending to Multi-Agent Systems

This chapter has presented two studies that have taken a step in quantitatively mapping out cognitive and behavioral factors for monolithic non-embodied agents (technological artifacts). The area of multi-agent systems is much less explored, particularly in the area of interacting with multi-agent swarms. This section concludes with a discussion of how the cognitive factors presented here would be relevant in multi-agent systems and swarms.

Multi-agent systems generally operate in objective task domains, where there is a tangible task such as surveillance that must be completed. The distribution of the agents is made to augment human decision-making capabilities according to concrete guidelines. Thus, the findings of the DD study will most strongly inform the discussion of this section. On occasion, multi-agent systems can be used to model the subjective preferences of humans to study their interactions. This is especially true in studies for marketing and crowd control (Humann & Madni, 2014; Kadyrova & Panasyuk, 2016). In these cases, the agents act as surrogates for human decision makers, in order to simulate the effects of system designs on the public.

With multiple interacting tangible agents, there is a high risk of cognitive overload. The greatest difficulty arises when interactions among agents are essential for the performance of the systems. In this case, the potential interactions grow quadratically with the number of agents in the system, quickly overwhelming a human's ability to track them all. Even just the sight of groups of interacting robots is enough to cause elevated stress levels in test subjects (Podevijn et al., 2016). Therefore, detailed explanation of an individual agent's decision process is almost always avoided. Instead, it may be necessary to generate monolithic explanations for multi-agent systems to reap the SAT and trust-related benefits. Moreover, multi-agent systems are often automated when gathering and summarizing data before it is presented to the user in raw form, reducing explanation. Repeated requests for attention from agents can quickly annoy and overwhelm a human controller.

Much research has gone into finding the fan-out of a human (i.e. the number of robots that one human can control) (Crandall, Goodrich, Olsen, & Nielsen, 2005; Humann & Pollard, 2019). A human must from time to time switch his attention among robots, and this imposes a cost in both time and situational awareness (Goodrich, Quigley, & Cosenzo, 2005). As a rough estimate, the fan-out of an operator can be calculated from the task switching metrics *interaction time* and *neglect time* (Chen, Barnes, & Harper-Sciarini, 2011b; Crandall et al., 2005). In the emerging field of swarm robotics, where the number of agents can reach into the thousands, there is no hope for human comprehension or control, or even mathematical prediction of behavior (Edmonds, 2004), so self-organizing algorithms, simulation (Humann, Khani, & Jin, 2014), and statistical testing are used to gain confidence in system performance.

The single-agent studies in this chapter show no strong correlation between cognitive load and decision optimality, but this may be because the users were not pushed to their cognitive limits by the monolithic agent. A multi-agent system which is prompting the user for control and decision making under time constraints represents a much more cognitively taxing scenario and may show a stronger relationship if it were studied in the same way.

Situation awareness is a design challenge both in the multi-agent system itself and in the user interface. Because most multi-agent systems are focused on objective tasks, global SA (analogous to SAG in the DD study) is the primary concern. Global SA is especially important when a human is used as a backup or troubleshooter for an automated system. In these cases, the automation is used for efficiency and precision during the majority of task execution, but when errors arise, the human is called in to restore functionality. If the human is not actively engaged with the system and maintaining global SA, it can be difficult to make accurate quick decisions when abruptly called back into duty (Ordoukhanian & Madni, 2017, 2018). Attempting to maintain SAT of large multi-agent systems is often irrelevant and could be harmful if it demands too much of the user's attention. The results of the DD study do not show an interaction between SAT and SAG, but we can speculate that as more cognitive effort is expended tracking the states of the individual agents, the user will have less ability to maintain awareness of the task and total system state. Therefore, global SA can be aided by design of low-explanation UIs that only present relevant task-centric information at the expense of more detailed SAT. Comprehension of the significance of system states could also be aided by UIs that alert the user when certain critical states are reached.

Projecting how multi-agent systems will behave in the future is perhaps the most critical research aspect, as this is an area where humans tend to falter (Tabibian et al., 2014). Systems can be made more predictable when designed with the human in mind, or UIs can present predictions directly to the user. If predictions are presented, it would be necessary to provide a limited volume of explanation, as they are based on simulations whose underpinnings would be difficult for the user to comprehend in real time.

Over-reliance on automation is even more of a risk in multi-agent systems, as the systems can quickly become so incomprehensible that the user's only viable choice to remain in control is to naively trust the recommendations of the expert system (Parasuraman & Riley, 1997). While trust propensity is only indirectly linked to adherence in these studies, it may need to be a part of the user's training with the system. Recommenders within multi-agent systems are built with the assumption that they can provide more optimal functionality than human judgment alone, so users may need to be taught to trust the system even in the face of errors and stress.

Unlike in monolithic systems, people may form a perception of each individual agent, so that the system perception may not be simply binary, as is indicated by Hassenzahl et al. (2008; Hassenzahl & Tractinsky, 2006). Here the concept of trust becomes more complicated, as a user may have differing levels of trust in each agent, and these may not all be easily predicted from that operator's trust propensity. Although control was not shown to strongly effect UXA, we can speculate that in a multi-agent system, having the ability to change or quarantine untrustworthy agents would have a positive effect on UXA. (But again, this may be counterproductive if the user substitutes his own erroneous judgment because he mistakenly distrusts an agent.)

Design of Intelligent Multi-Agent Systems and Future Work

The benefits and of multi-agent systems have made them an attractive design goal, but their complexity makes design and use very challenging. Traditional design processes can falter because of the unpredictability of multi-agent behavior, so advanced modeling, simulation, and optimization are often needed (Humann, 2015). These techniques include multi-objective and hierarchical optimization (Durand, Burgaud, Cooksey, & Mavris, 2017; Fisher, Cooksey, & Mavris, 2017), genetic algorithms to tune agent behavioral parameters (Humann & Jin, 2013; Humann et al., 2014), multi-agent simulation (including simulation of human behavior) (Gao & Cummings, 2012), and many others. The common theme is that computational tools are necessary to aid designers, as the complexity of systems is too great to draw them up fully formed from intuition or analytical methods.

Future multi-agent systems will require a more thorough understanding of human factors. This includes discovering the limitations of humans from the perspective of the agents. Agents will need to be able to sense when a human is overloaded or performing poorly and adapt. Executable models of humans will need to be developed in control and decision-making scenarios, so that designers can use realistic simulations of humans to test system autonomy levels prior to manufacture and deployment. Universal metrics for human interaction with multi-agent systems must be developed so that results of case studies can be interpreted and applied to new ideas.

Finally, emerging technology for human-systems interaction will also need to be deployed. This includes augmented reality, which can be used to summarize highlights of swarm states for situation awareness, and multi-modal (e.g. haptic) feedback to keep the user informed of the system status across multiple channels without overloading a single channel.

To reach these goals, we propose the following research agenda:

- Study the relationship between global SA and SAT: most UIs for multi-agent systems are designed under the assumption that users should be shielded from the internal details of each agent, only focusing on high-level tasks and system states. Is this assumption sound? Can a more detailed knowledge of SAT lead to inference of global SA?
- Pinpoint limits of human cognitive load: multi-agent environments will be much more cognitively taxing on the user, especially if he is forced to make decisions under time constraints. Effective design of systems must take the user's limits into account. How can workload be predicted by designers? How is workload related to decision optimality?
- Clarify relationships between levels of situation awareness and cognitive load: the "levels" of situation awareness (knowledge of states, comprehension of states, and prediction of future states) may not strictly build on one another, and any one could be tracked and summarized by a UI. Does attempting to maintain these different levels have a different effect on cognitive load? If one or more are presented to the user through a UI, making them easier to maintain, how does this affect cognitive load?
- Investigate forced vs. voluntary users: many multi-agent systems are designed out of necessity; it is unrealistic for users to complete the task without the aid of the system. In a work environment, employees can be forced to use the system. How does this affect trust in the system? Can users be trained to be more trusting? How does it affect UXA, and in the end does UXA only matter when trying to attract voluntary users?

SUMMARY

This chapter has investigated how human cognition reacts to the presence and configuration of monolithic agents. We have identified general system, user, and cognitive factors that predict decision behaviors related to interaction with systems, incorporate of system predictions (adherence), and domain decision success. We have presented surprising effects related to user traits and beliefs about systems that opens a door to future investigations. The analysis of multiple domains and the use of a common measurement methodology in two experiments has allowed us to better identify effects that should generalize well to other contexts. Furthermore, we have discussed these effects in the context of multi-agent systems and identified future research and challenges.

In the introduction, we posed the following research questions:

- How do a person's cognitive traits affect usage of an intelligent agent and resulting decision outcomes?
- Which cognitive or system factors explain variability in decision making (interaction, adherence, success) in the HAI system?
- What is the relationship between correct beliefs about agents, their use, and trust?

We provide the following answers to these research questions.

(1) In this work, we have quantitative evidence that suggests that self-reported experts are likely to be more trusting than the general population of users. These users not only interact less but also adhere to advice more often. True domain experts are more likely to have higher CRT. Intelligent agents could potentially adapt to users based on their personality, however, how to accommodate overconfident users remains an open research question.
(2) Despite the subjective (MR) vs. objective (DD) parameters of each task, we found that user experience and cognitive load were not as important as a user's understanding of task factors or understanding of the agent. Manipulation of the system's ECR profile also more strongly affected outcomes than subjective system perceptions.
(3) Situation-awareness based agent transparency (SAT) was an effective mediator of system explanation effects when trying to understand adherence to advice. Furthermore, there is strong evidence here to suggest that SAT and trust/system perceptions are discriminant, while SAT was found to be externally valid. This suggests that less trusting users might be convinced to use a system through effecting correct beliefs.

While this research has identified a number of HAI factors that transfer across domains and while we have provided expectations for their general relationships in a very limited scope, more research in other decision and task contexts, especially multi-agent tasks, is needed to develop a reliable, general theory about how intelligent agents affect human cognition and decision-making behavior. Additional factor modeling, especially task- and domain-specific factors, will be essential in achieving high levels of prediction about how human-machine systems evolve. This study has also not examined the longitudinal effects of repeated agent use on cognitive factors, nor how relationships between users and agents evolve over long periods of time. The domain knowledge and SAT metrics used in this task are exploratory and require further validation and study in each task domain where they are applied. Finally, the effects reported here warrant further and more detailed study where more variables are controlled.

In summary, we have discovered that cognitive traits and intermediate cognitive variables are crucial for understanding the effects of explanation, control, and error for HAI. Furthermore, we discovered that (1) the user profiling metrics trust propensity, cognitive reflection, reported knowledge, and domain knowledge increase the ability to predict decision-making behaviors in the presence of an agent, (2) objectively defined metrics such as situation awareness and domain knowledge are more indicative of outcomes than subjective system perceptions in the HAI system, and (3) correct user beliefs (SAT) about an agent mediate the effect of system explanation when predicting adherence to recommendations.

NOTES

All URLs last accessed June 2020.

1. www.vanityfair.com/news/business/2014/10/air-france-flight-447-crash
2. Our approach was nearly identical to http://grouplens.org/blog/similarity-functionsfor-user-user-collaborative-filtering/
3. www.qualtrics.com/
4. www.wired.com/2012/10/lance-armstrong-and-the-prisoners-dilemma-of-doping-in-professional-sports/
5. http://www.usabart.nl/QRMS/

REFERENCES

Ahuja, R. K., Mehlhorn, K., Orlin, J., & Tarjan, R. E. (1990). Faster algorithms for the shortest path problem. *Journal of the ACM (JACM)*, *37*(2), 213–223.

Ainslie, G. (2001). *Breakdown of will*. Cambridge: Cambridge University Press.

Arnold, V., Clark, N., Collier, P. A., Leech, S. A., & Sutton, S. G. (2006). The differential use and effect of knowledge-based system explanations in novice and expert judgment decisions. *Mis Quarterly*, 79–97.

Bainbridge, W. A., Hart, J., Kim, E. S., & Scassellati, B. (2008). The effect of presence on human-robot interaction. In *Robot and human interactive communication, 2008. RO-MAN 2008. The 17th IEEE international symposium on* (pp. 701–706). New York: IEEE.

Benbasat, I., & Wang, W. (2005). Trust in and adoption of online recommendation agents. *Journal of the Association for Information Systems*, *6*(3), 4.

Benjamini, Y., & Hochberg, Y. (1995). Controlling the false discovery rate: A practical and powerful approach to multiple testing. *Journal of the Royal Statistical Society. Series B (Methodological)*, 289–300.

Bilgic, M., & Mooney, R. J. (2005). Explaining recommendations: Satisfaction vs. promotion. In *Beyond personalization workshop, IUI* (Vol. 5).

Breazeal, C. (2004). Social interactions in HRI: The robot view. *IEEE Transactions on Systems, Man, and Cybernetics, Part C (Applications and Reviews)*, *34*(2), 181–186.

Breese, J. S., Heckerman, D., & Kadie, C. (1998). Empirical analysis of predictive algorithms for collaborative filtering. In *Proceedings of the fourteenth conference on uncertainty in artificial intelligence* (pp. 43–52). Burlington, MA: Morgan Kaufmann Publishers Inc.

Buhrmester, M., Kwang, T., & Gosling, S. D. (2011). Amazon's mechanical Turk a new source of inexpensive, yet high-quality, data? *Perspectives on Psychological Science*, *6*(1), 3–5.

Cafaro, A., Vilhjálmsson, H. H., Bickmore, T., Heylen, D., Jóhannsdóttir, K. R., & Valgarðsson, G. S. (2012). First impressions: Users' judgments of virtual agents' personality and interpersonal attitude in first encounters. In *International conference on intelligent virtual agents* (pp. 67–80). London: Springer.

Campbell, D. T., & Fiske, D. W. (1959). Convergent and discriminant validation by the multi-trait-multimethod matrix. *Psychological Bulletin*, *56*(2), 81.

Chen, J. Y., & Barnes, M. J. (2014). Human—agent teaming for multirobot control: A review of human factors issues. *IEEE Transactions on Human-Machine Systems*, *44*(1), 13–29.

Chen, J. Y., Barnes, M. J., & Harper-Sciarini, M. (2011a). Supervisory control of multiple robots: Human-performance issues and user-interface design. *IEEE Transactions on Systems, Man, and Cybernetics, Part C: Applications and Reviews*, *41*(4), 435–454.

Chen, J. Y., Barnes, M. J., & Harper-Sciarini, M. (2011b). Supervisory control of multiple robots: Human-performance issues and user-interface design. *IEEE Transactions on Systems, Man and Cybernetics, Part C: Applications and Reviews*, *41*(4), 435–454.

Chen, J. Y., Procci, K., Boyce, M., Wright, J., Garcia, A., & Barnes, M. (2014). Situation awareness-based agent transparency. *Technical report*, DTIC Document.

Chi, E. H. (2015). Blurring of the boundary between interactive search and recommendation. In *Proceedings of the 20th international conference on intelligent user interfaces* (pp. 2–2). New York: ACM.

Choi, A., de Melo, C. M., Khooshabeh, P., Woo, W., & Gratch, J. (2015). Physiological evidence for a dual process model of the social effects of emotion in computers. *International Journal of Human-Computer Studies*, *74*, 41–53.

Choi, S., & Clark, R. E. (2006). Cognitive and affective benefits of an animated pedagogical agent for learning English as a second language. *Journal of Educational Computing Research*, *34*(4), 441–466.

Colquitt, J. A., Scott, B. A., & LePine, J. A. (2007). Trust, trustworthiness, and trust propensity: A meta-analytic test of their unique relationships with risk taking and job performance. *Journal of Applied Psychology*, *92*(4), 909.

Crandall, J. W., Goodrich, M. A., Olsen, D. R., & Nielsen, C. W. (2005). Validating human-robot interaction schemes in multitasking environments. *IEEE Transactions on Systems, Man, and Cybernetics-Part A: Systems and Humans*, *35*(4), 438–449.

Cribbie, R. A. (2007). Multiplicity control in structural equation modeling. *Structural Equation Modeling*, *14*(1), 98–112.

Cummings, M. (2004). Automation bias in intelligent time critical decision support systems. In *AIAA 1st intelligent systems technical conference* (p. 6313). Reston, VA: AIAA.

Davidson, J. E., & Downing, C. (2000). Contemporary models of intelligence. In *Handbook of intelligence* (pp. 34–49). Cambridge, UK: Cambridge University Press.

Durand, J.-G. J., Burgaud, F., Cooksey, K. D., & Mavris, D. N. (2017). A design optimization technique for multi-robot systems. In *55th AIAA Aerospace sciences meeting* (p. 0690). Reston, VA: AIAA.

Edmonds, B. (2004). Using the experimental method to produce reliable self-organised systems. In *International workshop on engineering self-organising applications* (pp. 84–99). London: Springer.

Endsley, M. R. (1995a). Measurement of situation awareness in dynamic systems. *Human Factors: The Journal of the Human Factors and Ergonomics Society*, *37*(1), 65–84.

Endsley, M. R. (1995b). Toward a theory of situation awareness in dynamic systems. *Human Factors*, *37*(1), 32–64.

Endsley, M. R. (2000). Direct measurement of situation awareness: Validity and use of SAGAT. In *Situation awareness analysis and measurement* (p. 10). Boca Raton, FL: CRC Press.

Eppler, M. J., & Mengis, J. (2004). The concept of information overload: A review of literature from organization science, accounting, marketing, MIS, and related disciplines. *The Information Society*, *20*(5), 325–344.

Fisher, Z. C., Cooksey, K. D., & Mavris, D. (2017). A model-based systems engineering approach to design automation of SUAS. In *Aerospace conference, 2017 IEEE* (pp. 1–15). New York: IEEE.

Gao, F., & Cummings, M. (2012). Using discrete event simulation to model multi-robot multi-operator teamwork. In *Proceedings of the human factors and ergonomics society annual meeting* (Vol. 56, pp. 2093–2097). Los Angeles, CA: Sage Publications.

Gill, H., Boies, K., Finegan, J. E., & McNally, J. (2005). Antecedents of trust: Establishing a boundary condition for the relation between propensity to trust and intention to trust. *Journal of Business and Psychology*, *19*(3), 287–302.

Gomez-Uribe, C. A., & Hunt, N. (2016). The Netflix recommender system: Algorithms, business value, and innovation. *ACM Transactions on Management Information Systems (TMIS)*, *6*(4), 13.

Gonzalez, C., Ben-Asher, N., Martin, J., & Dutt, V. (2013). Emergence of cooperation with increased information: Explaining the process with instance-based learning models. *Unpublished manuscript under review.*

Goodrich, M. A., Quigley, M., & Cosenzo, K. (2005). Task switching and multi-robot teams. In *Multi-robot systems: From swarms to intelligent automata* (Vol. III, pp. 185–195). Netherlands: Springer.

Gregersen, H., & Sailer, L. (1993). Chaos theory and its implications for social science research. *Human Relations*, *46*(7), 777–802.

Gregor, S., & Benbasat, I. (1999). Explanations from intelligent systems: Theoretical foundations and implications for practice. *MIS Quarterly*, 497–530.

Gunning, D. (2017). Explainable artificial intelligence (XAI). *Defense Advanced Research Projects Agency (DARPA)*, nd Web.

Harman, J. L., O'Donovan, J., Abdelzaher, T., & Gonzalez, C. (2014). Dynamics of human trust in recommender systems. In *Proceedings of the 8th ACM conference on recommender systems* (pp. 305–308). New York: ACM.

Harper, F. M., & Konstan, J. A. (2016). The movielens datasets: History and context. *ACM Transactions on Interactive Intelligent Systems (TiiS)*, *5*(4), 19.

Hassenzahl, M. (2008). User experience (UX): Towards an experiential perspective on product quality. In *Proceedings of the 20th conference on l'Interaction HommeMachine* (pp. 11–15). New York: ACM.

Hassenzahl, M., & Tractinsky, N. (2006). User experience-a research agenda. *Behaviour & Information Technology*, *25*(2), 91–97.

Hauser, D. J., & Schwarz, N. (2016). Attentive Turkers: MTurk participants perform better on online attention checks than do subject pool participants. In *Behavior Research Methods* (Vol. 48, No. 1, pp. 400–407). Cham, Switzerland: Springer Nature.

Herlocker, J. L., Konstan, J. A., & Riedl, J. (2000). Explaining collaborative filtering recommendations. In *Proceedings of ACM CSCW'00 conference on computer-supported cooperative work* (pp. 241–250). New York: ACM.

Hertzum, M., Andersen, H. H., Andersen, V., & Hansen, C. B. (2002). Trust in information sources: seeking information from people, documents, and virtual agents. *Interacting with Computers*, *14*(5), 575–599.

Holmes, G., Donkin, A., & Witten, I. H. (1994). Weka: A machine learning workbench. In *Intelligent information systems, 1994: Proceedings of the 1994 second Australian and New Zealand conference on* (pp. 357–361). New York: IEEE.

Hoorens, V. (1993). Self-enhancement and superiority biases in social comparison. *European Review of Social Psychology*, *4*(1), 113–139.

Humann, J. (2015). *Behavioral modeling and computational synthesis of self-organizing systems.* Doctoral Thesis, University of Southern California, ProQuest, Ann Arbor, MI.

Humann, J., & Jin, Y. (2013). Evolutionary design of cellular self-organizing systems. In *ASME 2013 international design engineering technical conferences and computers and information in engineering conference* (pp. V03AT03A046–V03AT03A046). New York: American Society of Mechanical Engineers.

Humann, J., Khani, N., & Jin, Y. (2014). Evolutionary computational synthesis of self-organizing systems. *AI EDAM*, *28*(3), 259–275.

Humann, J., Khani, N., & Jin, Y. (2016). Adaptability tradeoffs in the design of self-organizing systems. In *ASME 2016 international design engineering technical conferences and computers and information in engineering conference* (pp. V007T06A016–V007T06A016). New York: American Society of Mechanical Engineers.

Humann, J., & Madni, A. M. (2014). Integrated agent-based modeling and optimization in complex systems analysis. *Procedia Computer Science*, *28*, 818–827.

Humann, J., & Pollard, K. A. (2019). Human factors in the scalability of multirobot operation: A review and simulation. In *2019 IEEE international conference on Systems, Man and Cybernetics (SMC)* (pp. 700–707). New York: IEEE.

Humann, J., & Spero, E. (2018). Modeling and simulation of multi-UAV, multi-operator surveillance systems. In *Systems conference (SysCon), 2018 annual IEEE international* (pp. 1–8). New York: IEEE.

Jin, Y., & Levitt, R. E. (1996). The virtual design team: A computational model of project organizations. *Computational Mathematical Organization Theory*, *2*(3), 171–195.

Jung, J. J. (2012). Attribute selection-based recommendation framework for shorthead user group: An empirical study by MovieLens and IMDB. *Expert Systems with Applications*, *39*(4), 4049–4054.

Kadyrova, L., & Panasyuk, M. (2016). Simulation modeling of consumer behavior in decision making about point of services purchase. *Academy of Marketing Studies Journal*, *20*, 70.

Kahneman, D. (1973). *Attention and effort*. Englewood Cliffs, NJ: Prentice-Hall.

Knijnenburg, B. P., Bostandjiev, S., O'Donovan, J., & Kobsa, A. (2012a). Inspectability and control in social recommenders. In *Proceedings of the sixth ACM conference on recommender systems* (pp. 43–50). New York: ACM.

Knijnenburg, B. P., Bostandjiev, S., O'Donovan, J., & Kobsa, A. (2012b). Inspectability and control in social recommenders. In P. Cunningham, N. J. Hurley, I. Guy, & S. S. Anand (Eds.), *RecSys* (pp. 43–50). New York: ACM.

Knijnenburg, B. P., Willemsen, M. C., Gantner, Z., Soncu, H., & Newell, C. (2012). Explaining the user experience of recommender systems. *User Modeling and User-Adapted Interaction*, *22*(4–5), 441–504.

Komiak, S. Y., & Benbasat, I. (2006). The effects of personalization and familiarity on trust and adoption of recommendation agents. *MIS Quarterly*, 941–960.

Koren, Y., & Bell, R. (2015). Advances in collaborative filtering. In *Recommender systems handbook* (pp. 77–118). London: Springer.

Kruger, J. (1999). Lake Wobegon be gone! the "below-average effect" and the egocentric nature of comparative ability judgments. *Journal of Personality and Social Psychology*, *77*(2), 221.

Kruger, J., & Dunning, D. (1999). Unskilled and unaware of it: how difficulties in recognizing one's own incompetence lead to inflated self-assessments. *Journal of Personality and Social Psychology*, *77*(6), 1121.

Lakkaraju, H., Kamar, E., Caruana, R., & Leskovec, J. (2017). Interpretable & explorable approximations of black box models. In *Proceedings of the KDD workshop on fairness, accountability, and transparency in machine learning*. New York: ACM.

Lee, M. K., & Turban, E. (2001). A trust model for consumer internet shopping. *International Journal of Electronic Commerce*, *6*(1), 75–91.

Marik, V., & McFarlane, D. (2005). Industrial adoption of agent-based technologies. *IEEE Intelligent Systems*, *20*(1), 27–35.

Martin, J. M., Juvina, I., Lebiere, C., & Gonzalez, C. (2011). The effects of individual and context on aggression in repeated social interaction. In *Engineering psychology and cognitive ergonomics* (pp. 442–451). London: Springer.

Mcknight, D. H., Carter, M., Thatcher, J. B., & Clay, P. F. (2011). Trust in a specific technology: An investigation of its components and measures. *ACM Transactions on Management Information Systems (TMIS)*, *2*(2), 12.

McKnight, H., Carter, M., & Clay, P. (2009). Trust in technology: Development of a set of constructs and measures. In *Digit 2009 proceedings* (p. 10). Atlanta, GA: Association for Information Systems.

Merritt, K., Smith, D., & Renzo, J. (2005). An investigation of self-reported computer literacy: Is it reliable. *Issues in Information Systems*, *6*(1), 289–295.

Mikhailov, A. S. (2011). *From swarms to societies: Origins of social organization* (pp. 367–380). London: Springer.

Miller, B. N., Albert, I., Lam, S. K., Konstan, J. A., & Riedl, J. (2003). MovieLens unplugged: Experiences with an occasionally connected recommender system. In *Proceedings of the 8th international conference on intelligent user interfaces* (pp. 263–266). New York: ACM.

Norman, D. A. (1986). Cognitive engineering. In *User centered system design: New perspectives on human-computer interaction* (p. 3161). Hillsdale, NJ: Erlbaum.

O'Donovan, J., & Smyth, B. (2005). Trust in recommender systems. In *Proceedings of the 10th international conference on intelligent user interfaces* (pp. 167–174). New York: ACM.

O'Donovan, J., Tintarev, N., Felfernig, A., Brusilovsky, P., Semeraro, G., & Lops, P. (2015). Joint workshop on interfaces and human decision making for recommender systems. In H. Werthner, M. Zanker, J. Golbeck, & G. Semeraro (Eds.), *RecSys* (pp. 347–348). New York: ACM.

Ordoukhanian, E., & Madni, A. M. (2017). Human-systems integration challenges in resilient multi-uav operation. In *International conference on applied human factors and ergonomics* (pp. 131–138). London: Springer.

Ordoukhanian, E., & Madni, A. M. (2018). *Introducing resilience into multi-UAV system-of-systems network* (pp. 27–40). London: Springer.

Paas, F., Tuovinen, J. E., Tabbers, H., & Van Gerven, P. W. (2003). Cognitive load measurement as a means to advance cognitive load theory. *Educational Psychologist*, *38*(1), 63–71.

Panait, L., & Luke, S. (2005). Cooperative multi-agent learning: The state of the art. *Autonomous Agents and Multi-agent Systems*, *11*(3), 387–434.

Parasuraman, R., & Riley, V. (1997). Humans and automation: Use, misuse, disuse, abuse. *Human Factors*, *39*(2), 230–253.

Parasuraman, R., Sheridan, T. B., & Wickens, C. D. (2008). Situation awareness, mental workload, and trust in automation: Viable, empirically supported cognitive engineering constructs. *Journal of Cognitive Engineering and Decision Making*, *2*(2), 140–160.

Payne, J. W., Bettman, J. R., & Johnson, E. J. (1993). *The adaptive decision maker.* Cambridge: Cambridge University Press.

Podevijn, G., O'grady, R., Mathews, N., Gilles, A., Fantini-Hauwel, C., & Dorigo, M. (2016). Investigating the effect of increasing robot group sizes on the human psychophysiological state in the context of human-swarm interaction. *Swarm Intelligence*, *10*(3), 193–210.

Prokopenko, M. (2013). *Advances in applied self-organizing systems.* London: Springer.

Pu, P., Chen, L., & Hu, R. (2011). A user-centric evaluation framework for recommender systems. In *Proceedings of the fifth ACM conference on recommender systems* (pp. 157–164). New York: ACM.

Raykov, T. (1992). Structural models for studying correlates and predictors of change. *Australian Journal of Psychology*, *44*(2), 101–112.

Ren, Z., Yang, F., Bouchlaghem, N., & Anumba, C. (2011). Multi-disciplinary collaborative building design—a comparative study between multi-agent systems and multi-disciplinary optimisation approaches. *Automation in Construction*, *20*(5), 537–549.

Requicha, A. (2013). *Swarms of self-organized nanorobots* (pp. 41–49). London: Springer.

Resnick, P., Iacovou, N., Suchak, M., Bergstrom, P., & Riedl, J. (1994). Grouplens: An open architecture for collaborative filtering of netnews. In *Proceedings of ACM CSCW'94 conference on computer-supported cooperative work* (pp. 175–186). New York: ACM.

Ribeiro, M. T., Singh, S., & Guestrin, C. (2016a). Model-agnostic interpretability of machine learning. *arXiv preprint arXiv:1606.05386*.

Ribeiro, M. T., Singh, S., & Guestrin, C. (2016b). Why should I trust you?: Explaining the predictions of any classifier. In *Proceedings of the 22nd ACM SIGKDD international conference on knowledge discovery and data mining* (pp. 1135–1144). New York: ACM.

Rosenfeld, A., Agmon, N., Maksimov, O., & Kraus, S. (2017). Intelligent agent supporting human—multi-robot team collaboration. *Artificial Intelligence*, *252*, 211–231.

Rubenstein, M., Ahler, C., & Nagpal, R. (2012). Kilobot: A low cost scalable robot system for collective behaviors. In *Robotics and Automation (ICRA), 2012 IEEE international conference on* (pp. 3293–3298). New York: IEEE.

Salem, M., Lakatos, G., Amirabdollahian, F., & Dautenhahn, K. (2015). Would you trust a (faulty) robot?: Effects of error, task type and personality on human-robot cooperation and trust. In *Proceedings of the tenth annual ACM/IEEE international conference on human-robot interaction* (pp. 141–148). New York: ACM.

Schaffer, J., O'Donovan, J., & Höllerer, T. (2018). Easy to please: Separating user experience from choice satisfaction. In *Proceedings of the 26th conference on user modeling, adaptation and personalization* (pp. 177–185). New York: ACM.

Shortliffe, E. H., Davis, R., Axline, S. G., Buchanan, B. G., Green, C. C., & Cohen, S. N. (1975). Computer-based consultations in clinical therapeutics: Explanation and rule acquisition capabilities of the MYCIN system. *Computers and Biomedical Research*, *8*(4), 303–320. Amsterdam: Elsevier.

Sinha, R., & Swearingen, K. (2002). The role of transparency in recommender systems. In *CHI'02 extended abstracts on Human factors in computing systems* (pp. 830–831). New York: ACM.

Snyder, M. G., Qu, Z., Chen, J. Y., & Barnes, M. J. (2010). Roboleader for reconnaissance by a team of robotic vehicles. In *Collaborative Technologies and Systems (CTS), 2010 international symposium on* (pp. 522–530). New York: IEEE.

Stephens, D. W., McLinn, C. M., & Stevens, J. R. (2002). Discounting and reciprocity in an iterated prisoner's dilemma. *Science*, *298*(5601), 2216–2218.

Stuart, D., Christensen, K., Chen, A., Cao, K.-C., Zeng, C., & Chen, Y. (2013). A framework for modeling and managing mass pedestrian evacuations involving individuals with disabilities: Networked Segways as mobile sensors and actuators. In *ASME 2013 international design engineering technical conferences and computers and information in engineering conference* (pp. V004T08A011–V004T08A011). New York: American Society of Mechanical Engineers.

Sweller, J. (1988). Cognitive load during problem solving: Effects on learning. *Cognitive Science*, *12*(2), 257–285.

Sweller, J. (1994). Cognitive load theory, learning difficulty, and instructional design. *Learning and Instruction*, *4*(4), 295–312.

Tabibian, B., Lewis, M., Lebiere, C., Chakraborty, N., Sycara, K., Bennati, S., & Oishi, M. (2014). Towards a cognitively-based analytic model of human control of swarms. In *2014 AAAI spring symposium series*. Palo Alto, CA: AAAI.

Tintarev, N., Kang, B., Höllerer, T., & O'Donovan, J. (2015). Inspection mechanisms for community-based content discovery in microblogs. In *IntRS@ RecSys* (pp. 21–28). New York: ACM.

Tintarev, N., & Masthoff, J. (2007). A survey of explanations in recommender systems. In *Data engineering workshop, 2007 IEEE 23rd international conference on* (pp. 801–810). New York: IEEE.

Tintarev, N., & Masthoff, J. (2011). Designing and evaluating explanations for recommender systems. In *Recommender Systems Handbook*, pages 479–510. Springer.

Tintarev, N., O'Donovan, J., Brusilovsky, P., Felfernig, A., Semeraro, G., & Lops, P. (2014). Recsys' 14 joint workshop on interfaces and human decision making for recommender systems. In *Proceedings of the 8th ACM conference on recommender systems* (pp. 383–384). New York: ACM.

Toplak, M. E., West, R. F., & Stanovich, K. E. (2011). The cognitive reflection test as a predictor of performance on heuristics-and-biases tasks. *Memory & Cognition, 39*(7), 1275–1289.

Ullman, J. B., & Bentler, P. M. (2003). *Structural equation modeling*. Hoboken, NJ: Wiley Online Library.

Varian, H. R., Bergstrom, T. C., & West, J. E. (1996). *Intermediate microeconomics* (Vol. 4). New York: Norton.

Veletsianos, G. (2007). Cognitive and affective benefits of an animated pedagogical agent: Considering contextual relevance and aesthetics. *Journal of Educational Computing Research, 36*(4), 373–377.

Wainer, J., Feil-Seifer, D. J., Shell, D. A., & Mataric, M. J. (2006). The role of physical embodiment in human-robot interaction. In *Robot and human interactive communication, 2006. ROMAN 2006. The 15th IEEE international symposium on* (pp. 117–122). New York: IEEE.

Welsh, M., Burns, N., & Delfabbro, P. (2013). The cognitive reflection test: How much more than numerical ability. In *Proceedings of the 35th annual conference of the cognitive science society* (pp. 1587–1592). Cognitive Science Society.

Yu, K., Berkovsky, S., Taib, R., Conway, D., Zhou, J., & Chen, F. (2017). User trust dynamics: An investigation driven by differences in system performance. In *Proceedings of the 22nd international conference on intelligent user interfaces* (pp. 307–317).New York: ACM.

8 Fuzzy Cognitive Maps for Modeling Human Factors in Systems

Karl Perusich

CONTENTS

INTRODUCTION

A security director of a large midwestern university was concerned about the possibility of a terrorist attack at her institution. Recognizing that how a response would be made and resources deployed would be very different if it were in fact a terrorist attack rather than a routine emergency, she wondered if it would be possible to develop an algorithm to analyze the available data in real time to assess its likeliness. To this end she canvassed a number of experts on campus on this posing the following question: How likely would it be that a terrorist attack was underway if there was a fire in a biology lab that contained lethal toxins, an accident at a strategic intersection on campus that stopped traffic, and an attack on personnel in the security office?

Since each individual had their own area of expertise, the director received a number of responses. One expert, the fire chief for the campus, was concerned about the fire in the biology lab. This individual hypothesized that a suspicious individual seen in the area might indicate that the fire was set. The fire being set definitely indicated the incident was non-routine, i.e. not an accident. Such a state of affairs might indicate that the attack was planned and might be part of a terrorist plot.

A second individual contacted by the director, a professor in the political science department with expertise in international terrorism, concentrated on whether there were simultaneous attacks present, believing that this increased the likeliness of the attack being planned, and any planned attacked likely indicated a terrorist plot. Good evidence of a terrorist plot and a catastrophic incident were a good indication of a local terrorist attack. For this individual, the presence of at least two of the three incidents described by the director indicated that a simultaneous attack was underway.

The campus director of emergency management services felt that any incident that resulted in fatalities should be classified as a catastrophic incident. Since fatalities were generally the end game for a terrorist attack, any catastrophic incident would likely indicate the possibility of a local terrorist attack. This individual believed that the attack on the security office would result in the loss of communication for emergency personnel on campus, in turn reducing their ability to coordinate a response. Without coordination the chances of fire or EMS resources reaching affected areas in time to reduce the threat of casualties would be reduced. The delay of the EMS personnel would increase the potential fatalities. Without department resources reaching the biology building, the fire would get out of control, likely increasing the chances lethal toxins would be released. The release of lethal toxins would definitely cause fatalities.

A graduate student in the policy department indicated that an accident at a strategic intersection on campus would block traffic, reducing the ability of fire department and EMS personnel to get to the place where the incidents were occurring. This in turn would allow the fire to get out of control, releasing deadly toxins that would result in potential fatalities.

These insights provided the director with a somewhat disjointed assessment of the potential situation she might encounter. She noted a few things about the information provided. Each subject matter expert provided detailed information and assessment about the domain they were most familiar with and did it by developing a chain of events. None provided a global, overarching picture of the situation. Many used common concepts in their assessment— simultaneous events, potential fatalities, attack was planned—in the chain of causality they developed. These could be used to connect the thinking of the different experts into an overall picture or model of the situation. What she needed was a modeling tool that could use these individual assessments to develop an overall picture of the situation.

Enter fuzzy cognitive maps. Fuzzy cognitive maps are a unique way to model multi-faceted, multi-disciplined systems that incorporate a variety of attributes. These attributes can range from "hard" concepts or values like the fuel remaining in an aircraft to very "soft" concepts like the attitudes of an adversary. Because the map captures causal relationships and requires only knowing the level of change that has occurred, a common numeric metric for the nodes is not necessary. In addition to this ability to compare apples to oranges, another key value of a fuzzy cognitive map is the ability to use multiple subject matter experts. A system, i.e. overall fuzzy cognitive map, is made by piecing individual maps of the subject matter experts together through common nodes. This allows each individual to develop their map using a language and concepts that they are familiar with.

A fuzzy cognitive map is a signed di-graph that captures the causality a subject matter expert or experts believe to define a problem space. The mapping concept has been used in a variety of different contexts and fields as diverse as economics, political science, human factors engineering, and others (Kosko, 1987; McNeese, Rentsch, & Perusich, 2000; McNeese & Perusich, 2000; Papageorigiou & Poczeta, 2015). As a modeling technique, fuzzy cognitive maps have a number of strengths that make them ideally suited for capturing the underlying relationships in a complex decision space that includes factors with a variety of attributes. Chief among these are the ability to compare "apples to oranges," the lack of a need for a common numerical metric, the identification of feedback loops, and the ability to construct a complete map from individual sub-maps. These strengths in this modeling technique come with a price. For the most part a fuzzy cognitive map can infer only qualitative changes, not quantitative changes. The best a fuzzy cognitive map can do is postulate a "large" increase in a concept, not an exact value for this increase (Kosko, 1986).

Once the map has been constructed, it can then be used in several ways for analyzing a problem space. In one the reachability matrix is calculated. This matrix is used to assess whether a particular node is one of the causes of another node. This can be used to identify only relevant nodes for a particular effect. In the second method certain nodes are designated inputs and assigned values that are then propagated through the map until some sort of equilibrium is reached. This infers a state of the system as represented by all of the nodes in the map.

CONSTRUCTING A FUZZY COGNITIVE MAP: NODES

A fuzzy cognitive map is composed of nodes connected by directed lines. A node in a fuzzy cognitive map represents a changeable concept in the map. Nodes are then connected by a directed line segment that infers a causality between the source and destination node. A change in the causal node is said to cause a change in the effect node, the definition of the type of node determined by the direction of the line segment connecting the two. Note that nodes are not strictly limited to being either a cause or an effect. It depends on the context at that point in the map. A particular node can be a cause for one pair of nodes and an effect for another pair.

Key to the construction of a valid fuzzy cognitive map is to recognize that the nodes must represent a changeable quantity, i.e. typically a characteristic of the underlying concept represented that can increase or decrease. A simple example from arms control theory can illustrate this. "Our battleships" causes "their battleships" would be an inappropriate pair of nodes for a fuzzy cognitive map. Neither the cause nor the effect in this relationship can change. But "an increase in the number of our battleships" would cause "an increase in the number of their battleships" meets the criteria for defining nodes in a fuzzy cognitive map because each node can now increase or decrease. In essence the map infers that an increase in the underlying concept of a node causes an increase (or decrease) of the concept of the underlying concept of the effect node. In addition to representing nodes that can capture an increase or decrease in an underlying concept it represents, they can also capture the presence or absence of some attribute associated with the space being modeled, termed binary nodes.

Let's return to the security director of the midwestern university. One of the experts that provided her information was concerned about the *fire in the biology lab*. Specifically, *this individual hypothesized that a <u>suspicious individual</u> seen in the area might indicate that the <u>fire was set</u>. The fire being set definitely indicated the <u>incident was non-routine</u>, i.e. not an accident. Such a state of affairs might indicate that the <u>attack was planned</u> and might be part of a <u>terrorist plot</u>.*

This individual has identified six changeable concepts that can be captured as nodes in a fuzzy cognitive map (underlined in the previous paragraph). Each of these nodes can be characterized as a binary node, which represents the presence or absence of the concept. Part of the challenge in constructing a fuzzy cognitive map is recasting or distilling the description given by an expert into some sort of causal chain. Keywords like *indicate*, *shows*, etc. mean that a causal relationship connects the concepts described. So for this description each of the changeable concepts identified can have two values: there was a fire in the biology lab or there wasn't, a suspicious individual was or wasn't seen in the area, the fire was set or it wasn't, the incident was routine or it wasn't, the attack was planned or not planned, and finally there was a terrorist plot or there wasn't one.

The causal connections are indicted by how an individual has described the relationships between the concepts. For example, *a suspicious individual seen in the area might indicate that the fire was set* indicates that the expert has identified a causal relationship between *suspicious individual* and *fire was set*, with a *suspicious individual* being a cause of the *fire being set.* A suspicious individual present at the biology building in their thinking would be a cause of the fire being set.

A completed fuzzy cognitive map for this expert's situation description is given in Figure 8.1.[1] Each of the nodes is a circle in the map with the concept it captures given by the label within it: *Fire in Biology Lab, Fire was Set*, etc. Causal relationships are indicated by a line connecting two nodes, the arrow showing the direction of "flow" of causality. The line segment starts on the cause and ends on the effect. Fuzzy cognitive maps for the descriptions provided by the other three experts contacted by the director are given in Figures 8.2 to 8.4.

This brings up another point. The line segments connecting a cause-effect pair are signed. If the line segment is positive, the change in the underlying concepts are of the same type; an increase in the causal node gives an increase in the effect node. Likewise, a decrease in the cause causes a decrease in the effect. For binary nodes the presence of the cause indicates the presence of the effect, with the absence of the cause indicating the absence of the effect when the edge connecting them is positive. The relationship can be inverse if the sign of the directed segment linking the nodes is negative. In this case an increase in the source causes a decrease in the effect and vice versa. For binary nodes with inverse causality, the presence of a cause would indicate the absence of the effect and vice versa.

To jump ahead a bit, inference is done in a fuzzy cognitive map by selecting a subset of nodes to define as inputs and assigning values to these. These input nodes represent the sources of causality in the map in the same way a voltage source represents sources of energy in an electric circuit. Keep in mind that the node represents a change, increase or decrease, presence or absence, in the underlying concept being captured by it. Although it is possible to define fuzzy values for this change for the node using linguistic qualifiers, for example *somewhat increases* or *significantly decreases*, it is more standard to simply map the node to crisp values of increase, decrease, or no change, or in the case of binary nodes, presence or absence (Perusich & McNeese, 1998). This allows the nodes to be modeled with simple numerical values: 1 for increase, –1 for decrease, and 0 for no change, or 1 for presence and 0 for absence for binary nodes.

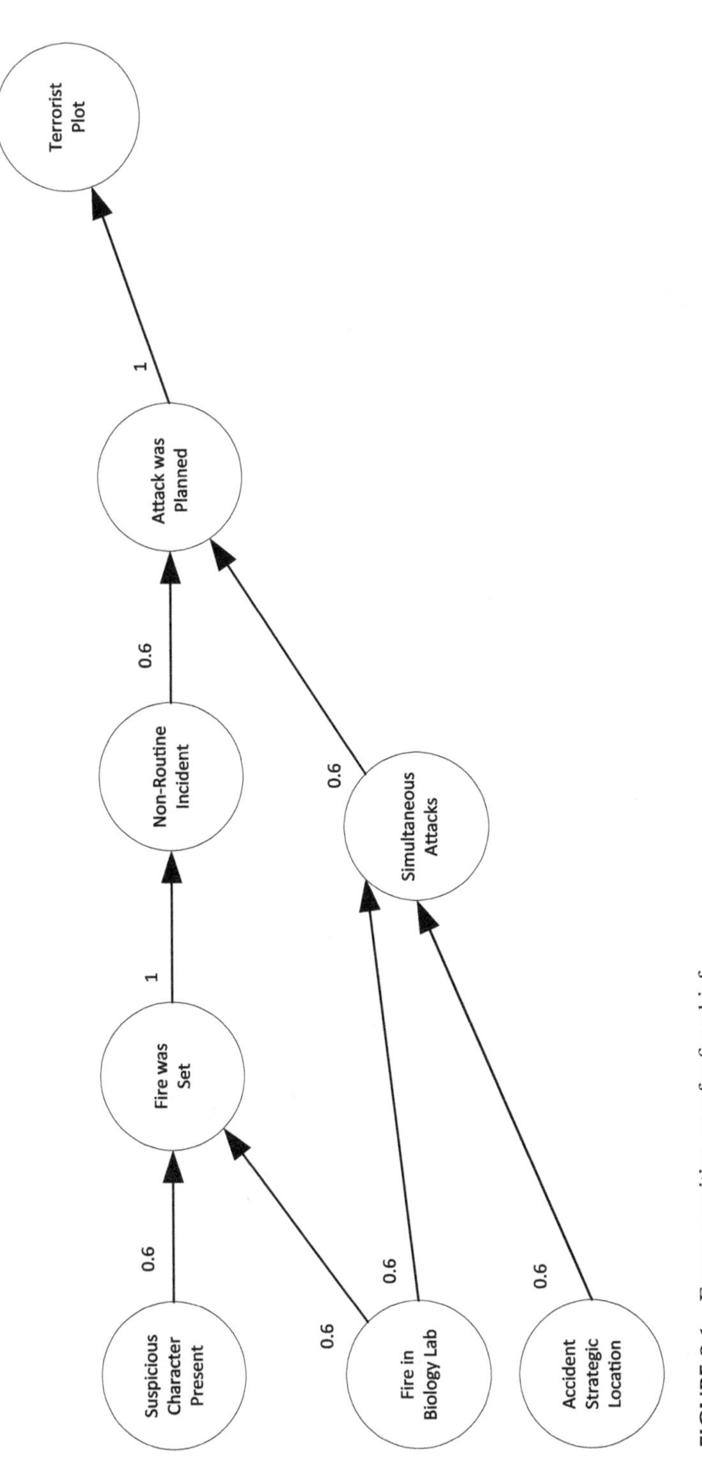

FIGURE 8.1 Fuzzy cognitive map for fire chief.

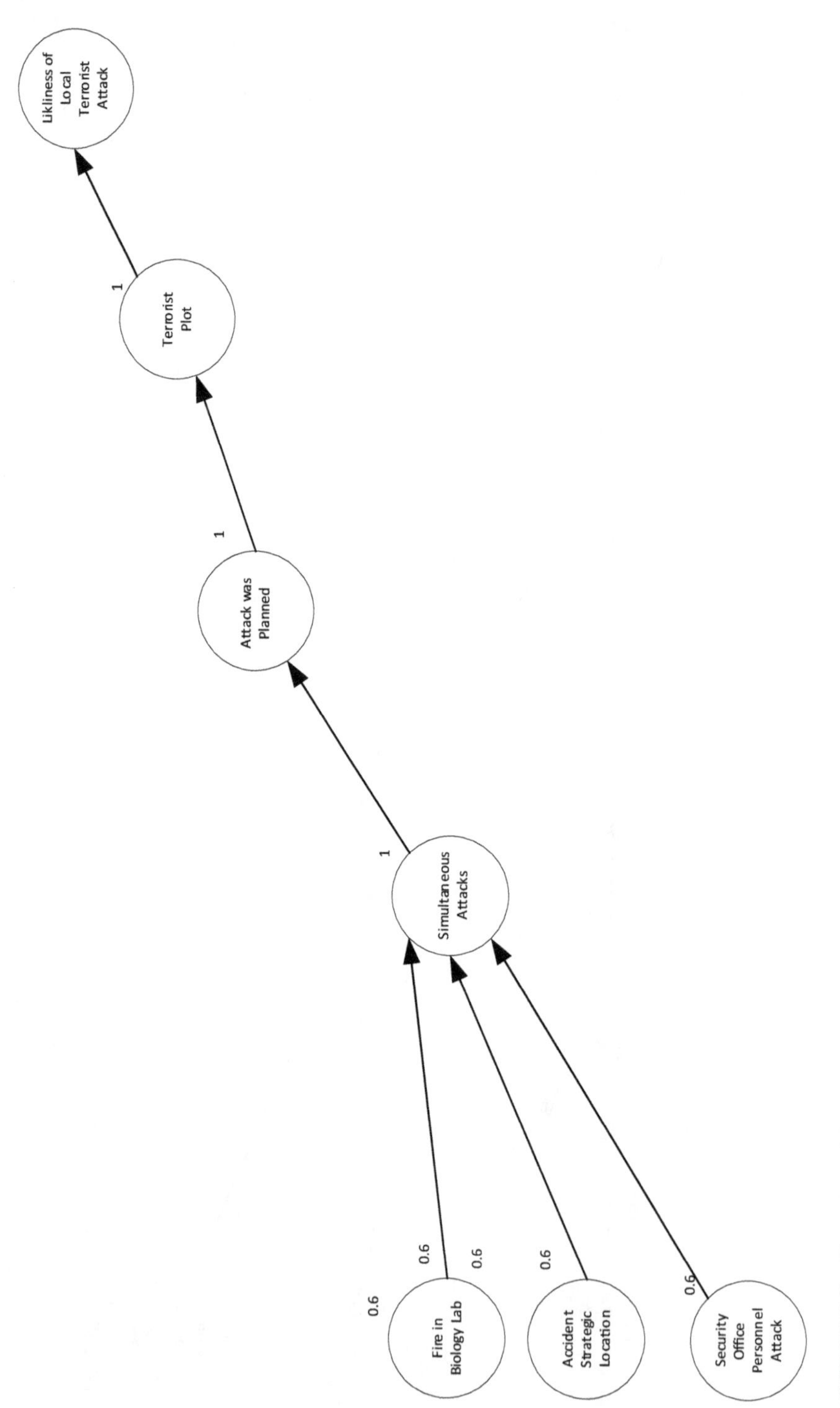

FIGURE 8.2 Fuzzy cognitive map for political science professor.

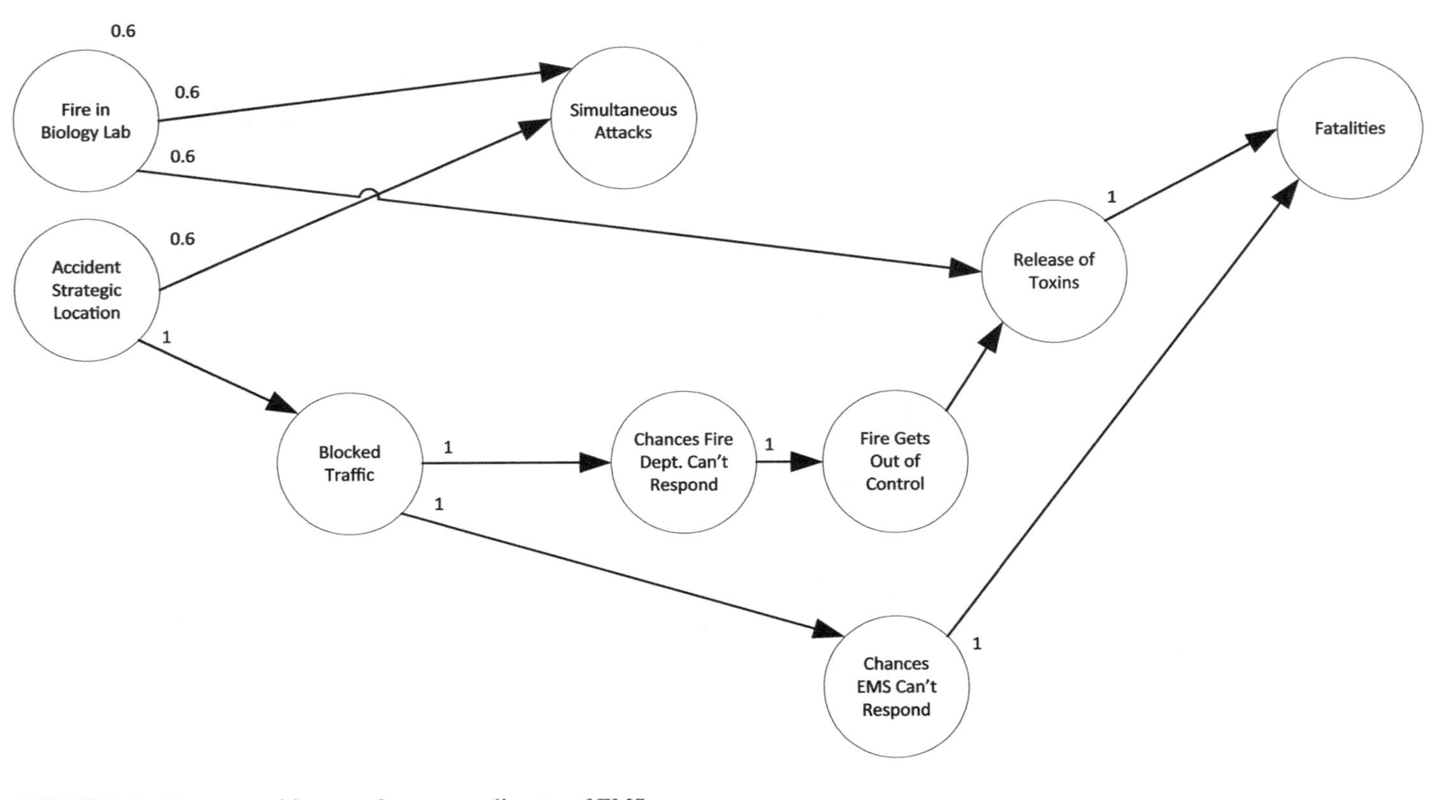

FIGURE 8.3 Fuzzy cognitive map for campus director of EMS.

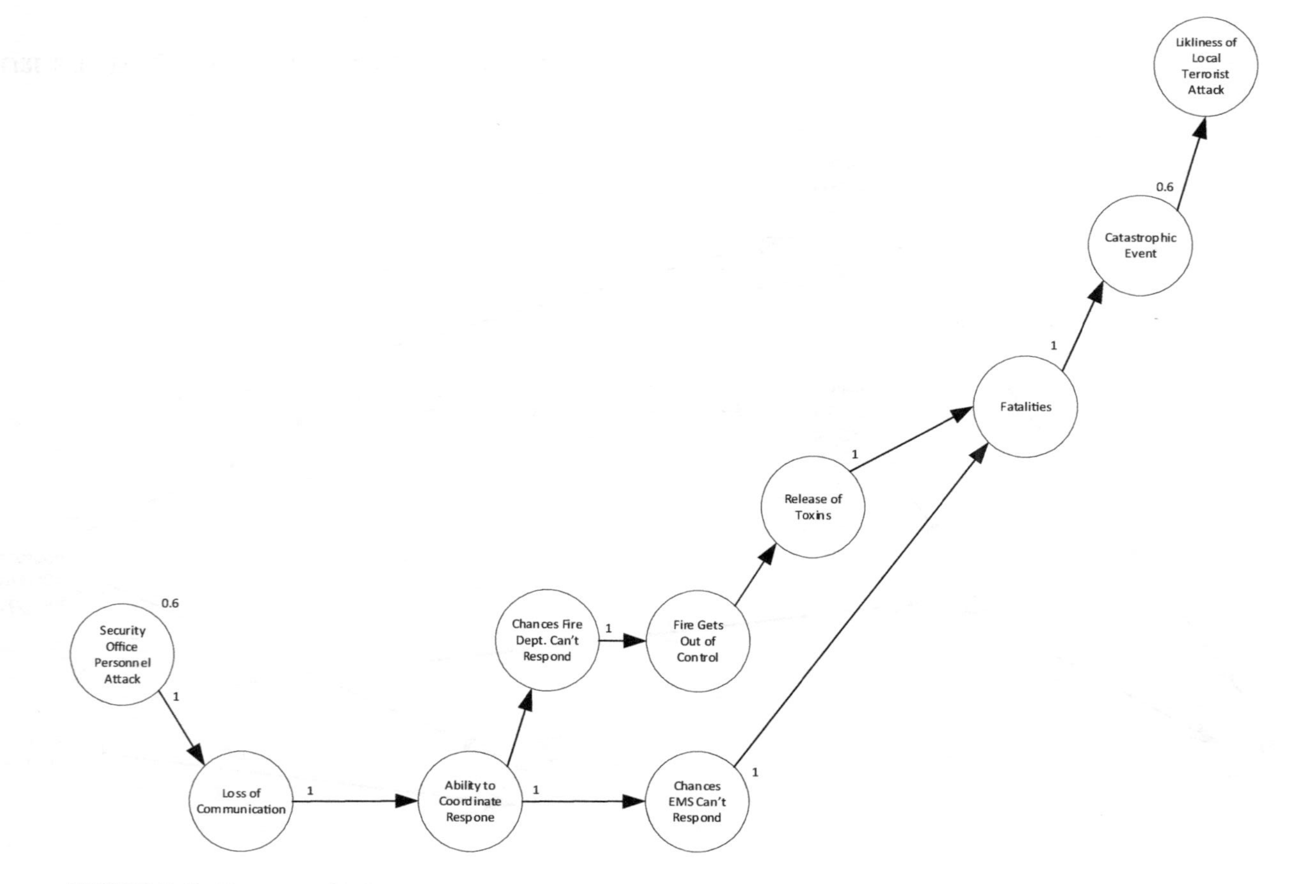

FIGURE 8.4 Fuzzy cognitive map for graduate student in the policy department.

TABLE 8.1
Common Nodes for Experts (Partial List)

Node	Fire Chief	Political Science Professor	Director of EMS	Policy Department Graduate Student
Fire in biology lab	X	X	X	
Accident at strategic location	X	X	X	
Security office personnel attack		X		
Simultaneous attack	X	X	X	
Attack is planned	X	X		
Terrorist plot	X	X		
Potential fatalities			X	X
Catastrophic incident		X		X

Nodes common to the indicated expert are denoted by X

One of the advantages of using fuzzy cognitive maps is that they can be constructed from sub-maps develop by multiple experts through common nodes. Each expert can describe the space in which they are most fluent using a language that they are most comfortable with. Very often these experts will only describe (i.e. map) a part of the overall problem space under examination. Fuzzy cognitive mapping allows the build up of a high-fidelity model of a problem from multiple experts.

Returning to the problem of identifying a terrorist attack on a university, the director received expert information from four individuals, each with a different viewpoint. None completely described all facets of the problem. Instead, each relied on their own expertise and described what they knew best. These individual sub-maps can be merged through common nodes to develop a composite, global map of the entire problem. A partial list of common nodes for the four experts are given in Table 8.1 with the composite map given in Figure 8.5.

CONSTRUCTING A FUZZY COGNITIVE MAP: EDGES

The fuzziness in the map comes in primarily through the strength of the line segments connecting a cause-effect pair. Take for example a simple two-node pair where A is the cause and B is the effect, as shown in Figure 8.5. As can be seen in the figure, the line segment, directed from A to B, has a strength of "a little." The idea is here is that an increase in A causes a little increase in B. The small increase in B may or may not be important to the system as modeled by the totality of values of the nodes in the map.

As previously stated, one of the chief strengths of a fuzzy cognitive map is its ability to compare apples to oranges. Since the map infers changes in nodes from changes in the underlying concepts, a common numerical metric is not needed. Values for the nodes do not need to be graphed to a common numerical measure, for example money, many times artificially, to be used in the map. The only knowledge that is needed is whether the underlying concept the node represents has changed and how qualitatively that change affects the effects that it is deemed to cause. Each

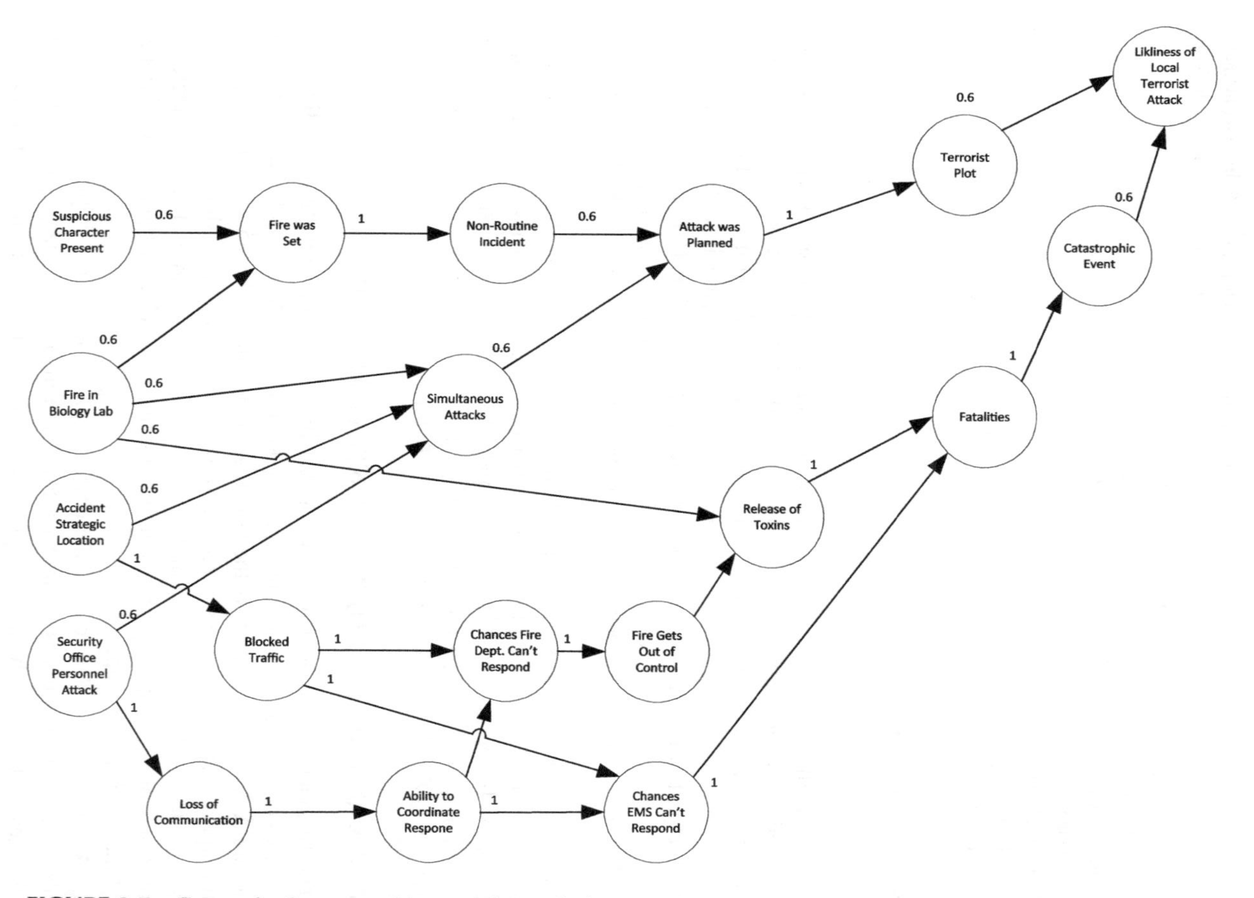

FIGURE 8.5 Composite fuzzy cognitive map for analyzing terrorist attack against a university.

subject matter expert that may have insight into part of the problem space can construct maps using their own "language." This allows the individual to incorporate a variety of attributes into the map, from hard details about the system like fuel level, remaining weapons, and altitude, to soft characteristics and decisions, like the cognitive state of the pilot and the rules of engagement they are adhering to.

This highlights another important value of fuzzy cognitive maps as a modeling tool. Very often problem spaces can be large, complex, and cut across many disciplines and areas of expertise. Rarely can a single individual have enough insight to model the entire problem space themselves, yet, it is the meta-space that is most interesting. Since common numeric metrics are not needed in a fuzzy cognitive map, a group of subject matter experts can be used to model this meta-space. Each develops a small map (sub-map) that incorporates their understanding of the causal relationships that exist from their viewpoint as a relevant subject matter expert. This viewpoint typically does not incorporate or attempt to model the entire problem space. The meta-model is then contracted by piecing these individual sub-maps together through common nodes.

For example, assume that some problem beset an aircraft as it was making a final approach for a landing in storm. In this case one might identify three subject matter experts relevant to understanding the situation of the aircraft at the time of the mishap: the pilot, the air traffic controller, and the meteorologist at the weather service. Each would have a particular vantage point with relevant information, but most likely none of the three has a complete picture of all of the relevant information important for understanding the situation, if for no other reason than that each only has access to data through their instruments. The pilot can see firsthand through the cockpit window prevailing weather conditions in the immediate surroundings of the aircraft with additional information about the status of the aircraft available through their flight instruments. The air traffic controller would have data about flight paths, location of other aircraft in the area, etc. The weather officer would have real-time data about winds, changing weather conditions along a flight path, location of thunderstorms and their movements, etc.

Each has a particular vantage to the problem space that encompasses all of this data and each would make decisions based on what is available to them. Ideally, information sharing would occur, but one actor might not know that data they have in real time is of value to another actor unless asked. Likewise, this actor may not know the importance of a particular datum, and hence ask for it, given the developing situation, or may not know it's available. When reconstructing the problem space using the data and vantage point of each actor and then piecing it together to form a meta-map of the situation at the time of the mishap, the importance of certain data may become apparent. Future similar situations can then be avoided by providing this relevant data in real time and educating the participants of its importance.

One of the interesting results that can happen from the process of constructing a meta-map from individual sub-maps is the development of feedback loops. It may turn out that as a change in a concept is propagated through the meta-map, there may be a lengthy causal connection that returns to the originating node, i.e. there is a causal feedback loop through the map to this initial cause. If positive feedback is present the causal loop tends to reinforce the initial change increasing its effect, sometimes without bounds. If negative feedback is present the initial change will decrease and decay away, normally the desired state of affairs.

Identification of feedback loops in a map is one of the most desirable features of a fuzzy cognitive map when it is constructed from multiple subject matter experts. One can argue that since an expert "sees" only a piece of the problem space they tend to be unaware of the global effects of their decisions and designs, i.e. they are missing the forest for the trees. They are missing the big picture not necessarily because they want to but more normally because they have to. They do not have the cognitive bandwidth or the knowledge or access to relevant information to fully understand other aspects of the problem space in which their expertise is embedded. Feedback loops are one of those unintended consequences that are only apparent when expertise from multiple individuals is combined to form a meta-model and can be used to explain negative effects when a "big picture" approach is taken to understanding a problem.

Once constructed, a fuzzy cognitive map can be used in two major ways. In the first, the map can be used as a guide in future data assessment by looking at the causal chain between nodes. In this case a user is interested in what causes have a direct connection to a particular effect. In the second, the map is used as a meta-model of the problem space for predictive purposes. Certain nodes are assigned values that are then propagated through the map to assess the resulting changes in other nodes. In this case the totality of nodes about a problem space is the assessment of the output given the input.

CONSTRUCTING A FUZZY COGNITIVE MAP: ASSIGNING VALUES EDGES

Inference in a fuzzy cognitive map will typically involve the propagation of nodal values through the map. As stated previously, nodes can be given fuzzy values through linguistic qualifiers or crisp values that strictly represent an increase or decrease in the concept it represents. Let's concentrate on the latter situation where nodes are assigned crisp values of –1 for decrease, 0 for no change, or 1 for increase. A standard technique for assigning nodal values explained in detail in the next section is to sum all of the causal nodes weighted by their edge strengths that are connected to a particular node. This value is then mapped to one of the three possible nodal values. The fuzziness, then, in both the map and the inference process is captured by the strength of the edge connecting two nodes (Osoba & Kosko, 2017).

The values of these edges can be assigned directly by defining fractional values to particular linguistic terms used in the description of the causal chain being modeled. Typically this involves first sorting the terms in a prescribed way, for example, from weakest to strongest. Fractional values are then assigned for each on the interval [0,1]. Since there is normally not a compelling reason to do otherwise, the interval is divided into equally spaced intervals with each value then assigned to one of the terms. For example, the description has identified five terms such that the interval [0,1] would be divided into five increments: 0.2, 0.4, 0.6, 0.8, and 1.0. Each term would then be assigned to one of these values based on where it is in the ranking from weakest to strongest. An example of these assignments is given in Table 8.1.

TABLE 8.2
Numerical Values for Fuzzy Edges

Linguistic Qualifier for an Edge in a Map	Numerical Value Assigned
Very little	0.2
A little	0.4
Somewhat	0.6
Greatly	0.8
Extremely	1.0

These numeric values are always then checked by inferring nodal values in a map to ensure that they give the desired results and adjusted if necessary.

Let us return to the descriptions provided to the director of security by the political science professor. This expert was most concerned about whether a simultaneous attack was present. *If there were simultaneous attacks present, then the likeness of the attack being planned increased, and any planned attacked likely indicated a terrorist plot. For this individual the presence of at least two of the three incidents described by the director*[2] *were present.* Again the causal concepts captured by nodes in the map are underlined. Except for the likeliness of the attack being planned, each of the causal concepts is binary in nature. Rather than an increase or decrease being captured, the presence or absence of the concept is captured by the node. The map for this expert's reasoning is given in Figure 8.2.

Assessing the strength of the causal connections in this map is done partly with reference to the inference process, described in the next section. For this map, inference will be done by summing causal nodes for a particular effect weighted by the strength of the edge connecting them. The sum is then mapped to the nodal value of 1 (increase or presence), –1 (decrease), or 0 (no change). As stated in the description by the professor, the presence of *simultaneous attacks* "likely" indicates that the *attack is planned.* This is not direct causality. *Simultaneous attacks* do not a priori indicate the *attack was planned*, but there is a strong relationship present. One might assume given the strong causal relationship that few if any other causes would need to be present to give the effect *attack is planned*, so a value of 0.6 is planned.[3]

For the node *simultaneous attacks*, a condition defined by the professor is that if any two of the three incidents, *attack on the security office*, *accident at critical location*, and *fire in the biology lab*, then the node *simultaneous attacks* will fire. Values of 0.6 are assigned for the edge strengths in this case. In this way the sum of the causes weighted by their strengths for the node simultaneous attacks will always be greater than 1 whenever any two of the causes are present.

INFERENCE IN A FUZZY COGNITIVE MAP

There are several ways in which a fuzzy cognitive map can be used to understand a problem space. In one the fuzzy cognitive map is a model with predictive powers. Given an initial set of nodal values it can be used to infer other values in the map.

The idea is to keep these input nodes constant and propagate the causality through the map until it equilibrates to static values or a limit cycle, a repeating fluctuation in nodal values (dynamic equilibrium). The state or solution of the problem at hand is the value of all the nodal values once initial conditions are applied and the map reaches static or dynamic equilibrium.

To begin the process a subset of nodes in the map are identified as "inputs." One of the interesting things about a fuzzy cognitive map is that this subset can change depending on the context of the problem. In some cases less information is available so a smaller set of nodes is used. In other cases more information is available so a larger subset is used. In still others, an entirely different set of nodes is used that may or may not overlap with the original set.

Once a set of input nodes is identified, they must be assigned initial values. These will represent the sources of causality for the inference process and will be held constant throughout the process. In some cases the initial value is set purely subjectively by the user, for example, a "large increase" in the adversary's willingness to attack. In other cases the nodal value (keeping in mind that it must represent an increase or decrease in the underlying concept represented by the node) can be mapped from actual data from the attribute that the node represents. This is generally done by breaking the range of actual values associated with the node into mutually exclusive intervals that are then mapped to a linguistic modifier. Again there is no a priori method for assessing these intervals. They are determined by the judgment of the subject matter experts. Very often an initial set of intervals is assigned and then adjusted based on the inference process to give known outcomes.

An alternative approach for nodal values is to assign each one of three possible values: 1 to indicate an increase in the underlying concept captured by the node, –1 to indicate a decrease, and 0 to indicate no change. For binary nodes in the map 1 would indicate the presence of the concept and 0 its absence. Which format is used for the definition of nodal values determines which inference technique is used. Both will be examined here.

Let us first examine the case where nodal values are defined with linguistic qualifiers. An example fuzzy cognitive map is given in Figure 8.6. In this case node A is to capture changes in the weight associated with a component in the system being modeled by the fuzzy cognitive map. The exact ranges of increases in weight for each linguistic qualifier would be determined by the system component that the node is capturing in the map and the context in which it is in. For this example, if the exact increase was 245 kgs then the node would be given a fuzzy value of "big increase." If this node was considered an input during the inference process, then it would retain the value "big increase" regardless of changes the map might attempt to make to it because of its connections to other nodes. It is a source of causality so its value is fixed in the same way that a voltage sources value is fixed in an electric circuit.

A basic fuzzy cognitive map is defined by its connection matrix, **A**, where the A_{ij} represents the causal link from node i to node j, sign and strength. An entry of 0 indicates that no causal link exists from node i to node j. So, with this definition the rows of **A** represent causes and the columns represent effects. The values of the entry capture the fuzziness in some way, with a positive value indicating direct causality and a negative sign indicating inverse causality. Fractional values capture

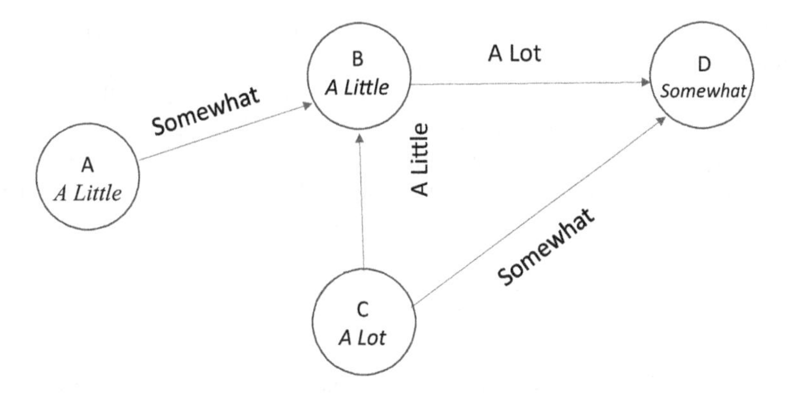

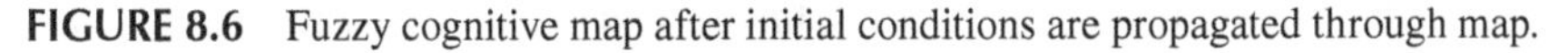

FIGURE 8.6 Fuzzy cognitive map after initial conditions are propagated through map.

TABLE 8.3
Example Interval Mapping to Linguistic Qualifier

Change in Weight	Linguistic Qualifier
0–200 kgs	A little increase
100–200 kgs	Somewhat increases
201–300 kgs	A lot of increase
301–400 kgs	Very big increase
401–500 kgs	Extremely big increase

the fuzziness in strength of the causality representing modifiers like "somewhat," "a little," "a lot," etc.

Inference proceeds in the following way once the input nodes have been defined and their initial values have been assigned. A vector of all nodes with the input nodes set to their assigned values is multiplied by the connection matrix to propagate the causality one step through the map, yielding an updated vector of nodal values. Should they change because of connections in the map, the input nodes, which represent sources of causality, are reset to their initial values. This process is then repeated with this updated vector being multiplied by the connection matrix, with input nodes reset if necessary, to give an updated vector. Mathematically the update process is given by the following equation:

$$\mathbf{V}_{N+1} = \mathbf{A}\mathbf{V}_N$$

where $\mathbf{V}_{N+1}$ is the updated vector of nodal values, $\mathbf{A}$ is the connection matrix, and $\mathbf{V}_N$ is the initial vector of nodal values. This process is repeated until the nodal vector $\mathbf{V}_N$ equilibrates to a static set of values or a limit cycle. With a limit cycle the nodal values repeat in a particular pattern indicating an instability in the system. The final set of nodal values reached in the map represents the state of the system it models. All nodal values are part of this inference.

For maps that use a linguistic qualifier for nodal definitions methodology that mimics fuzzy logic, a max-min approach is used. In this approach each cause is measured against the link with the minimum value of the cause and the strength of its link used as its contribution to the effect node. This is done for each causal node and the maximum of these is chosen as the value of the effect node.

$$E_j = \sum_{0}^{K} \max\left[\min\left[C_i, A_{ij}\right]\right]$$

where E_j is the effect node, K is the number of causal nodes that have non-zero links to E_j, C_j is the fuzzy value of the ith causal node, and A_{ij} is the strength of the connection from C_i to E_j.

As an example of the max-min inference process consider the simple fuzzy cognitive map given in Figure 8.6. It contains four nodes, with four causal connections. The nodal connections have linguistic strengths summarized in Table 8.2, where "somewhat" is considered stronger than "a little", "a lot" is stronger than "somewhat" and so on. So, for example, the causal link from A to B can be read as "an increase in A somewhat causes an increase in B".

To see how max-min inference works, assume that nodes A and C are designated as inputs, with linguistic values of "a little" and "a lot" respectively. Since there is no feedback present in the map, it is entirely feed forward, so the inference process is straightforward in the sense that it takes only one pass through the map to equilibrate. The values for A and C are propagated through the map until all remaining nodes have values. Should feedback have been present, the inference process would have to have been repeated multiple times, with nodes A and C reset to their initial conditions as inputs should they have been changed.

Node B is affected by the two input nodes, A and C. To determine a value for node B, the values of nodes A and C are evaluated against the strength of the connection between them and B, with the minimum value, as given in Table 8.2, as the contribution of each node to B. Node A has a value of "a little" while its connection strength to node B is "somewhat." Choosing the minimum of the two, "a little" becomes the contribution of A to causing B.

Node D in turn is caused by nodes B and C. (Note this is why the value of node B must be calculated before node D.) The contribution of node C to node D is the

TABLE 8.4
Sample Linguistic Values for a Fuzzy Cognitive Map

Strength	Linguistic Value
Weakest	A little
	Somewhat
	A lot
Strongest	A great deal

minimum of "a lot," the nodal value of C, and "somewhat," the strength of connection from C to D, thus giving a value of "somewhat." The minimum of "a little" (the value of node B) and "a lot" (the strength of the connection from B to D) is the contribution of B to D, in this case "a little." The value of the node D, then, is the maximum of "a little" and "somewhat," or "somewhat." Remembering that the system is the totality of nodes, its final state is given as in Figure 8.6. Although only some nodes may be considered outputs and of immediate interest, the state of the system is the inferred values for all nodes.

If the nodes are not defined using linguistic qualifiers an arithmetic approach can be used. In this case, numerical values are assigned to the strength of the nodes and the links between them. Nodal values are limited to 1, –1, and 0, as defined previously. A positive value of 1 for a node indicates an increase, while a value of –1 indicates a decrease in the underlying concept of the node. A node is updated by multiplying all causes connected to it by the strength of the link. These are then summed to give a value for the effect node. This value can then be mapped to a linguistic scale. In this approach causality accumulates. Mathematically:

$$E_j = \sum_{0}^{K} C_i A_{ij}$$

where E_j is value of the effect node, C_i is the ith node that is a cause of E_j, and A_{ij} is the strength of the connection between E_j and C_i. The sum is over the K nodes that are causes. Once the value of E_j is determined it can then be mapped back to a fractional interval and its associated linguistic value. This method is most often used when the nodal values are defined by 1, –1, and 0. A common mapping used is:

If $E_j \geq 1$, then the node = 1
If $E_j \leq -1$, then the node = –1
Otherwise the node = 0

Returning to the composite map prepared by the director of assessing the likeliness of a terrorist attack, any set of its nodes can be used as inputs. Let's assume that information is available that there was a fire in the biology lab, there was no suspicious character seen in the area but there was also an accident at a key intersection on campus that blocked traffic. For this case nodal values for initial conditions would be as given in Table 8.5.

TABLE 8.5
Initial Conditions for Scenario

Node	Nodal Value	Numerical Nodal Value
Suspicious character present	Absence	0
Fire in biology lab	Presence	+1
Accident at strategic location	Presence	+1
Blocked traffic	Presence	+1

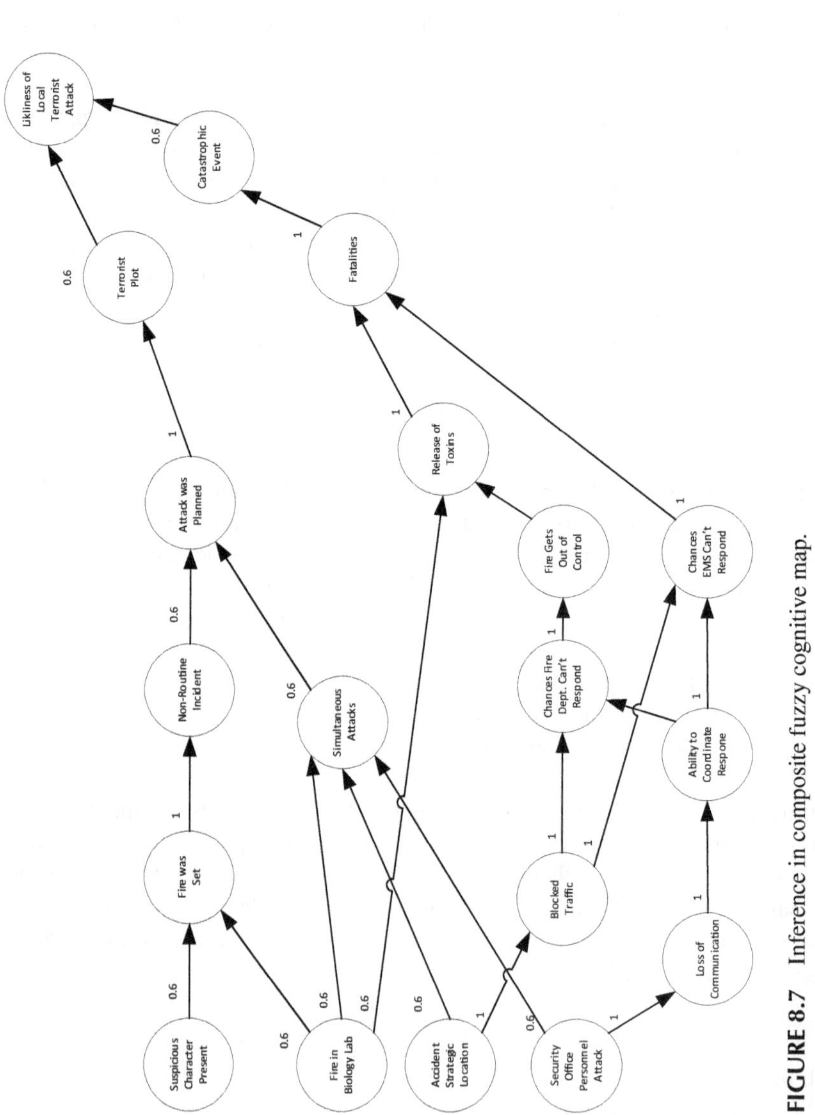

FIGURE 8.7 Inference in composite fuzzy cognitive map.

Note: Initial inputs are: No suspicious person present. There is a fire in the biology lab and there is an accident at a strategic location on campus.

To illustrate how the inference process works let's examine it for the node *Simultaneous Attacks*. In this case the node has three causes, *Fire in Biology Lab*, *Accident Strategic Location*, and *Security Office Personnel Attack*. As stated in the scenario description *Fire in the Biology Lab* and *Accident at Strategic Location* are initial conditions so their values are set at +1 and fixed during the inference process. No information is available about *Security Office Personnel Attack* so it is given a value of 0. The initial sum for the node *Simultaneous Attacks* is then calculated:

$$
\begin{aligned}
\text{NodeValue}_{\textit{Simultaneous Attacks}} &= \text{NodeValue}_{\textit{Fire in Biology Lab}}{}^{*}0.6 + \\
&\quad \text{NodeValue}_{\textit{AccidentStrategicLocation}}{}^{*}0.6 + \\
&\quad \text{NodeValue}_{\textit{SecurityOfficePersonnelAttack}}{}^{*}0.6 \\
\text{NodeValue}_{\textit{Simultaneous Attacks}} &= 1{*}0.6 + 1{*}0.6 + 0{*}0.6 = 1.2
\end{aligned}
$$

This value is then mapped to give a value of 1 for the node *Simultaneous Attacks*. The remaining initial conditions are propagated through the map to give the values in Figure 8.7.

OTHER USES OF FUZZY COGNITIVE MAPS: THE REACHABILITY MATRIX

One of the important ways in which a fuzzy cognitive map can be used in the assessment of a problem space is to determine the causal links between nodes. In this instance a user is trying ty identify which nodes will ultimately affect a node or nodes of interest, that is, the user is trying to find causal paths from nodes that are defined as inputs to nodes assigned as outputs. Note that depending on the context, the set of nodes defined "inputs" and "outputs" can change (Perusich & McNeese, 2005).

Such information can be used in two different ways. In the first way a user is trying to identify what input nodes affect output nodes of interest. Large maps constructed from multiple subject matter experts may contain tens, hundreds, even thousands of nodes cutting across multiple expert domains. It is not always obvious what and how nodes will ultimately affect another node. Developing this information will help the user to concentrate on only nodes that do in fact affect the output node of interest. This can be useful when looking for causes when the output node is some type of disaster, steering the user to only those nodes, and the concepts that they are modeling, that do in fact cause the output. Sometimes nodes that would seem initially to be contributing to the output can be eliminated. More importantly, nodes that might not be considered initially could become highlighted through this process, especially when multiple sub-maps are pieced together in developing the final map. It might not be obvious in a sub-map that a particular node could be a contributing trigger to some event, but when the sub-maps are combined a chain of causality is identified that individual subject matter experts failed to understand.

Once the causal link has been determined from a cause to an effect it can be used to identify and assess "interventions" that can be used to change the outcome. Interventions are add-ons that be incorporated into the system being modeled. These add-ons then become new nodes in the map that change the causal nodes present in it in some way. In some instances it may be possible to break the link by eliminating a node, for example, blocking an intersection if the traffic at that point is contributing to a problem downstream. A second possibility is to change the map by adding additional nodes at key points in it that mitigate or change the effects of already-present nodes. For example, it may be the case that a particular node is significantly contributing to the overheating of the system modeled by the map. In this case it may be possible to add nodes that contribute to the cooling of other nodes in the map. If this is the case then the technology modeled by the node would need to be physically built into the system. Using the map in this way, though, would help user identify practical and effective ways to design interventions that would give the desired results before time and resources are committed to changing the actual system.

Another way to design interventions is to force and eliminate feedback in the map. As stated previously, one of the interesting but often not recognized attributes of a fully constructed map is the presence of feedback loops. These loops of causality reinforce or mitigate certain nodes. Identification of these loops can suggest ways in which they can be broken if their net result is undesirable. It may also be possible by careful examination of the causal chain leading back to the node to change its characteristic. Many times these loops give positive feedback, reinforcing a node. Just as in a control circuit this can be undesirable, causing a particular attribute to grow without bounds and saturate. If this loop can be flipped to negative feedback then the "shock" to the system will be mitigated and decay away.

To identify whether one node ultimately will affect another node, the reachability matrix for the map is calculated. To calculate the reachability matrix entries in the connection matrix **A** are first replaced with a 1 if they are non-zero. Since for this calculation knowing only that a causal connection exists is necessary, knowing the strength is not necessary. The reachability matrix **R** is calculated by repeated multiplication of **A**. After each multiplication the entries in **R** are also mapped to 1: non-zero values in the matrix are given values of 1 with values of 0 remaining unchanged. This process is repeated until the entries in **R** are static. In theory this could take N multiplications, where N is the number of nodes in the map. In practice the map equilibrates well before N multiplications.

$$R = \mathbf{A}^N$$

A non-zero entry in the reachability matrix at $\mathbf{R}_{ij}$ indicates that there is a causal path from node i to node j. Note that there may be multiple paths through the nodes in the map that connect node i to node j. A visual inspection of the map is then generally used to determine the path or paths from node i to node j.

CONCLUSION

Fuzzy cognitive maps are an effective way to model multi-faceted problem spaces with many different types of attributes present. Nodes in the map represent changes in the underlying concept so a common numeric metric is not needed to compare the different concepts being modeled. Another advantage of this technique is that the map can be constructed from sub-maps, each prepared by an expert with knowledge about only part of the problem through common nodes. The map itself is a true model in the sense that it can predict outcomes. A set of nodes is chosen as inputs and their causal effects are allowed to propagate through the map. The resulting state of the system is then the values of all the nodes in the map viewed together. In addition to predictive capabilities a fuzzy cognitive map can be used to understand the relationships between concepts incorporated in the model. This can be used to examine the effectiveness of proposed changes to aspects in it.

NOTES

1. Each line segment has a value associated with it, indicating the strength of the causal relationship between the two nodes. A description of the meaning of these values and how they are assigned is given in the next section.
2. The incidents were a fire in a critical biology lab, an accident at a strategic location, and an attack on personnel in the security office.
3. The mapping used later for inference in the map is that if the sum is greater than 1, then the node maps to 1. So by choosing an edge strength of 0.6, only two of the causes for this node need to fire for it be 1.

REFERENCES

Kosko, B. (1986). Fuzzy cognitive maps. *International Journal of Man-Machine Studies, 24*, 65–75.

Kosko, B. (1987). Adaptive inference in fuzzy knowledge networks. *IEEE International Conference on Neural Networks*, June (II-261-268).

McNeese, M. D., & Perusich, K. (2000). Constructing a battlespace to understand macro-ergonomic factors in team situational awareness. In *Proceedings of the Industrial Ergonomics Association/Human Factors and Ergonomics Society (IEA/HFES) 2000 Congress* (pp. 2-618–2-621). Santa Monica, CA: Human Factors and Ergonomics Society.

McNeese, M. D., Rentsch, J. R., & Perusich, K. (2000). Modeling, measuring, and mediating teamwork: The use of fuzzy cognitive maps and team member schema similarity to enhance BMC3I decision making. in *Proceedings of the IEEE international conference on systems, man, and cybernetics*. New York: Institute of Electrical and Electronic Engineers (pp. 1081–1086).

Osoba, O., & Kosko, B. (2017). Fuzzy cognitive maps of public support for insurgency and terrorism. *Journal of Defense Modeling and Simulation: Applications, Methodology, Technology, 14*(1), pp. 17–32.

Papageorigiou, E., & Poczeta, K. (2015). Application of fuzzy cognitive maps to electricity consumption prediction. *2015 annual conference of the North American Fuzzy information processing society*. 10.1109/NAFIPS-WConSC.2015.7284139

Perusich, K., & McNeese, M. D. (1998). *Understanding and modeling information dominance in battle management: Applications of fuzzy cognitive maps*. Air Force Research Laboratory. AFRL-HE-WP-TR-1998-0040.

Perusich, K., & McNeese, M. D. (2005). Using fuzzy cognitive maps as an intelligent analyst. *Proceedings of the 2005 IEEE international conference on computational intelligence for homeland security and personal safety* (pp. 9–15). 10.1109/CIHSPS.2005.1500602

9 Understanding Human-Machine Teaming through Interdependence Analysis

Matthew Johnson, Micael Vignatti, and Daniel Duran

CONTENTS

MOTIVATION

As any craftsman knows, having the right tool for the job makes all the difference. Sadly, adequate formative tools are lacking for the design of human-machine teaming. The vast majority of formative design tools are technology-centric in which the human is not even considered part of the system (e.g. MATLAB, LabVIEW).

Human factors research has highlighted the need for consideration of the human, but this is often accomplished using summative assessments to evaluate existing systems at the end of development (e.g. NASA TLX). User-centered design has pushed for such evaluations to happen earlier in the development process. Apple, for example, has a reputation for early user evaluations translating into successful products. However, this remains the exception rather than the norm for most system development efforts. Moreover, while user evaluations are undoubtedly valuable, designers need to be able to account for such considerations long before there is a system to evaluate. To assist designers in accomplishing this, they need formative tools available in the design process that include the human as part of the system from the beginning.

Another major impediment to designing human-machine teaming is that most existing tools focus on the taskwork, in particular the physical activity. Taskwork is only a small piece of the teaming puzzle. More advanced efforts from cognitive systems engineers extend this with consideration for cognitive activity (e.g. Crandall & Klein (2006)). This is a valuable contribution, but still tends to be taskwork focused. While these traditional task-based approaches are important, they do not sufficiently capture human-machine teaming requirements. Moreover, the language, concepts, and products of those who focus on cognitive aspects are often far removed from those who design and implement working systems and do not translate "into a language that helps a product be built or coded" (Hoffman & Deal, 2008).

The key to understanding human-machine teaming, and in fact teaming in general, is understanding interdependence. Teaming, both human-human and human-machine, comes in many forms with varying characteristics and properties. The one concept that is consistent in every case is the importance of interdependence. This truth is invariant across all domains. This paper begins with an explanation of interdependence. To help designers reorient their minds and view the problem space through the lens of interdependence, we propose several design principles. These principles are intended to help designers shift their perspective and reframe the design challenges to focus on the core elements of teaming, specifically the interdependence needed to support it. Lastly, we discuss a formative tool called the Interdependence Analysis tool.

UNDERSTANDING INTERDEPENDENCE

In order to analyze a human-machine system's interdependence, it is necessary to understand the concept of interdependence. It is often simply equated to dependence, where one entity relies on another because it lacks some capability provided by the other. However, this definition of the concept is too simplistic to capture the nuances observed in interdependence relationships among teams engaged in joint activity. Our definition of interdependence is as follows:

> "Interdependence" describes the set of complementary relationships that two or more parties rely on to manage required (hard) or opportunistic (soft) dependencies in joint activity.
>
> **(Johnson et al., 2014)**

Interdependence is about relationships, not taskwork. It is true that the taskwork influences the potential relationships, however, designing human-machine teaming is about designing the interdependence relationships, not the taskwork functions. These relationships are used to manage the interdependence in the joint activity. Relationships can be required, but a significant portion of teaming is about exploiting opportunistic relationships.

To better intuit the concept of interdependence, consider an example of playing the same sheet of music as a solo versus a duet. Although the music is the same, the processes involved are very different (Clark, 1996). The difference is that the process of a duet requires ways to support the interdependence among the players. Understanding the nature of the interdependencies among team members provides insight into the kinds of coordination or teaming that will be required, such as the management of timing, tempo, and volume. Supporting these relationships through agreed signaling and exploiting these interdependence relationships at run-time is what teaming is all about. Success in a duet requires not only execution of the musical score (i.e. individual taskwork competency), but also the extra work of coordinating with others via interdependence relationships to produce effective performance. Flawless execution of the musical score individually is not enough for the duet to be successful, nor is exceptional individual competence sufficient for effective teaming.

INTERDEPENDENCE DESIGN PRINCIPLES

While having the right tools is valuable, having the right mindset or perspective on the problem is also essential to effectively leveraging the tools. This turns out to be particularly challenging for human-machine teaming largely because of the pervasiveness of today's function allocation-based paradigms (Johnson, Bradshaw, Feltovich, Hoffman et al., 2011). In general, such approaches date back to Sheridan and Verplank's work on supervisory control (Sheridan & Verplank, 1978). Follow-up research on dynamic and adaptive function allocation has led to numerous proposals for dynamic adjustment of autonomy. Such approaches have been variously called adjustable autonomy, dynamic task allocation, sliding autonomy, flexible autonomy, and adaptive automation. In each case, the system must decide at runtime which functions to automate and to what level of autonomy (Parasuraman, Sheridan, & Wickens, 2000), thus we refer to them as function allocation-based approaches. A restrictive outcome of such approaches is that automation choices dictate interaction possibilities. A simple example of this can be seen with the Roomba vacuum. The original bump-and-go exploration strategy of the Roomba prevented the ability to pause and resume work. Newer models actually map the room and plan efficient routes, enabling such interaction. A more recent example can be found in machine learning. This automation choice has demonstrated sophisticated capabilities that have raised interest, while at the same time the DARPA XAI project is examining how to redesign these types of technologies to be more explainable, so they can be more useful to the humans with whom they will need to work. These are just a few of the examples where automation choices limited interaction potential.

Our approach inverts this paradigm, suggesting that desired teaming interactions should shape the automation design (Johnson, Bradshaw, Feltovich, Jonker et al.,

2011). Thus, our focus is on the interdependence relationships that enable interaction. Focusing on relationships instead of functions is challenging for most people raised under the traditional task decomposition paradigm. Task decomposition is the staple approach to tackling hard problems by breaking them down into smaller ones. Understanding interdependence requires an additional step of considering how those pieces will fit back together again when distributed between people and machines. As Peter Drucker noted long ago, "when it comes to the job itself, however, the problem is not to dissect it into parts or motions but to put together an integrated whole" (Drucker, 1954). While his statement was about business management, it summarizes the true challenge of human-machine teaming nicely.

To help designers reorient their minds and view the problem space through the lens of interdependence, we propose several design principles. The first principle is about the value of teaming. While there is certainly a cost to teaming, and that cost must be controlled (Klein, Woods, Bradshaw, Hoffman, & Feltovich, 2004), there is a value to teaming that must be balanced against the cost. Historically, traditional approaches appear almost surprised by interdependence. For example, an early task decomposition approach to planning noted that "the expansion of each node produces child nodes. Each child node contains a more detailed model of the action it represents. The individual subplan for each node will be correct, but there is as yet no guarantee that the new plan, taken as a whole, will be correct. There may be interactions between the new, detailed steps that render the overall plan invalid" (Sacerdoti, 1975). Similarly, another approach noted "because the relevant constraints have been shared during the planning process, the expectation is that few, if any, conflicts will appear during plan merging. However, because of the complexity of planning dependencies, conflicts can arise" (desJardins & Wolverton, 1999). These references from early planning systems unknowingly highlighted the need to handle interdependence. Their solution was to try to supplement their task decomposition approach with capabilities such as critics used to resolve conflicts and backtracking techniques to handle cases that could not be resolved. Today's function allocation approaches are no different from these early planning approaches because they focus on what to automate and who to assign it to (Parasuraman et al., 2000). More recent work still notes that "'conventional wisdom' is often an over-simplification, and will be modified and sometimes reversed by a host of contextual factors" (Wickens, Li, Santamaria, Sebok, & Sarter, 2010). The main problem with traditional approaches was noted by Stefik long ago who observed "Subproblems interact. This observation is central to problem solving" (Stefik, 1981). This is a critical insight, however he concluded, "a key step in design is to minimize the interactions between separate subsystems" (Stefik, 1981). While traditional approaches view interdependence as an unexpected or unfortunate problem to be resolved, avoided, or minimized, we view it as the core design element. It is something to be leveraged opportunistically. The main reason is because the value of teaming comes not from avoiding, ignoring, limiting, or minimizing teaming, but from exploiting it. For example, searching a building can be more effectively done as a team. However, to attain the benefits, the team must exploit their interdependencies by coordinating search activity and sharing information about their individual search efforts. Thus, our first principle:

Principle 1: the value of teaming comes from exploiting it not avoiding it.

In order to exploit teaming, it is important to understand teaming. Historically, engineers have designed work to be done by an individual. Staying with our building search example, engineers have created robotic algorithms for an individual robot to search a building. However, if two robots with such an algorithm were put into a building, the result would not be teaming, but two individual efforts. The reason is the algorithm designers did not view the work as joint work. In contrast, we propose that all behaviors should be designed from the beginning to be joint work (Johnson, Bradshaw, Feltovich, Jonker et al., 2011). This means that as any behavior or algorithm is created, designers should consider the potential teaming associated with the activity. If this is done, the single agent case is just a degenerate case and is achieved for free. However, the converse is not true. Thus, our second principle is:

Principle 2: All work should be designed as joint activity (coactive), with independent work being the degenerate case.

The implication of this principle is that designers should be designing collaborative algorithms for distributed and decentralized systems. This is important because human-machine systems are inherently distributed and decentralized.

In order to achieve the second principle, we need to be able to describe joint activity. This requires a theoretical understanding of how work can be performed jointly, which is absent in existing approaches. Currently, engineers decide where to divide work into sub-tasks or actions. This can be aided by approaches like hierarchical task analysis (Annett, 2003), but few teams employ such approaches formally in practice. Even if task analysis approaches are used, they provide no logic for the division choices. Thus work decomposition is done with limited understanding of the consequences or impact of these choices, resulting in well-known issues like the substitution myth (Christoffersen & Woods, 2002). It is not just the automation, but the design of the automation that affects performance. Because of this, there is a need for understanding the work itself and the interdependence created by decomposition and distribution. Some examples of this can be found in research on group processes and productivity which provides a taxonomy of group tasks (Steiner, 1972). It includes knowing if a task is divisible or not, whether the task requires maximizing or optimizing, and whether the task is additive, disjunctive, or conjunctive just to name a few. Continuing with the building search example, moving debris is an additive task allowing for joint activity, while pushing an elevator button is disjunctive, with little value if done jointly. This leads us to our third principle:

Principle 3: Any work has inherent potential for jointness and limitations to jointness.

This principle is about uncovering and identifying the potential interdependencies in joint work. One particularly human capability is our amazing ability to team. People are thrown into teams without training and sometimes even without domain knowledge and yet are capable of at least rudimentary teaming. Our life experiences have enabled us to understand what aspects of work require synchronization, when acknowledgment is useful, what information is relevant to share, and how to request assistance. While people do make mistakes and some people are more naturally skilled than others, this generalizability across domains—regularly exhibited by people—provides the intuition that there are common patterns in joint work which support our third principle.

In order to model joint work, we must first address what counts as work. Traditional engineering and planning target physical actions that affect the world, but can ignore cognitive aspects such as sensing, perception, memory, reasoning, and understanding. This oversight is demonstrated by the development of an entirely separate field known as cognitive engineering (Norman, 1986) to address issues ignored by traditional engineering. While teamwork does involve coordinating physical actions, a significant role of teamwork is coordinating cognitive activities as well (Fong, 2001). Examples include monitoring the state of the world, drawing attention to significant events, assessing the progress of team members, and reasoning over current circumstances, all of which can be enhanced through effective teaming. Designs that do not include these aspects will not produce effective team players (Klein et al., 2004). Thus, our fourth principle:

Principle 4: Teaming occurs not just on the physical level, but also on the cognitive level.

As we get into more specific discussions on modeling and representation, we can draw upon good design principles that cross domain boundaries. Our fifth principle is just good design practice in general and will be applied in several ways:

Principle 5: Separate the "what" from the "how."

As an example, consider the goal of making sure a teammate does not fall into a hole while working. This could be accomplished by specifying a route for the teammate that avoids the hole, informing the teammate about the hole and letting them avoid it, or positioning oneself or an object by the hole to force the teammate to avoid both the obstacle and the hole. All of these achieve the same purpose (the what) but through different means (the how).

There are many factors that go into effective teaming. These include the work itself, the team members involved in the work, and the environmental factors in which the work is performed. The challenge is that all of these factors interact. In accordance with principle five, we propose our sixth principle:

Principle 6: Joint work can and should be modeled in an agent-agnostic manner.

By this we mean that the description of the work should not include considerations for specific team compositions or team member capabilities. Using the building search example, the work would involve moving around the building and identifying people. It would not include walking versus flying, or infrared versus visible light detection. This does not mean team composition will not be accounted for, just that the initial work description is not the appropriate place to do this. This is indeed a foreign concept in robotics, as most developers strive to customize their solutions to the specific targeted hardware, though recent efforts like ROS (robot operating system; Quigley et al., 2009) have shown the value of abstraction in the robotics domain. This concept is less foreign in domains such as cross-platform application development. In such domains, layers of abstraction are used to enable generalization. Similarly, our principle is trying to emphasize that you may not know the characteristics of the teammates ahead of time, so designing your joint work on top of a reasonable abstraction of the work itself will allow for broader applicability and reuse.

If work should be modeled as joint work, how is this different than regular work? We propose it is the inclusion of interdependence in the model that captures the potential jointness. Malone and Crowston define coordination as "managing dependencies between activities" (Malone & Crowston, 1994). Studying human teams and team effectiveness, researchers have identified team member interdependence as a critical feature defining the essence of a team (Salas, Rosen, Burke, & Goodwin, 2009). From human-machine research, Feltovich et al. propose interdependence is the essence of joint activity (Feltovich, Bradshaw, Clancey, & Johnson, 2007). Interdependence focuses on how the decomposed and potentially distributed work remains interdependent. This provides exactly what is needed to understand joint work, to understand the implications of different decomposition choices, and to specify requirements based on teaming (Johnson et al., 2014). Thus, our seventh principle:

Principle 7: Interdependence provides the basis for understanding potential jointness.

Some interdependencies are obvious. For example, resource constraints on a shared resource or sequencing constraints on dependent tasks. These are hard requirements that must be coordinated through teaming. As hard requirements, they are unavoidable and therefore obvious. However, a significant amount of normal human teaming involves opportunistic (i.e., soft) interdependence relationships. Soft interdependence does not stem from a hard constraint or a lack of capability. It arises from recognizing opportunities to be more effective, more efficient, and/or more robust by working jointly. Soft interdependence is less obvious because it is optional and opportunistic rather than strictly required. It includes a wide range of helpful things that a participant may do to enhance team performance. Examples include progress appraisals ("I'm running late"), warnings ("Watch your step"), helpful adjuncts ("Can I get the door for you?"), and observations about relevant unexpected events ("It has started to rain"). Many aspects of teamwork are best described as soft interdependencies. Our observations to date suggest that good teams can often be distinguished from great ones by how well they manage soft interdependencies. Thus, our eighth principle:

Principle 8: Teaming involves both required (hard) and opportunistic (soft) interdependencies.

So how does a designer identify interdependencies, particularly the less obvious soft interdependencies? Coactive design proposed three essential interdependence relations: observability, predictability and directability (Johnson et al., 2014). Observability means making pertinent aspects of one's status, as well as one's knowledge of the team, task, and environment observable to others. Predictability means one's actions should be predictable enough that others can reasonably rely on them when considering their own actions. Directability means one's ability to influence the behavior of others and complementarily be influenced by others. There are certainly additional types of interdependence relationships, such as explainability and trust, but we view these three as foundational to the others. This leads to principle nine:

Principle 9: Observability, predictability, and directability are compulsory interdependencies in teamwork.

One characteristic of teaming that provides compelling value is the ability for teams to be robust to individual failures. This does not happen by accident or without any effort. Teams monitor and assess the state of each other in order to achieve this advantage. Underlying this skill is the understanding that failure is always an option. Human failings and limitations are well known and include issues like experience, motivation, and attention. These issues can vary over time and across individuals. Human failings are often the motivation for more automation (Johnson & Vera, 2019). Yet automation has its own failings and limitations. Automation can have blind spots, it can be brittle, and it often lacks contextual awareness just to name a few limitations. Neither human failings nor automation failings are a problem as long as the team is attentive to the potential for failure. More directly, that the human-machine team is designed to address it. Thus, principle ten:

Principle 10: Failure is always an option and teaming should be designed to help the team be robust to failure of both people and machines.

This means that designers should not only be considering if performing some task is possible, but should be considering the risks and frailties of a task with respect to all team members. This provides insight into the importance and potential necessity of different teaming options in order to prevent any single team member from being a critical point of failure. Supporting this principle involves designing appropriate observability and predictability relationships to support monitoring, backup behaviors, and other teaming competencies.

Our last principle is another based on principle five—separating the what from the how. It addresses an aspect of teaming that is important, namely teaming strategy. Teaming strategy is the means by which a team chooses to exploit available interdependence options. In traditional systems that tackle multi-agent teaming, the teaming strategy is often tightly coupled to the overall system implementation, making it difficult if not impossible to change teaming strategies (e.g. (Dias, Zlot, Kalra, & Stentz, 2006)). It also confounds the scientific analysis of teamwork by not separating the teaming capability within the work from the teaming strategy employed by the team members. Learning to understand the teaming capability of the work

TABLE 9.1
Summary of Interdependence Design Principles

	Interdependence Design Principles
1	The value of teaming comes from exploiting it not avoiding it.
2	All work should be designed as joint activity (coactive), with independent work being the degenerate case.
3	Any work has inherent potential for jointness and limitations to jointness.
4	Teaming occurs not just on the physical level, but also on the cognitive level.
5	Separate the "what" from the "how."
6	Joint work can and should be modeled in an agent agnostic manner.
7	Interdependence provides the basis for understanding potential jointness.
8	Teaming involves both required (hard) and opportunistic (soft) interdependencies.
9	Observability, predictability, and directability are compulsory interdependencies in teamwork.
10	Failure is always an option and teaming should be designed to help the team be robust to failure of both people and machines.
11	The jointness of work and the teaming strategy can and should be considered separately, though they need to be designed to work together.

separate from the teaming strategy of the team members helps designers comprehend the influence of each on overall system performance. Thus, principle eleven is:

> *Principle 11: The jointness of work and the teaming strategy can and should be considered separately, though they need to be designed to work together.*

The purpose of these principles is to help reshape a designer's mindset in order to effectively employ the Interdependence Analysis tool. Table 9.1 summarizes the interdependence design principles.

INTERDEPENDENCE ANALYSIS TOOL

The purpose of interdependence analysis (IA) is understanding how people and automation can effectively team by identifying and providing insight into the potential interdependence relationships used to support one another throughout an activity. The Interdependence Analysis tool was developed to assist with IA. The tool can be used to analyze existing systems, but one of the tool's strengths is that it can also be used formatively to guide the initial design process. The need for formative tools is consistent with Kirlik et al., who emphasize "the importance of understanding why cognitive demands are present, prior to determining a strategy for aiding the operator in meeting these demands" (Kirlik, Miller, & Jagacinski, 1993, p. 950). Understanding and designing for interdependence can provide this type of guidance. IA provides insight into when the information requirements needed for specific interdependencies are adequately supported and when they are not. It can inform the designer of what is and is not needed, what is critical, and what is optional. Most importantly, it can indicate how changes in capabilities affect relationships.

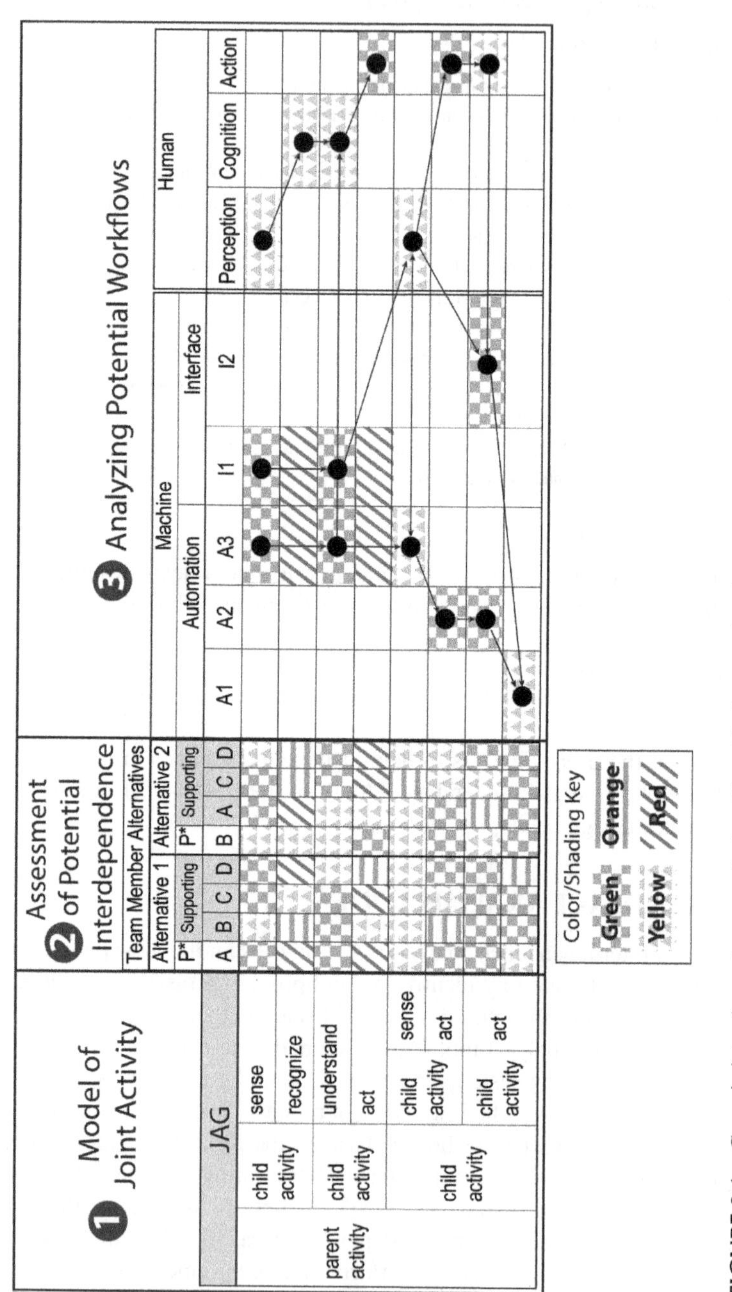

FIGURE 9.1 Generic interdependence analysis table with three main section labeled.

Note: Section 1 helps designers model the joint activity, section 2 helps them identify potential interdependencies in the activity, and section 3 helps analyze the potential workflows to better understand the flexibility and risk in the human-machine system.

As systems develop and improve, understanding the impact of how these changes impact human-machine teaming is critical to ensuring acceptance and utility of new technology.

The IA tool is in the form of a table, as shown in Figure 9.1, with three main sections: joint activity modeling, assessment of potential interdependence, and analysis of potential workflows.

Modeling of Joint Activity

The first section of the IA tool focuses on modeling the joint activity. In accordance with principle two, this modeling should model all work as joint activity. In order to accomplish this, it is important to understand what is unique and important about joint activity and how these aspects can be modeled.

What Is Joint Activity?

So, what does it mean to model some task or function as joint activity? Our view of joint activity comes from work on joint activity theory (Feltovich et al., 2007; Klein et al., 2004), a generalization of Herbert Clark's work in linguistics (Clark, 1996). Joint activity has important characteristics with respect to structure, process, and the potential for interaction (Bradshaw, Feltovich, & Johnson, 2011).

The first distinction is in the overall structure of joint activity. Joint activity is sets of nested actions. When viewed as a task, traditional engineering practice suggests making the given work an atomic isolated module that performs the specified function when called. The internals of the standalone module are typically hidden through encapsulation. While this is good programming practice for software development and integration, it turns out to be problematic for teaming. This is because tasks or functions are actually part of sets of nested actions, in other words an activity. The function may have several sets of nested actions within it, or it may itself be nested within a set of actions. Though team members may be working on functions that can be represented individually, it is important, from a teaming perspective, to understand the overall team activity context provided by the activity structure.

The second distinction is that joint activity is a process, one that extends in space and time. When viewed as a task or function, this is often overlooked, as if no time transpires and the world stands still. This has many implications from a teaming perspective. The first is that events of the past, current status within an activity and intentions for the future are all potentially relevant for effective teaming. Additionally, having a process implies that there is additional work necessary to compose the hierarchical structure. This additional work, often referred to as coordination, is also important for effective teaming.

The last distinction important to this discussion is that joint activity has the potential for interaction. The previous distinctions (considering work as sets of nested actions that are part of a process) shift thinking from individual tasks to activities. However, it is consideration of the potential for interaction that enables the activity to be considered joint. If there is not substantive interaction, then the work is parallel—not joint (Bradshaw et al., 2011). This need to support interaction drives what it means to model joint work. It means to decompose joint activity into a set of nested actions and to instrument those actions to properly support interactions needed for distributed team members to recompose the final solution. This is where designing for interdependence opposes traditional information-hiding practices. Functional encapsulation and opaque automation are often at odds with effective teaming. Understanding the nature of joint activity also plays a role in a more nuanced understanding of interaction. Since activity involves nested sets of activities, interaction can happen across a variety of levels of abstraction. Since it is a process, interaction can involve the past, the future, and ongoing progress during an activity. Defining a system as a human-machine team is defining it as joint activity, but defining it alone is not enough to make it a reality; the system must be designed and built.

How to Model Joint Activity—The Joint Activity Graph

While there are conceivably numerous ways to model joint activity, we propose one solution called the Joint Activity Graph (JAG). The purpose of the JAG is to capture the key elements of joint activity: structure, process and potential interactions.

The data structure underlying a JAG, is a simple graph. It is a tree of unit height—a single activity with zero or more children. More complex activity can be constructed by assembling multiple JAGs into a larger structure that is itself a JAG. Any high-level JAG can be understood by recursively expanding all the children to provide a tree-like view on the overall activity. This recursive data structure has intrinsic benefits for composition and reuse. The JAG provides a description of all the nested activities that are necessary to the proper execution of the activity it describes. As a hierarchical structure, it is similar to hierarchical task networks (HTNs) (Erol, Hendler, & Nau, 1994; Georgievski & Aiello, 2014). As with HTNs, JAGs can be directly executable or conceptual like a goal, sometimes referred to as primitive and non-primitive tasks respectively. However, JAGs do not require explicit declaration as such. Whereas planners are built on top of existing systems with known capabilities, human-machine teaming is constantly evolving and the potential variation in team composition means rigid assumptions on the type of actions that are executable by one agent will likely be invalid for some other agent. Thus, JAGs were designed to postpone the decision on what is directly executable allowing teams flexibility to resolve this at runtime.

If structure alone was all that was needed to understand teaming, then JAGs would be unnecessary as a hierarchical tree would suffice. However, the assembly of a JAG into a more complex activity also requires a process description. JAGs contain a process model whose purpose is modeling how to assemble its immediate descendants into the parent. As with structure, flexibility is an important aspect of the teaming process. The goal of modeling the process with a JAG is not to dictate the "right way" to complete an activity, but to model the permissible ways to do so. The JAG aims to depict the roadmap of options not just one viable road. In other words, it is not meant to define the one solution, but a solution space. The re-composition process can be individually defined for each unit JAG and is fully customizable. While these can be designed from scratch to fit specific needs, we natively support a set of common useful alternatives. These alternatives are combinations of two broad concepts: sequencing and logic. Sequencing provides for sequential or parallel execution. Logic provides the operators "And" and "Or" for specifying conjunctive and disjunctive tasks. Together these provide four useful combinations:

1. Sequential-And: All activities need to be completed successfully in sequence.
2. Sequential-Or: Activities are executed in sequence until one is completed successfully.
3. Parallel-And: All activities can be executed in parallel and need to be completed successfully.
4. Parallel-Or: All activities can be executed in parallel but are racing each other until one completes successfully.

These four processes provide sufficient mechanisms to model many different activities. However, because it is unlikely that we can cover all types of processes imaginable, we left the process definition open for future extensions.

Structure and process dictate how an activity can be decomposed and distributed and how it can be recomposed to achieve the goals of the activity, however neither defines the potential for interactions that determine the jointness. This is accomplished by identifying the interdependence relationships that underlie and define the needed interactions which in turn enable teamwork. Some interdependencies come from the structure. For example, if a JAG's children are distributed across team members, to know that the parent JAG is successful may require information sharing. Interdependence can also stem from the process. For example, if sequential activities are distributed across team members then they will need to coordinate their sequencing. However, many of the important teamwork dimensions such as adaptability, situation awareness, performance monitoring and feedback, and decision making (Driskell, Goodwin, Salas, & O'Shea, 2006) involve interdependencies beyond those associated with structure or process. The purpose of the IA tool is to help designers identify potential interdependencies. Specifically, observability, predictability and directability (OPD) requirements explain what kind of interactions are potentially valuable to achieve effective teaming (principle 9). More importantly they depict precisely how a given activity must be instrumented to expose and ingest appropriate information.

Assessing Potential Interdependence

After modeling the joint activity, the next step is to assess the potential interdependence. This involves enumerating the viable team role alternatives, assessing the capacity to perform the work, assessing the capacity to support another team member as they perform the work, identifying potential interdependencies, and then determining the requirements to support the interdependence relationships of interest.

Enumerating Viable Team Role Alternatives

While the joint activity is modeled in an agent-agnostic manner, the assessment of interdependence is in general not. This is because interdependence is about relationships. Therefore, any team alternative needs at a minimum two team members, though larger teams are permitted. Having specific individual team members is not required, though it does permit greater specificity. The team alternatives section of the tool captures the fact that team composition has an impact on teaming. It also permits analysis and comparison of how changes to team composition impact teaming.

To aid interdependence analysis, the IA tool makes a distinction between a performer and supporting team members. The performer is the individual primarily doing the given aspect of the work. The supporting team members are then viewed from the perspective of assisting the performer. It is the supporting team member columns that are key to identifying interdependencies. For a given alternative, only one entity is assigned as the performer, labeled P* in Figure 9.1. This is not to say others cannot do the work, but is simply a mechanism to aid the designer in considering a certain perspective.

In general, IA tables will have a minimum of two alternatives. The first with a given performer and supporter and the second with the roles reversed. This allows the designer to consider all of the joint work from both perspectives. This permutation is another key element to identifying all potential interdependencies, which in turn guides the design of effective teaming. In addition to considering any party performing any part of the work, this also forces consideration for any party assisting in any part of the work. If a team consisted of two identical team members, then having two team alternatives would be redundant since they would be identical. The main reason for a minimum of two alternatives is that people and machines are inherently different. This asymmetry needs to be understood and accounted for when designing human-machine teaming.

The columns in each alternative can represent specific individuals (existing or planned). If the team has more than two members then additional columns can be used, as represented by columns A, B, C, and D in Figure 9.1. For larger teams this can become unwieldy, but categories and roles can be used to keep it manageable. For example, consider a single operator managing four unmanned aerial vehicles (UAVs) and two unmanned ground vehicles (UGVs). Assuming the vehicles are of the same type then categories can be used. The team alternative would be three columns: one human operator, one for UAVs, and one for UGVs. Multiple people are also permitted and can be simplified with roles. Consider extending the previous example to be two such units being managed by a commander. These 15 entities can be captured by four columns: commander, operator, UAV, UGV. There is a limit to the feasibility of extending the table to large teams, but such teams are more like organizations than teams.

In general, a permutation of team members in which each is assigned the performer is how team alternatives are created. While permutations across a team seems a bit daunting and suggest visions of exponential explosion, in practice we have found that humans generally team in small numbers and larger teams use categories and roles to manage scale. Some may question whether such consideration for the team composition is necessary for effective design. Consider a simple example of providing a driver instructions on how to get to your house with some road construction blocking the obvious route. If the driver you are teaming with is someone familiar with your neighborhood versus someone unfamiliar, you probably need different coordination mechanisms. If the driver is a teenager learning to drive, a young adult, or a senior citizen you might also imagine different teaming mechanisms. There is no "one size fits all" when designing teaming, especially when considering team composition. This is not a limitation of our approach, but merely a reflection of reality. There is a tradeoff to be made with respect to how much detail to include, but having no representation of team composition is unlikely to be effective.

Assessing Capacity to Perform and Capacity to Support

After the team alternatives are determined, the next step is an assessment of each column in a team alternative to each row in the joint activity model. For formative tasks, the assessment is necessarily subjective, but when assessing existing systems, it is possible to use empirical data to inform the assessment (Johnson et al., 2017). The assessment process uses a color coding scheme, as shown in Figure 9.2. The color scheme is dependent on the type of column being assessed.

Under the "performer" columns, the colors are used to assess the individual's capacity to perform the activity specified by the row. It complies with principle 10—considering the potential for failure. The color green in the "performer" column indicates that the performer can do the task. For example, a robot may have the capacity to navigate around an office without any assistance. Yellow indicates less than perfect reliability. For example, a robot may not be able to reliably recognize a coffee mug all the time. Orange indicates some capacity, but not enough for the task. For example, a robot may have a 50-pound lifting capacity, but would need assistance lifting anything over 50 pounds. Another use of orange is to indicate hard-coded assumptions that limit the performance to very specific contexts. For example, a robot may be able to pick up a coffee mug from a table, but it may not be able to do so from the floor or cluttered cabinet. The color red indicates no capacity, for example, a robot may have no means to open a door.

Under the "supporting team member" columns, the colors are an assessment of that team member's potential to support the performer for the activity specified by the row. The color red indicates no potential for interdependence, thus independent operation is the only viable option for the task. Orange indicates a hard constraint, such as providing supplemental lifting capacity when objects are too heavy. Another example of orange is when a machine needs human authorization to perform the activity. Yellow is used to represent improvements to reliability. For example, a

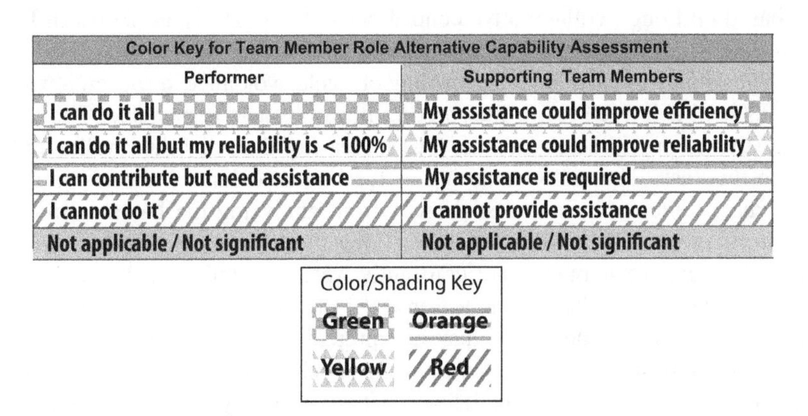

Color Key for Team Member Role Alternative Capability Assessment	
Performer	Supporting Team Members
I can do it all	My assistance could improve efficiency
I can do it all but my reliability is < 100%	My assistance could improve reliability
I can contribute but need assistance	My assistance is required
I cannot do it	I cannot provide assistance
Not applicable / Not significant	Not applicable / Not significant

FIGURE 9.2 Color key for team member role alternative capability assessment.

human could provide recognition assistance to a robot and increase the reliability in identifying coffee mugs. Green is used to indicate assistance that may improve efficiency. For example, a robot may be able to determine the shortest route much faster than a human or could assist in cleaning up a room.

One last note on color coding. Some relationships are not significant and so either the performer or the supporting team members' assessment may not be applicable. In these cases, use of gray is suggested to minimize the attention drawn by such cases (Beierl & Tschirley, 2017).

Identifying Potential Interdependence Issues

Once the assessment process is finished, the color pattern can be analyzed. The color pattern characterizes the nature of the interdependence within a team for the given joint activity. Colors other than green in the "performer" column indicate some limitation of the performer, such as potential brittleness due to reliability (yellow) or hard constraints due to lack of capacity (orange). The hard constraints in the performer column indicate a need to team to accomplish the work and are usually fairly obvious. The more interesting situation is the potential brittleness which is often less obvious. Teaming is not required in this circumstance, but doing so can make the team more resilient.

Colors other than red in the "supporting team member" columns indicate required (orange) or opportunistic (yellow and green) interdependence relationships between team members. Again, it is the opportunistic cases that tend to be the most interesting, less obvious, and contribute to resilience.

Determining System Requirements

With the assessment complete, the IA tool can now help designers extract clear-cut design requirements needed to enable and support specific interdependence relationships. For each relationship of interest, the design considers the compulsory interdependencies of observability, predictability, and directability (OPD) in accordance with principle 9. In other words, identifying who needs to observe what from whom, who needs to be able to predict what, and how members need to be able to direct each other for a given aspect of the work. As an example, we have created a small IA table based on Fong's collaborative control work (Fong, 2001), as shown in Figure 9.3. In Fong's example there was one teaming alternative: the human assisting the robot. The robot was capable of performing obstacle avoidance; however, it was less than 100% reliable (yellow) in interpreting if an obstacle is passable. The human was capable of providing assistance, thus increasing the reliability (yellow) of the robot in this task. The yellow coloring of the human column indicates soft interdependence. Requirements can be derived from analyzing the IA table in Figure 9.3. The robot must be predictable in notifying the human when unsure about an obstacle because the human is not constantly watching (red). The human's ability to interpret depends on being able to sense the obstacle, so there is an observability requirement. Once the human has interpreted if the obstacle is passable, this information must have a way to alter the robot's behavior, so there is a directability requirement. These requirements define what is needed by both the algorithm and the human interface to support this interaction. Note that the IA tool's purpose is to identify what the requirements are, not how to meet them (principle 5). That is an implementation choice. These particular OPD requirements are based on the desire

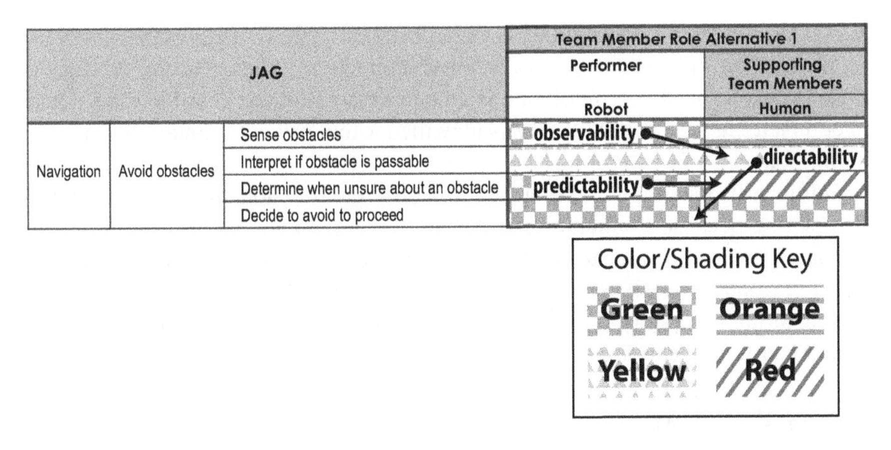

FIGURE 9.3 Interdependence analysis example from Fong's (2001) collaborative control work, showing observability, predictability, and directability requirements based on choosing to allow the human to provide interpretation assistance to the robot during navigation.

to support a particular interdependence relationship: the human assisting in interpretation of whether an obstacle is passable. This example demonstrates how OPD requirements derive from the role alternatives the designer chooses to support, their associated interdependence relationships, and the required capacities. This example does not include the reciprocal teaming alternative with the human assisted by the robot because this was outside the scope of the original work.

Analyzing Potential Workflows

The original IA tool was introduced as a way to help designers understand and design for the interdependence in human-machine teams (Johnson et al., 2014). The assessment of potential interdependence provided this, but demanded a fair bit of imagination to envision the implementation alternatives. As the tool was applied in the development process within a team of engineers, it became clear that it would be valuable to provide an improved visualization of the teaming alternatives to aid the team in developing a unified understanding (Johnson et al., 2017). Specifically, we needed a way to connect the theoretical understanding of interdependence to the physical instantiation of what had been developed thus far (i.e., the implementation) and what was planned to be developed.

Establishing Workflows

To achieve this, we associated leaf JAGs to particular team member's capability of supporting them. We also expanded the team members to allow more detailed mapping to specific algorithms, interface elements, or human abilities. Across the top of section 3 in Figure 9.1, there are column headings for each algorithm, interface element, or human capability used to accomplish the task. Below each heading is a black dot to indicate where in the activity that particular component has a role. We then connect the dots with arrows to indicate potential workflows to accomplish the goal. The resulting graph structure is a visual description of all existing and potential workflows. Therefore, it is a depiction of the flexibility in the system (Johnson et al.,

2015). The graph can be thought of as the adjustment options in adjustable autonomy or the initiative options in mixed-initiative interaction (Allen, Guinn, & Horvitz, 1999). However, what they really are is an enumeration of the possible ways a team can complete the joint activity. This is how the IA tool provides not a solution, but a solution space.

Additionally, the graph makes it clear that discrete function allocation is not what is happening because the human is informed by automation and display elements, and automation can be assisted by the human as indicated by the numerous horizontal and diagonal lines shown in section 3 in Figure 9.1. Most importantly, it shows where teaming is supported and where it is not. This allows grounded debate on the value of additional teaming support versus the cost of implementation.

Assessing Workflows

The color coding in the workflow section of the IA tool, section 3 of Figure 9.1, is indicative of risk for a given pathway. The color for a given cell is determined by the performer column of the assessment section. Moving vertically the risk is compounded across different functions. However, horizontal pathways indicate potential mitigations through teaming.

The potential workflows section of the IA tool is valuable for several reasons. First, it ensures the joint activity model grounds out in actionable terms and does not fall prey to wishful mnemonics (McDermott, 1976) or suitcase words (Brooks, 2017). Delineating the roadmap of possibilities also helps to guide a design team in their development. They can easily see what pathways are supported and which are not, but more importantly they can be guided to potentially opportunistic pathways they may not have found otherwise. The IA tool structure is also well suited to collection of empirical data to help inform design choices throughout the design iteration process (e.g. (Johnson et al., 2017)). Information about reliability, performance time, and frequency of use associated with specific pathways can help shape design decisions and provide credence to implementation directions.

INTERDEPENDENCE ANALYSIS EXAMPLE

As an example of how to use the IA tool we will use it to analyze manned aircraft collision avoidance. Figure 9.4 depicts an abbreviated IA of this challenge. The JAG, shown in the left side of Figure 9.4, depicts the agent agnostic joint activity (principles 3, 5, and 6). The work includes the cognitive aspects (principle 4) to sense and interpret traffic and decide when there is a conflict and what to do about it, as well as the physical task to take the necessary actions. We assume the work is jointly shared (principle 2) by some automation working together with a human pilot. Our hypothetical system includes automation, such as a Traffic Collision Avoidance System (TCAS) and a Detect and Avoid System (DAA), shown as headings in the workflow section of Figure 9.4. The figure depicts two TCAS options for discussion purposes. In teaming alternative one (TCAS-I), assessing the automation's ability to perform, we see that TCAS can reliably sense and interpret traffic, but only cooperative traffic. TCAS has no ability to detect non-cooperative traffic (i.e. non-transponder

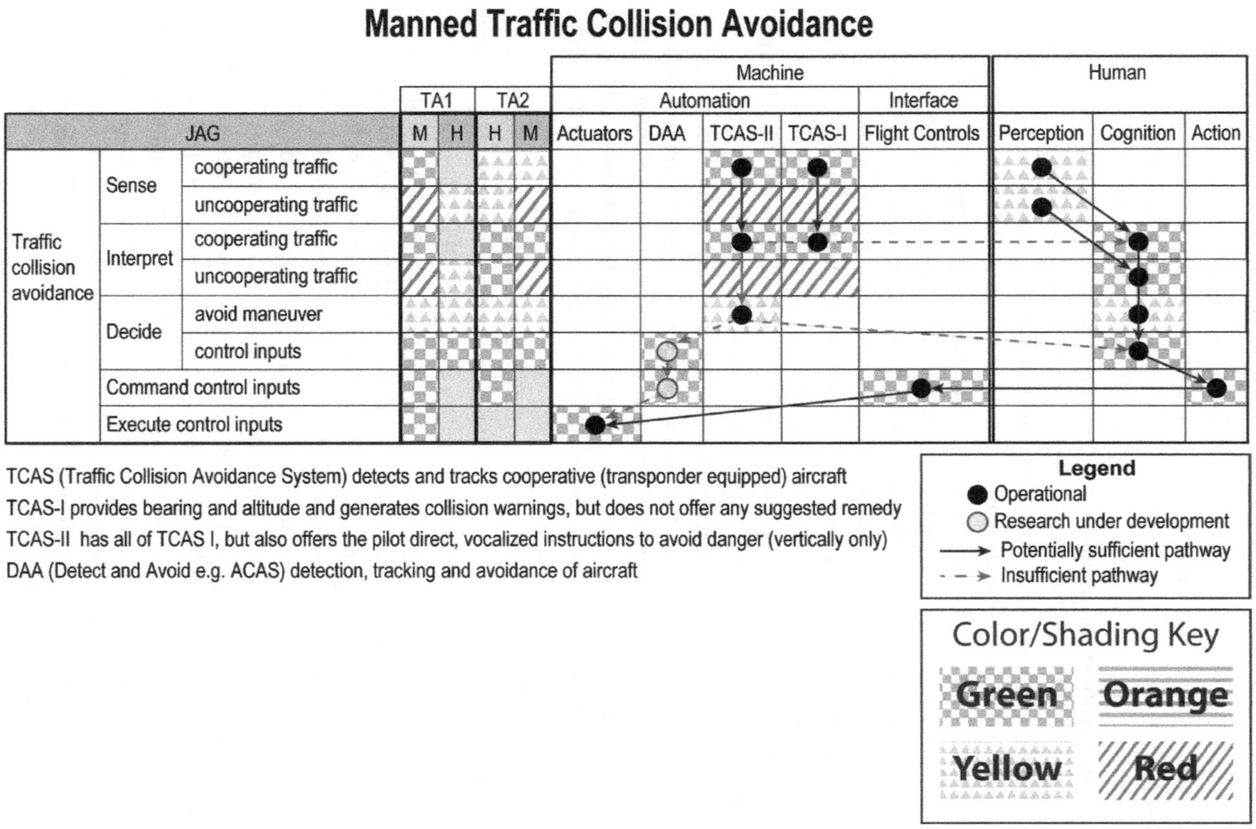

Manned Traffic Collision Avoidance

Machine | Human
TA1 | TA2 | Automation | Interface
JAG | M | H | H | M | Actuators | DAA | TCAS-II | TCAS-I | Flight Controls | Perception | Cognition | Action

Traffic collision avoidance
Sense: cooperating traffic; uncooperating traffic
Interpret: cooperating traffic; uncooperating traffic
Decide: avoid maneuver; control inputs
Command control inputs
Execute control inputs

TCAS (Traffic Collision Avoidance System) detects and tracks cooperative (transponder equipped) aircraft
TCAS-I provides bearing and altitude and generates collision warnings, but does not offer any suggested remedy
TCAS-II has all of TCAS I, but also offers the pilot direct, vocalized instructions to avoid danger (vertically only)
DAA (Detect and Avoid e.g. ACAS) detection, tracking and avoidance of aircraft

Legend
Operational
Research under development
Potentially sufficient pathway
Insufficient pathway

Color/Shading Key
Green
Orange
Yellow
Red

FIGURE 9.4 Interdependence analysis of manned traffic collision avoidance.

equipped). While TCAS-I only warns about traffic, TCAS-II can decide what to do and will provide vocalized instructions, but this is based on the limited sensing capability of TCAS-I. While DAA systems are not fully operationalized yet, here we assume we are designing a system to have such capability. Our envisioned DAA system leverages TCAS decisions to determine control inputs. We have assumed perfect execution (green) but left open the possibility that the automated system might misdiagnose on rare occasion (yellow) accounting for potential failure (principle 10). By considering that the human pilot has capabilities that could assist the automation (yellow) such as recognizing uncooperative traffic and making the avoidance decision we have identified potential soft interdependencies (principles 7 and 8). TCAS-II shows the human is capable of everything except directly executing the controls with errors most likely occurring in sensing due to attention or decision making (slow reaction time). Automation can assist the pilot in sensing cooperative traffic and making the avoidance decision (yellow) and generating control commands (green).

The potential workflows are depicted in the right half of Figure 9.4. One can see the potential automated solution with TCAS identifying traffic and DAA commanding avoidance maneuvers. Though the final execution of commands is highly reliable by today's automation (green), it is dependent on sensing that lacks some context (red) and decision making that may not be perfect (yellow). This provides designers with an overall risk assessment for the automated solution path. The manual solution is also possible with the pilot detecting traffic, deciding what to do, and taking action. There is risk in this solution, but it is different. Pilots have limited attention and may miss traffic and misjudge a decision. These are just two solutions, neither of which exploits teaming. By considering the supporting team member columns in each teaming alternative, designers are guided toward teamwork options throughout the activity.

Teaming alternative 2 (TA2) considers how the automation can help the pilot. Each option can be represented by horizontal lines that cross any boundary between teammates. Each relationship will have its own associated OPD requirements (principle 9). For example, TCAS-I can sense and interpret cooperating traffic for the pilot. This interdependence relationship is indicated by the upper horizontal red line connecting the TCAS-I interpretation to the human pilot cognition. To support this teaming pattern, the system must make potential traffic conflicts it detects observable to the pilot, in this case through a warning display. This activity can be done in parallel and thus both the pilot and the automation can contribute. The workflow includes both sense and interpret pathways because the automation is not replacing, but supplementing the pilot's ability, as most flight manuals will warn. It also means the risks of each individual pathway, missing context for automation and lack of attention for pilots, can potentially be mitigated by the other. Alternatively, TCAS-II can provide a suggested course of action. This interdependence relationship is indicated by the lower horizontal red line connecting the TCAS-II decision to the human pilot decision. It has different OPD requirements than the first. It requires a mechanism to direct pilot action, in this case voice instructions. Note that although the DAA has the potential to determine control inputs, it currently does not support providing such assistance directly to the pilot, only through the fully automated solution.

Teaming alternative 1 (TA1) considers how the human can help the automation. Although the assessment column for the human in TA1 indicates the potential to help recognize uncooperative traffic or decide on a course of action, no support is provided. There are no horizontal arrows from the human to the machine in the workflow, except the control input command. If the pilot performs these activities, they are done outside of any teaming and are not sharable with the automation in the current design.

Figure 9.4 represents the existing or proposed interaction possibilities. By describing the interdependence landscape in such a manner, it is easy for the designer to see where interdependencies exist and where they do not. This can guide designers toward new interaction concepts. For example, is it possible for the human pilot to supplement the TCAS automation of sensing and interpretation and if so, what kind of OPD requirements would need to be supported? How can the automation and pilot jointly engage in decision making, instead of having the either-or situation as in Figure 9.4? The IA tool is useful in both understanding one's design and helping designers identify novel design alternatives.

This example demonstrates the process of interdependence analysis and how it helps designers understand human-machine teaming. We connected the design principles to the analysis story to show how they relate to teaming and how the IA table guides the designer through these considerations. It should be clear that the potential for teaming and the requirements for supporting it will change as design changes are considered, such as different types of automation (e.g. TCAS-I vs. TCAS-II) or having the pilot be remote. This is how the IA tool aids designers in diagnosing the teaming implications of design choices.

Summary of IA Tool Capabilities

The IA tool is unique in its capability to aid designers in assessing interdependence in human-machine teams. Currently, it is the only design tool that specifically shows interdependence. It is also novel in that it does not try to determine which team member, the human or the machine, should be allocated to what aspects of work. Instead its focus is capitalizing on the opportunities for synergy in the team. It enables designers to more effectively find teaming opportunities and helps them understand the requirements needed to exploit those opportunities (principle 1).

The IA tool is also novel in how it includes both people and machines in the design of a system. Many approaches do not include people as part of the system, while others assess human performance without the context of the automation and interfaces. The novel approach to modeling all work as joint work in an agent agnostic way allows for complete flexibility of teaming. The support for consideration of teaming alternatives helps designers to see the problem from multiple perspectives. Providing workflow visualization that includes people, algorithms, and interfaces grounds the theory into actionable practice.

The IA tool is also unique in how it captures soft interdependencies. These types of capabilities are often overlooked, but their importance to teaming should not be. Support for such opportunistic teaming options can lead to improved flexibility and better overall system resilience (Johnson et al., 2017). All too often the purpose of human-machine teaming is viewed as a constraint due to automation not being able to do something. IA is about capitalizing on opportunities for synergy, not just filling required slots. It is about mutual enhancement, not replacement or substitution.

The IA tool is also unique in providing not just a single solution, but a roadmap of design alternatives. This type of view is critical for development of advanced technologies which involve highly iterative development. Supporting multiple pathways on the roadmap provides flexibility and contributes to system resilience.

It is worth noting that teaming strategy is not addressed by the IA tool, but it is included in the discussion because strategy is an important part of teaming. The IA tool provides insight into the tools and options available to teaming strategies (e.g. state, structure, and skills; Johnson & Vera, 2019), but not the strategy itself.

A critique of the IA process is that it is potentially time consuming to develop joint activity models and consider the different teaming alternatives. This is a valid concern, but building systems that fail or struggle to meet people's needs is also costly. While interdependence analysis will definitely take more time than simple task decomposition, it is unlikely to take more time than the first iterative development of any complex system. Our experience is that its value throughout the life of the project greatly exceeds the initial time investment.

Another concern is that IA requires a high level of expertise. In some sense this is true. The main issue is shifting away from the traditional paradigm. Shifting from a function allocation mindset is not easy, just as shifting to object-oriented programming or functional programming is not easy for someone trained in procedural programming. It also requires two skill sets not often found in combination: technology expertise and human expertise. However, ignoring either technology or the human because dealing with both is difficult is not a viable answer. Future educational tracks that emphasize both could alleviate this problem. We have proposed several guiding principles to help designers reorient their perspective, but as in any field, there is no substitute for practice and experience.

CONCLUSION

The purpose of interdependence analysis (IA) is understanding how people and automation can effectively team by identifying and providing insight into the potential interdependence relationships used to support one another throughout an activity. These interdependence relationships enumerate the potential human-machine interactions that comprise teaming. In this way, IA provides a roadmap of the opportunities afforded by a given system design. Interdependence relationships help to understand both the human factors and the technological factors that enhance or inhibit effective teaming. IA can be used to derive specific design requirements, both algorithmic and interface, that determine human-machine interactions for each alternative. The analysis also provides a risk assessment associated with each teaming option. Having alternatives provides flexibility, while understanding the risk associated with each option assists operational decision making.

As we build more intelligent machines to tackle more sophisticated challenges, designers need formative tools for creating effective human-machine teaming. IA meets this need by explicitly modeling the machine, the human, the work, and the interplay of all three. This enables developers to architect effective teaming by designing support for interdependence. IA should be viewed as an approach to determining how and what automated capabilities are built such that intelligent systems are imbued with teaming competence.

REFERENCES

Allen, J. E., Guinn, C. I., & Horvitz, E. (1999). Mixed-initiative interaction. *IEEE Intelligent Systems*, *14*(5), 14–23. https://doi.org/http://dx.doi.org/10.1109/5254.796083

Annett, J. (2003). Hierarchical task analysis. In *Handbook of cognitive task design* (pp. 17–35). London: Lawrence Erlbaum Associates.

Beierl, C., & Tschirley, D. (2017). *Unmanned tactical autonomous control and collaboration situation awareness*. Retrieved from www.dtic.mil/dtic/tr/fulltext/u2/1046299.pdf

Bradshaw, J. M., Feltovich, P. J., & Johnson, M. (2011). Human-agent interaction. In *The handbook of human-machine interaction: A human-centered design approach*. Burlington, VT: CRC Press, Ashgate Publishing Company.

Brooks, R. (2017). *The seven deadly sins of AI predictions*. Retrieved November 17, 2017, from MIT Technology Review website: www.technologyreview.com/s/609048/the-seven-deadly-sins-of-ai-predictions/?utm_campaign=add_this&utm_source=email&utm_medium=post

Christoffersen, K., & Woods, D. D. (2002). How to make automated systems team players. *Advances in Human Performance and Cognitive Engineering Research*, *2*, 1–12.

Clark, H. H. (1996). *Using language*. Retrieved from www.loc.gov/catdir/toc/cam023/95038401.html

Crandall, B., & Klein, G. (2006). *Working minds: A practitioner's guide to cognitive task analysis*. Cambridge, MA: The MIT Press.

desJardins, M., & Wolverton, M. (1999). Coordinating a distributed planning system. *AI Magazine*, *20*(4), 45. https://doi.org/10.1609/aimag.v20i4.1478

Dias, M., Zlot, R., Kalra, N., & Stentz, A. (2006). Market-based multirobot coordination: A survey and analysis. In *Proceedings of the IEEE* (pp. 1257–1270). Retrieved from http://robotics.cse.tamu.edu/dshell/cs689/papers/dias06market.pdf

Driskell, J. E., Goodwin, G. F., Salas, E., & O'Shea, P. G. (2006). What makes a good team player? Personality and team effectiveness. *Group Dynamics: Theory, Research, and Practice*, *10*(4), 249–271. https://doi.org/10.1037/1089-2699.10.4.249

Drucker, P. F. (1954). *The practice of management*. Retrieved from www.amazon.com/The-Practice-Management-Peter-Drucker/dp/0060878975

Erol, K., Hendler, J., & Nau, D. S. (1994). HTN planning: Complexity and expresivity. In *AAAI* (pp. 1123–1128). Retrieved from www.aaai.org/Papers/AAAI/1994/AAAI94-173.pdf

Feltovich, P. J., Bradshaw, J. M., Clancey, W. J., & Johnson, M. (2007). Toward an ontology of regulation: Socially-based support for coordination in human and machine joint activity. In G. O'Hare, M. O'Grady, A. Ricci, & O. Dikenelli (Eds.), *Engineering societies in the agents world VII: Vol. lecture no* (pp. 175–192). Heidelberg, Germany: Springer.

Fong, T. W. (2001). *Collaborative control: A robot-centric model for vehicle teleoperation*. Pittsburgh, PA: Robotics Institute, Carnegie Mellon University.

Georgievski, I., & Aiello, M. (2014). *An overview of hierarchical task network planning*. Retrieved from http://arxiv.org/abs/1403.7426

Hoffman, R. R., & Deal, S. V. (2008). Influencing versus informing design, part 1: A gap analysis. *IEEE Intelligent Systems*, *23*(5), 78–81.

Johnson, M., Bradshaw, J. M., Feltovich, P. J., Hoffman, R. R., Jonker, C., Riemsdijk, B. V., & Sierhuis, M. (2011). Beyond cooperative robotics: The central role of interdependence in coactive design. *IEEE Intelligent Systems, 26*(3). https://doi.org/10.1109/MIS.2011.47

Johnson, M., Bradshaw, J. M., Feltovich, P. J., Jonker, C., van Riemsdijk, B., & Sierhuis, M. (2011). The fundamental principle of coactive design: Interdependence must shape autonomy. In M. De Vos, N. Fornara, J. Pitt, & G. Vouros (Eds.), *Coordination, organizations, institutions, and norms in agent systems VI* (Vol. 6541, pp. 172–191). Berlin: Springer-Verlag. https://doi.org/10.1007/978-3-642-21268-0_10

Johnson, M., Bradshaw, J. M., Feltovich, P. J., Jonker, C. M., van Riemsdijk, B. M., & Sierhuis, M. (2014). Coactive design: Designing support for interdependence in joint activity. *Journal of Human-Robot Interaction, 3*(1), 43–69.

Johnson, M., Shrewsbury, B., Bertrand, S., Calvert, D., Wu, T., Duran, D., . . . Pratt, J. (2017). Team IHMC's lessons learned from the DARPA robotics challenge: Finding data in the rubble. *Journal of Field Robotics, 34*(2), 241–261. https://doi.org/10.1002/rob.21674

Johnson, M., Shrewsbury, B., Bertrand, S., Wu, T., Duran, D., Floyd, M., . . . Pratt, J. (2015). Team IHMC's lessons learned from the DARPA robotics challenge trials. *Journal of Field Robotics, 32*(2), 192–208. https://doi.org/10.1002/rob.21571

Johnson, M., & Vera, A. H. (2019, Spring). No AI is an Island: The case for teaming intelligence. *AI Magazine*, pp. 16–28.

Kirlik, A., Miller, R. A., & Jagacinski, R. J. (1993). Supervisory control in a dynamic and uncertain environment II: A process model of skilled human environment interaction. *IEEE Transactions on Systems, Man, and Cybernetics, 23*, 929–952. https://doi.org/10.1109/21.247880

Klein, G., Woods, D. D., Bradshaw, J. M., Hoffman, R. R., & Feltovich, P. J. (2004). Ten challenges for making automation a "team player" in joint human-agent activity. *IEEE Intelligent Systems, 19*(6), 91–95. https://doi.org/http://dx.doi.org/10.1109/MIS.2004.74

Malone, T. W., & Crowston, K. (1994). The interdisciplinary study of coordination. *ACM Computing Surveys, 26*(1), 87–119. https://doi.org/http://doi.acm.org/10.1145/174666.174668

McDermott, D. (1976). Artificial intelligence meets natural stupidity. *ACM SIGART Bulletin*. Retrieved from http://dl.acm.org/citation.cfm?id=1045340

Norman, D. A. (1986). Cognitive engineering. *User-centered system design* (pp. 31–61). Retrieved from https://pdfs.semanticscholar.org/57f1/76992f92ae559d9c110211d7f04c5143cb44.pdf

Parasuraman, R., Sheridan, T., & Wickens, C. (2000). A model for types and levels of human interaction with automation. *Systems, Man and Cybernetics, Part A, IEEE Transactions On, 30*(3), 286–297. http://doi.org/10.1109/3468.844354

Quigley, M., Conley, K., Gerkey, B., Faust, J., Foote, T., Leibs, J., . . . Mg, A. (2009). ROS: An open-source robot operating system. *Icra, 3*(Figure 1), 5. Retrieved from www.willowgarage.com/papers/ros-open-source-robot-operating-system

Sacerdoti, E. D. (1975). The nonlinear nature of plans. *Proceedings of the 4th International Joint Conference on Artificial Intelligence, 1*, 206–214. https://doi.org/10.1017/CBO9781107415324.004

Salas, E., Rosen, M. A., Burke, C. S., & Goodwin, G. F. (2009). The wisdom of collectives in organizations: An update of the teamwork competencies. In *Team effectiveness in complex organizations: Cross-disciplinary perspectives and approaches* (pp. 39–79). New York: Routledge/Taylor & Francis Group.

Sheridan, T. B., & Verplank, W. (1978). *Human and computer control of undersea teleoperators*. Cambridge, MA: Man-Machine Systems Laboratory, Department of Mechanical Engineering, MIT.

Stefik, M. (1981). Planning with constraints (MOLGEN: Part 1). *Artificial Intelligence*, *16*(2), 111–139. https://doi.org/10.1016/0004-3702(81)90007-2

Steiner, I. D. (1972). Group process and productivity. *New York*. Retrieved from http://books.google.com/books?id=20S3AAAAIAAJ&pgis=1

Wickens, C. D., Li, H., Santamaria, A., Sebok, A., & Sarter, N. B. (2010). Stages and levels of automation: An integrated meta-analysis. *Proceedings of the Human Factors and Ergonomics Society Annual Meeting*, *54*(4), 389–393. https://doi.org/10.1177/154193121005400425

10 Using Conceptual Recurrence Analysis to Decompose Team Conversations

Michael T. Tolston, Gregory J. Funke, Michael A. Riley, Vincent Mancuso, and Victor Finomore

CONTENTS

Teams form the essential substrate of most modern organizations, from small businesses to large government agencies. In order for teams to perform effectively, team members have to coordinate behaviors and tasks and interact in prosocial ways. This requires team members to possess a common understanding of the team's resources, long-term goals, immediate objectives, and the constraints under which the team works (Salas, Sims, & Burke, 2005). In other words, effective teamwork requires the establishment and alignment of shared conceptual understanding between teammates. Examples include team member knowledge, beliefs, and attitudes regarding entities, processes, and strategies relevant to setting and achieving team objectives and goals (DeChurch & Mesmer-Magnus, 2010). The overlap of these cognitive structures across teammates has broadly been referred to as "shared mental models" (Cannon-Bowers, Salas, & Converse, 1993; Cannon-Bowers & Salas, 2001; Mohammed, Klimoski, & Rentsch, 2000), and alignment of these models is thought to be an essential coordinative mechanism that underwrites successful team outcomes (Salas et al., 2005). However, some authors argue that conventional operationalizations of this mechanism only partly explain how teams successfully interact (Cooke, Gorman, Myers, & Duran, 2013; Gorman, Dunbar, Grimm, & Gipson, 2017).

While the concept of shared mental models was initially formed to explain why teams can efficiently coordinate behaviors implicitly by relying on shared understanding (Cannon-Bowers et al., 1993), the term has taken on a narrower meaning largely synonymous with similarity in long-term memory structures (Mohammed, Hamilton, Sanchez-Manzanares, & Rico, 2017). As such, the "mental models" aspect of shared mental models has been conceptualized as having existence prior to and independent of the task or process the models would serve. In this framework, measuring shared mental models equates to assessing pre-existing cognitive structure individually for each team member and aggregating or comparing them within teams to understand the degree to which conceptual knowledge structures overlap (DeChurch & Mesmer-Magnus, 2010). However, some have argued that the static long-term knowledge structures participants have in place prior to engaging a task, while important, are insufficient predictors of performance outcomes (Cooke et al., 2013; Gorman, Dunbar, Grimm, & Gipson, 2017). Accordingly, the dynamic nature of teams and team tasks may benefit from alternative approaches to understanding how teams solve problems.

Teams can be characterized as dynamic systems with emergent properties (Kozlowski & Ilgen, 2006). Importantly, both teams and circumstances frequently change in many ways; teams expand and dissolve, task objectives are met and replaced with new ones, team goals are revised, and task constraints shift, intensify, or disappear. As such, many important processes and properties of teams exist in the high-order dynamic relationships team members establish with each other and with their environment (Cooke, Gorman, & Rowe, 2009). From this perspective, interactive and dynamic alignments at the conceptual, perceptual, and behavioral levels underlie cognitive similarities in teammates. This process of dynamic alignment has been called team cognition (Cooke et al., 2013).

In order to assess higher-order relational properties inherent in team cognition, higher-order observables are needed; measuring team cognition requires measuring the team interacting in a context specific to the goals and objectives relevant to the motivating research question. Cooke et al. (2013) argued that team effectiveness and subsequent successes are dependent upon the ways information and knowledge are employed by the team, i.e., through the emergent information processing that implicitly relies on knowledge structures and which is embodied in the interactions between teammates as they uncover, share, and negotiate the context-specific meaning of goal-relevant information. As such, many efforts to understand and enhance team performance have been directed toward analysis of team communication patterns (Gorman, Cooke, Amazeen, & Fouse, 2012; Gorman et al., 2019; Russell, Funke, Knott, & Strang, 2012; Wiltshire, Butner, & Fiore, 2018).

Although team communications provide a potential window to the underlying cognitive processes that enable team performance, there are many challenges associated with operationalizing and measuring cognition (Mohammed, Ferzandi, & Hamilton, 2010). Any method aimed at doing so must address three fundamental aspects of cognitive measurement (DeChurch & Mesmer-Magnus, 2010): elicitation—how cognitive content is observed; structural aspects—how the elements of cognitive content are related to each other; and emergence—how team-level understanding relates to individual understanding. Importantly, the inferred content and relational structure of cognition is sensitive to the techniques employed in its measurement (DeChurch & Mesmer-Magnus, 2010).

In regard to these three challenges, it is important to remark that team settings allow cognitive behaviors that are normally implicit or hidden in observations of individual cognitive performance to be readily observed as teammates interact and communicate. Team members often interact through naturalistic communication to establish and operate on common ground understanding—shared mental models—that supports descriptions, explanations, and predictions of team tasks and efforts (Mohammed et al., 2010). In other words, teams verbally share and operate on knowledge during their interactions, rendering the act of information processing into an observable operation. Thus, a natural language solution to team-level cognitive measurement that takes advantage of team interactions to provide answers to all three aspects outlined by DeChurch and Mesmer-Magnus (2010) is to present a collaborative problem-solving task to a group and record their communicative interactions to determine the content and structure of their cognitive processing (cf., Ericsson & Simon, 1998). As outlined in this chapter, this approach, combined with natural language processing (NLP) of communication analysis, allows the observation of concepts defining a problem space and important relationships between them, and also provides a way to directly capture emergent team cognition.

Despite its potential for revealing insights into team cognitive processes, team communication analysis presents a number of special challenges to researchers. Largely, these challenges arise because semantic information is often ambiguous and difficult to quantify. Simple approaches such as frequency counts from hand-coded communication data can provide meaningful descriptions of communicative exchanges (e.g., Mancuso, Finomore, Rahill, Blair, & Funke, 2014), but these methods are time consuming and limited to pre-defined dictionaries that may not make use of idiosyncratic information available in a particular communication dataset. Other approaches, like latent-semantic analysis (Foltz, 1996), and, more recently, word2vec (Mikolov, Chen, Corrado, & Dean, 2013), make use of powerful mathematical approaches that can uncover statistical regularities in text data. Though such techniques have been used to meaningfully quantify team communications (e.g., Gorman et al., 2016; Gorman, Foltz, Kiekel, Martin, & Cooke, 2003), these approaches can produce abstract mathematical spaces whose dimensions can be difficult to understand (Iliev, Dehghani, & Sagi, 2014). In this chapter, we discuss the utility of conceptual recurrence analysis (CRA; Angus, 2019; Angus, Smith, & Wiles, 2012a) and show how CRA conducted in the Discursis software package (Angus et al., 2012a; Angus, Smith, & Wiles, 2012b) can be extended using network analysis to provide insights into the structure of team communications.

Specifically, in this chapter we show: (1) how CRA can be used to identify key concepts that form the basis of meaningful task-specific discourse; (2) how CRA can then be used to evaluate the similarity of verbal exchanges in team process; (3) how dimensionality reduction of the matrix encoding conceptual similarity between utterances can be combined with network analysis to facilitate clustering of utterances and coding according to semantic themes; (4) how this classification then forms the basis of generating a categorical time series from team utterances; (5) how the categorical time series can be used to assess transitions between conceptual themes; and (6) how network approaches can again be used to construct a graph from the transition matrix to show the relationships between themes.

We view the final product as a representation of an aggregated team mental model of task-relevant concepts and relationships between them obtained by measuring team cognition. In other words, by observing temporal relations between concept transitions, we can observe how teams are processing conceptual information. We present the following methods as an introduction that we believe will be useful to team researchers.

CONCEPTUAL RECURRENCE ANALYSIS

CRA is an extension of recurrence quantification analysis (RQA; Webber & Zbilut, 1994), a nonlinear technique used to quantify structure in complex time series data. RQA quantifies structure in terms of the recurrence (i.e., repetition) of states of the system (i.e., of values in the time series). CRA quantifies communication data, such as a conversation, by measuring the extent to which individuals' utterances—complete conversational turns—are semantically similar. Importantly, in addition to providing a measure of semantic similarity between utterances, CRA also provides a set of concepts over which similarity is measured. This approach, as implemented in the program Discursis (Angus et al., 2012a, 2012b), has been used to quantify interpersonal communication data, including those observed in doctor-patient interactions (Angus, Watson, Smith, Gallois, & Wiles, 2012; Atay et al., 2015; Watson, Angus, Gore, & Farmer, 2015), talk show interviews (Angus et al., 2012a; Angus & Wiles, 2018), and team communications (Tolston, Riley, Mancuso, Finomore, & Funke, 2019).

Details underlying CRA have been outlined in previous work (Angus et al., 2012a, 2012b), but here we will broadly discuss essential aspects of the technique. CRA as implemented in Discursis first conducts pre-processing of text data by removing common words and optionally stemming words to remove prefixes and suffixes. The resultant words are referred to as concepts. Important words—referred to as key concepts—are automatically chosen from this list, partly based on their prevalence in the corpus. This has been conceptualized as a "bottom up" approach to revealing conceptual content, in contrast with a "top down" approach in which important concepts are identified a priori (Tolston et al., 2019). Discursis then constructs a semantic vector space from these key concepts by assessing concepts for similarity to key concepts using statistical techniques that depend in part on word co-occurrence. This forms the mathematical structure by which each utterance can be represented in a vector space in terms how prevalent each key concept is in that utterance. Specifically, the conceptual content of each utterance is computed as the sum total of the projections of all individual concepts in that utterance onto the key concept basis. In other words, the semantic content of an utterance is computed by the vector addition of the vectorized representation of the concepts in the utterance. This resultant vector codes the extent to which each key concept is invoked in each utterance. A similarity value is then computed for each pair of utterances, yielding a number that captures the degree to which key concepts occur in similar proportions between each utterance. This value is similar to a pointwise correlation; for each pair of utterances, the dimension-wise (i.e., key-concept) projections of the two utterances in the semantic space are multiplied and summed (akin to a covariance) and then normalized by the products of the lengths of the two projections (akin to the product of the variances). The result is a matrix that captures the degree of semantic similarity between all utterances. Readers interested in a more comprehensive discussion of CRA and Discursis are referred to Angus, Watson et al. (2012) and to Tolston et al. (2019).

RECURRENCE QUANTIFICATION ANALYSIS AND NETWORKS

As we will demonstrate in this chapter, the utility of CRA can be expanded using network-based analyses. This expansion is derived from techniques developed to extend the parent analysis (RQA) to generate recurrence networks—complex networks generated from an adjacency matrix obtained from RQA of time-series data (Donner et al., 2011; Donner, Donges, Zou, & Feldhoff, 2015; Donner, Zou, Donges, Marwan, & Kurths, 2010). Such an approach requires a similarity threshold that discretizes the distance matrix into a recurrence matrix that contains values of 0 (for non-recurrent states) and 1 (for recurrent states), which can then be visualized in a recurrence plot with "on" pixels indicating recurrent states and "off" pixels indicating non-recurrent states. In other words, the threshold sets the minimum degree of computed conceptual similarity that utterances must have to be defined as recurrent, i.e., as having the same conceptual content. The resultant recurrence matrix can be used as an adjacency matrix to create a complex network, with nodes in the network corresponding to utterances, and edges linking utterances that have sufficiently similar content. Donner et al. (2010) described how this network-based approach can reveal geometric properties of time series in terms of linking together similar system states. Importantly, such network constructions provide a set of powerful metrics that can quantify geometric and topological aspects of the system under study (e.g., Zhang & Small, 2006).

Networks are often characterized in terms of both local and global characteristics (e.g., Donner et al., 2015), where local characteristics capture information relating to each node or edge and global characteristics describe the network as a whole. However, there are also intermediate, or meso-scale, structures that may be informative for quantifying complex networks (Saggar et al., 2018). For example, *community structure* captures how well a network can be partitioned into meaningful subgraphs, or communities, with a quality metric indicating the proportion of edges that join nodes within a community to edges that join nodes outside of the community. This metric can capture intermediate level structures that exist when highly similar observations are clustered together within the larger network. This community detection can be seen as a form of clustering, in which nodes in a network are classified according to which community they belong.

With respect to communication analysis of team exchanges, community structure can capture the degree to which teams are producing heterogenous utterances. In other words, the degree of community structure in a semantic network indicates the degree to which utterances are focused on particular combinations of key terms, where utterances belonging to communities dominated by particular combinations of key terms have less similarity to other communities than within the community. Traditionally, the analysis of similarities between concepts to create themes or kernels—clusters of related concepts—has been a central endeavor in the concept mapping approach to studying cognition (McNeese & Reddy, 2015; Zaff, McNeese, & Snyder, 1993). In the present work, we show how clusters of highly related concepts can be obtained automatically by combining CRA with network analyses.

In previous research, CRA has largely been conducted utilizing similarity matrices without a threshold applied (e.g., Angus & Wiles, 2018; Tolston et al., 2019). In this chapter, we evaluated the effect that thresholding has on the structure of semantic networks derived from CRA applied to communication data collected from teams performing a collaborative consensus building task. Specifically, we combined thresholding with a dimensionality reduction technique to determine their effects on a resulting network's capacity to capture the semantic relationships between utterances. Our main research question was whether there was specific structure in the utterances created by teams in a collaborative consensus-building task that could lend insight into the ways that teams were processing semantic information. We used a task with known expert logic linking items being reasoned about to external concepts. We asked whether our decomposition using CRA and network analyses could be used to create a categorical time series to show how conceptual clusters may be associated during team conversations about a consensus-building task in ways that reveal how teams reason about the task.

APPROACH

Lost in the Desert Task

"Lost in the Desert" is a problem-solving task in which groups or teams of individuals are presented with a desert survival scenario (Lafferty, Eady, & Elmers, 1974). Teams are told to imagine that they are the sole survivors of a passenger plane crash-landing in a desert. They are told that they are all uninjured, that they should stay where they are until help arrives, and that there are 15 items that can be salvaged from the plane wreckage. Team members are then asked to rank those items according to their importance for survival, first individually and then as a team via consensus building discussion (see Table 10.1 for a list of the items, ranked by subject matter experts in order of most to least important for survival).

This task was chosen as an ice breaker for a subsequent distributed team decision-making task (results for the subsequent task are reported in Tolston et al., 2017). This and similar tasks have been shown to generate rich conversations and are often used

TABLE 10.1
Items Presented to Teams in the Lost in the Desert Scenario

Item	Rank	Reason
A cosmetic mirror	1	Visual signaling
Overcoat (for everyone)	2	Helps ration sweat by slowing evaporation
Two liters of water per person	3	Drinking water (a person requires a gallon of water a day in the desert)
Flashlight with four battery cells	4	Nighttime signaling
Parachute (red and white)	5	Shelter
Folding knife	6	For cutting rope, preparing food, etc.
Plastic raincoat (large size)	7	To collect condensation
45-caliber pistol (loaded)	8	Defense and signaling
Sunglasses (for everyone)	9	Protection against sun glare
First-aid kit	10	Emergency use and ad hoc rope (nobody on the team is injured in the crash)
Magnetic compass	11	Reflective signaling device
Air map of the area	12	Paper for fire and environmental protection
A book entitled "Desert Animals That Can Be Eaten"	13	Food is less important than water in the desert; digestion consumes water
Two liters of 80-proof vodka	14	Useful as an antiseptic, firestarter, etc.; will cause dehydration if consumed
Bottle of 1,000 salt tablets	15	Of no use in desert

to assess factors that affect team behaviors and decision making (e.g., Citera, 1998; Ferrin & Dirks, 2003; Liu & Li, 2017).

METHOD

PARTICIPANTS

Data were collected from 64 participants (29 men, 35 women) recruited from the Dayton, Ohio area. The range of participants' ages was 18 to 33 years ($M = 23.05$, $SD = 3.76$). Participants completed the experiment as members of four-person teams, resulting in 16 experimental teams in our sample. Participants were compensated $15/hour for their time. All participants had normal or corrected to normal vision.

PROCEDURE

Upon arrival, participants completed an informed consent document and were provided instructions regarding the task. During the task, participants sat at individual computer workstations that were visually isolated from each other. Team members received instructions about the task and were asked to individually rank the 15 items. They were then given an unlimited amount of time to discuss their ranking preferences with their teammates. The goal of this phase was for the team to reach a consensus regarding the rank for each item. During the experiment, participants were asked to communicate exclusively by typing in a chat interface. Participants did not receive any guidance regarding how to coordinate or strategize. After the team ranked all items the task ended.

Analyses

CRA: Data Preparation

Data were preprocessed by correcting spelling errors and merging semantically identical terms (e.g., "rain coat" was changed to "raincoat"). The data for each team was then concatenated into a long-form file containing all team communications. This file was then analyzed with Discursis. The default parameter was used for the upper limit on the number of concepts extracted (100) and the merge word variants option was turned on, which is a word-stemming option that identifies and merges words with similar roots. These settings were chosen from an evaluation of parameter settings in CRA that was conducted in a previous study (Tolston et al., 2019). The default stop-list from Discursis—which includes 431 common words, such as "a" and "and"—was expanded to include additional terms that were prevalent in team communications (e.g., "important" and "use"). This stop-list was used to remove common words, including pronouns and prepositions, which provide little discriminating information and that would otherwise saturate the semantic space. The output similarity matrix from Discursis was read into MATLAB for further processing.

Dimensionality Reduction: Multidimensional Scaling

In the present analyses, we used multidimensional scaling (MDS) to assess the impact of dimensionality reduction on community structures in the complex networks obtained from similarity matrices outputted from Discursis. MDS is similar to principal components analysis (PCA), but MDS is a more general dimensionality reduction technique that can be used on either Euclidean (i.e., classical MDS) or non-Euclidean distance matrices (e.g., generalized MDS), rather than only Euclidean vector spaces, as in PCA. Dimensionality reduction techniques like PCA can improve clustering results (Ding & He, 2004), and MDS has been used specifically to evaluate clusters in a semantic space to patterns in sentiment analysis (Cambria, Song, Wang, & Howard, 2014). We used classical MDS to capture the regularities in the relationships between utterances in the space spanned by the important topics identified by CRA, and then projected back through the original space spanned by key concepts to identify interpretable combinations of data.

Classical MDS—hereafter referred to as MDS—requires a Euclidean distance matrix upon which to operate. Such a distance matrix can be calculated from the similarity matrix obtained from CRA using the fact that cosine similarity is related to Euclidean distance by the following relationship:

$$d_{ij} = (2 - 2s_{ij})^{\frac{1}{2}}, \tag{1}$$

(Seber, 2004), where d_{ij} is a Euclidean metric distance and s_{ij} identifies pairwise cosine similarities between utterances. The output from MDS is a lower-dimensional representation of the metric space that best preserves the distances between points in the original space. We used MDS to create a vector space representation of the utterance-utterance similarity matrix such that the maximum variance in utterance similarity was accounted for with the fewest number of dimensions. Prior to analyzing the similarity matrix with MDS, all utterances with no similarity to other utterances were removed from the matrix.

Thresholding and Network Creation

The output of MDS is a projection of the data onto the coordinates of the vector space reconstructed from the distance matrix, with a component for each observation in the distance matrix. If the space is Euclidean and there is enough data, then eigenvalues will be non-zero and positive up to the number of dimensions that define the coordinate axes of the original space from which the distance matrix was derived, and zero after. A subset of these components can be used in subsequent analyses, for example by creating new distance matrices that use only the first few principal components that account for the majority of variance in the distance matrix. For our analyses, we created a range of such matrices, using an increasing number of components to identify the number that resulted in the clearest community structure. For each component of the MDS vector space, from 2 to 96—the number of key concepts identified in our CRA analyses of the team communications (see the section "Results" for further information regarding key concepts extracted)—a new distance matrix was created from reconstructions of the vector space using components from one up to that number, and then thresholded using the connectivity efficiency metric introduced in Huang, Xu, Wang, Zhang, and Liu (2018). The connectivity efficiency metric identifies the threshold that results in a network that has the least number of disconnected components (e.g., the fraction of the coverage of the network is maximal) while penalizing for network density. It is defined as

$$CE = F - D, \tag{2}$$

where F is the fraction of the coverage of the network—the proportion of nodes connected to the largest component of the network—and D is the network density—the number of edges observed in the network divided by the number of possible edges. Connectivity efficiency has been shown to have a convex shape as a function of similarity threshold, meaning that a single optimal value is often obtained that balances network connectivity with sparsity (Huang et al., 2018).

Community Detection

For each of the thresholded networks, the community structure of that network was assessed using the Louvain method for community detection (Blondel, Guillaume, Lambiotte, & Lefebvre, 2008), which results in a quality metric (Q) that captures the proportion of edges that link nodes within a community to edges that link outside of a community. Such quality metrics are used to assess the relative communities identified by different algorithms and also as an objective function (Blondel et al., 2008), which is how it is implemented here. After finding the community quality for each network, the best network (e.g., the network that had the largest Q) was chosen for subsequent evaluation.

Calculating Representative Conceptual Content

Once the final graph was found, the vectorized representations of the conceptual content of the utterances were normalized to unit length such that the sum of the squared components of each vector equaled one. These were averaged together in each community to create a representative vector capturing the average conceptual content of utterances in that community. These were then used to identify prominent conceptual themes.

Identifying Conceptual Transitions

Once the conceptual network was constructed and community detection was conducted, each node in the network (e.g., each utterance) was coded according to its community. This was used to create a categorical time series for each team where each utterance took on the label of the community to which it belonged. This time series was then used to form a transition matrix of first-order transitions between conceptual communities. For example, if a categorical time series 1-2-3-2-1 was observed, the transition matrix would be

	1	2	3
1		1.00	
2	0.50		0.50
3		1.00	

where the rows correspond to the initial state and the columns correspond to the state that is transitioned to, and the entries are the empirical probabilities of going to a state given an initial state. This approach leverages the fact that structure between concepts can be observed in the temporal interdependencies of transitions between concepts. Specifically, the probabilities of transitions between categories in a time-ordered symbolic sequence can be considered a measure of similarity between them (Cooke, Gorman, & Kiekel, 2008; Cooke, Neville, & Rowe, 1996). Such transition matrices have been used to quantify the structure of hand-coded categorical transitions in individual and dyadic brainstorming activities (Brown, Tumeo, Larey, & Paulus, 1998), as well as interaction patterns in team communications (Cooke et al., 2008).

RESULTS

On average, teams spent 33.50 minutes (*SD* = 14.62) and produced 242.50 utterances (*SD* = 78.66) discussing the Lost in the Desert scenario. A word-cloud depicting the relative frequencies of terms in team utterances can be seen in Figure 10.1. As can be seen in the figure, items that the team had to rank, such as "knife" and "mirror," make up the majority of prominent terms.

The similarity matrix output from Discursis consisted of a semantic space consisting of 96 dimensions (key concepts). Of the 3,880 total utterances observed from

FIGURE 10.1 Word cloud showing relative frequency of words in team communications after removing stop words.

Note: In the figure, increases in font size correlate with increases in the frequency of occurrence of the word in team communications. The words highlighted in orange (color online) are those which occur most frequently.

teams discussing the Lost in the Desert scenario, 2,281 were similar to at least one other utterance and were retained in MDS analyses—the high number of discarded utterances can be largely accounted for by the fact that short utterances consisting entirely of common stop words, like "yes" and "uh huh," were removed during preprocessing.

The eigenvalues of the components extracted from MDS of the distance matrix obtained by transforming the similarity matrix output by Discursis can be seen in Figure 10.2. A visual inspection shows that the first 20 components were highly important, accounting for about 65% of the variance in the 96-dimensional space. The eigenvalues decrease sharply to zero after 96 components, meaning that the implicit dimensions that determine the distances in the matrix were fully recovered.

A representative example of the results from calculating the connectivity efficiency of networks obtained by thresholding the distance matrix can be seen in Figure 10.3, where the number of dimensions used to construct the similarity matrix was 57 (the number of dimensions identified as optimal for community detection; see the section "Results" for further information). The convex curve around the peak value shows a single point that optimally balances network connectivity with network sparsity.

The quality results of the community detection algorithm as a function of the number of components retained from MDS can be seen in Figure 10.4. There is an obvious peak of .87 at 57 dimensions in the thresholded distance matrix. These results show the value of applying a similarity threshold during CRA, since it leads

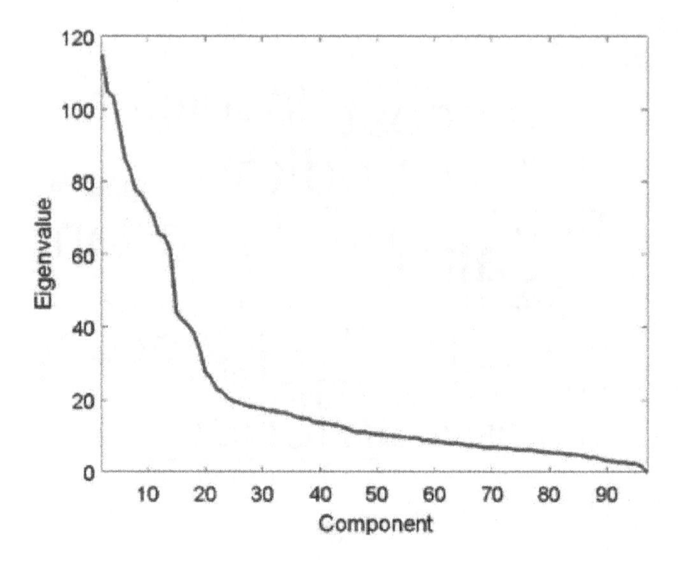

FIGURE 10.2 Eigenvalues of metric multidimensional scaling: Components 1 through 96 all have non-zero positive eigenvalues, while components greater than 96 have eigenvalues of zero (values past 97 are not shown).

Note: The implicit dimensions of the conceptual space constraining the values in the distance matrix were fully recovered using MDS.

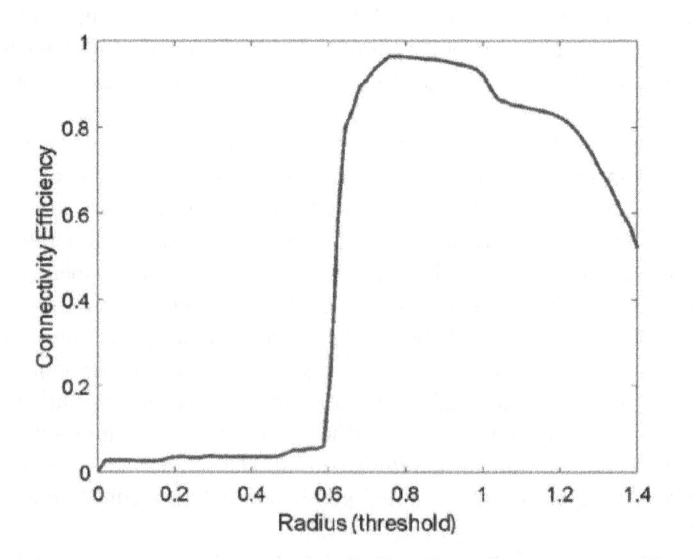

FIGURE 10.3 The connectivity efficiency of the graph obtained using 57 components from the MDS analysis of the distance matrix computed from the Discursis similarity matrix.

Note: The convex curve shows a single best answer balancing network connectivity with sparsity.

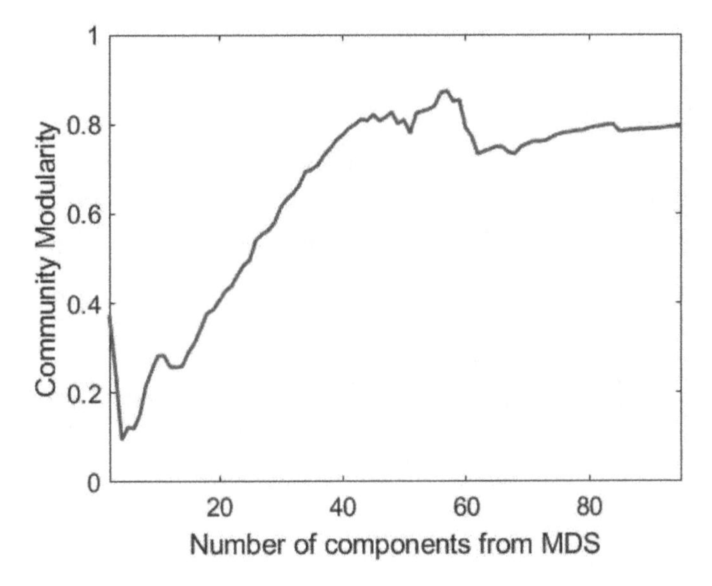

FIGURE 10.4 Number of components retained from MDS and the resultant community quality for both thresholded and unthresholded distance matrices.

Note: Since using the distance matrix from the full span of non-zero eigenvalues is equivalent to the distance matrix from Discursis, the highest value for the non-thresholded matrix is equal to that from the raw Discursis output.

to clearer identification of similar utterances and a better separation of dissimilar utterances.

A simplified version of the final network obtained from using the optimal parameters estimated from the connectivity efficiency and community detection procedures can be seen in Figure 10.5. The graph was plotted using a force-directed algorithm (Fruchterman & Reingold, 1991). Each community is uniquely colored to aid visual inspection of the graph (colored figures available online). The high degree of modularity in the graph is apparent in that communities with high intra-community edge density and low inter-community edge density dominate the graph. To increase legibility of the figure, only well-populated communities (with more than 21 utterances) and densely connected nodes (with at least 20 edges) are shown. The full graph has similar visual characteristics as the presented reduced graph. The distribution of utterances by community can be seen in Figure 10.6.

After the community structure was calculated, each utterance was categorized by the community to which it belonged, and a transition matrix was created for each team. These were then aggregated into a single matrix to obtain empirical transition probabilities between conceptual clusters (see Figure 10.7). To reduce the dimensionality of the transition matrix, transitions to communities that had fewer than 21 observations were relabeled as transitions to a single ground state that represented all of these less frequently occurring states (Fu, Shi, Zou, & Xuan, 2006). The number 21 was chosen as a cutoff from a visual inspection of the graphs for intelligibility,

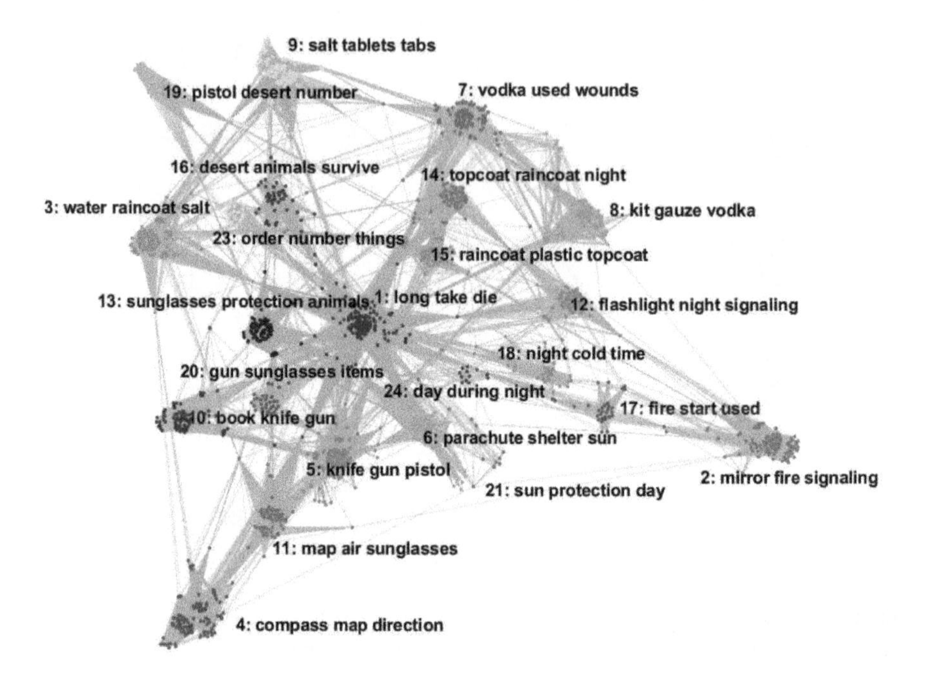

FIGURE 10.5 Final network with communities numbered, highlighted, and labeled; labels correspond to the top three concepts of each community.

Note: To increase legibility of the figure, only well-populated communities (with more than 21 utterances) and densely connected nodes (with at least 20 edges) are shown.

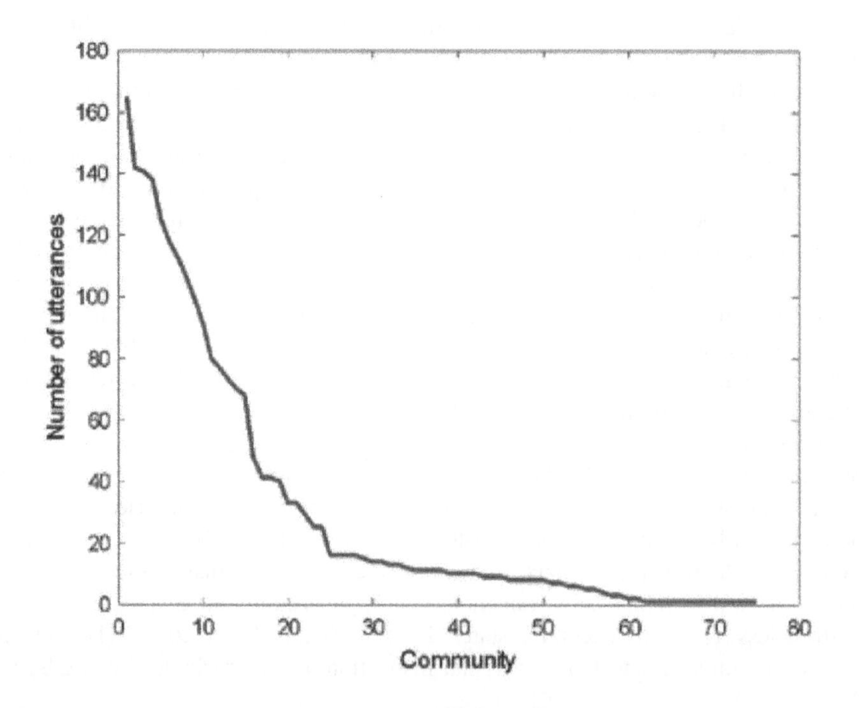

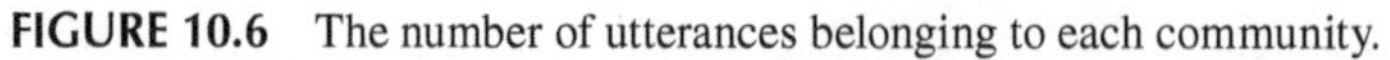
FIGURE 10.6 The number of utterances belonging to each community.

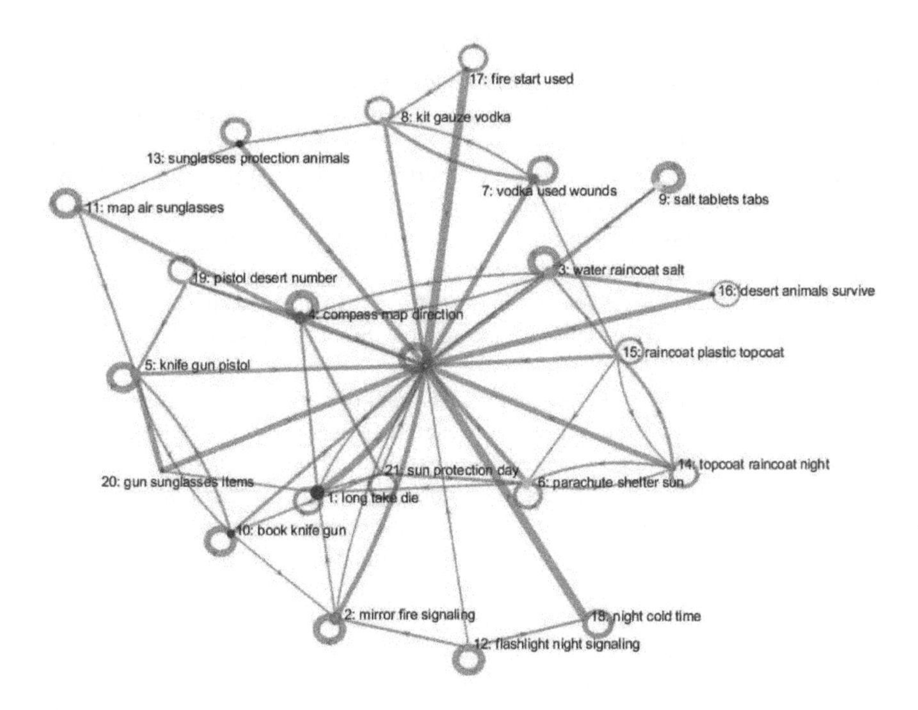

FIGURE 10.7 Network showing directed transitions between communities.

Note: Nodes are labeled with the top three concepts, in order, observed in each community. Edges represent transitions between nodes with probabilities of at least .075. The width of edge lines in the figure are proportional to the probability of a transition and node size is proportional to the number of utterances belonging to a community. The central node is a ground node that represents transitions to communities with fewer than 21 observations.

sparsity, and connectivity. There were 22 communities that had at least 21 observations. Finally, the transition matrix was thresholded to highlight important transitions and increase interpretability. From visual inspections of the graphs obtained by thresholding, a transition probability of .075 was chosen as a cutoff. Further decreasing this value led to a saturation of connections, while increasing it removed informative structure from the graph.

DISCUSSION

Measuring the cognitive structures underlying team reasoning and decision making is a critical endeavor with many important practical implications. In the present chapter, we introduced a methodological approach to mapping team cognition using a naturalistic language problem-solving task combined with CRA and network analyses. Our results showed the utility of generating networks from conceptual recurrence matrices, how MDS can improve clustering of similar utterances, how community structure in the networks reveals thematic patterning in the conceptual content of utterances, and how transitions between communities reveals structural aspects of team cognition. In doing so, we showed how this method can be used to leverage the overt nature of cognition in team interactions to map the conceptual content, structure, and emergent patterns underlying how teams think about a problem. By using bottom-up statistical approaches based on NLP, we uncovered conceptual themes that fit well with expectations based on expert reasoning in a problem-solving task without a priori concept selection that some other approaches to measuring cognitive content rely upon. Taking this a step further, we then showed how researchers can make use of the temporal order of team communications to reveal the links between thematic conceptual clusters that are important for team decision making. Below, we review our specific findings and provide several directions researchers may pursue with the given techniques.

The task we used, Lost in the Desert, has a set of items that teams must reason about with respect to their utility for desert survival. Our motivating question was whether the approach to measuring team cognition outlined in this chapter provides insight into the conceptual content and structure that constrains team reasoning in a decision making task. We believe it does. We were further interested in whether team-level cognitive structure would be in line with expert reasoning in that task. Our results reveal several ways in which it was. Analyses showed that utterances in the Lost in the Desert scenario tended to be about specific thematic combinations of key concepts. This was evidenced by our findings showing that the connectivity efficiency of thresholded distance matrices was a useful criterion for generating semantic networks, and that the best network had a high-quality community structure (0.84). Many of the themes uncovered in our analyses revolved around groups of nouns which have similar utility. Several examples include that the compass was often discussed along with the map and direction, the parachute was associated with shelter and the sun, the book identifying edible desert animals was discussed in conjunction with the gun and knife, the flashlight and night were discussed along with fire mirror and signaling, and nighttime was associated with cold. Teams also discussed vodka in conjunction with the first aid kit (i.e., as an antiseptic) and as a means of starting a fire. Many of these themes highlight conceptual connections that are in line with expectations based on expert reasoning (see Table 10.1). By next looking at transitions between themes, our analyses showed some of the reasoning behind team decision making that also lined up with expert analysis.

The transition matrix based on community structure in the network of utterances shows that semantically similar items were often discussed along trajectories exploring the meaning and utility of similar or complementary items in ways that lined up with reasoning by experts. For instance, discussions of needing protection against the sun (community 21 in Figure 10.7) are connected to items that provide shelter (e.g., the parachute, community 6) and protection from extreme temperatures (communities 14 and 15). Below is example data showing a transition from an utterance mainly about the desert (community 16) to one mainly about water (community 3):

> "How much desert experience do you have from your deployment, Mike?"
> "Um none really but I know water will be important."

These connections and their overlap with expert reasoning illustrates the utility of the method for generating meaningful representations of cognitive structure from bottom-up analysis of team communications.

Interestingly, the network analysis shows that while the terms teams used routinely clustered into semantically coherent groups, the specific way that teams created and traversed the conceptual landscape was largely unique, with a substantial portion of the utterances teams made being rather idiosyncratic combinations of topics (represented as the central ground node). Specifically, transitions to the central node in the network identified in Figure 10.7 account for about 19% of the total number of utterances analyzed, with the other transitions accounting for 7% or less each. However, the network provides insight into the concerns of a large proportion of the teams.

Using the approaches outlined in this chapter, we were able to gain insight into the motivations and concerns of the teams that guided their decision making. However, there are important links between syntactic, semantic, behavioral coordination, and communication that go beyond the current analysis (e.g., Dale, Fusaroli, Duran, & Richardson, 2013; Dale & Kello, 2018). Future work can leverage the present techniques, along with other measures of linguistic and behavioral alignment, to evaluate the interactive and hierarchical nature of team interactions that embody information processing. For instance, measures of syntactic alignment as well as measures of similarity of postural dynamics can both be obtained with RQA (i.e., Dale & Spivey, 2006; Shockley, Santana, & Fowler, 2003). Networks can then be generated from these analyses which can be assessed using multilayer network approaches to determine how well the structure across these modalities can predict team effectiveness and performance outcomes (Kivelä et al., 2014).

Additionally, it would be possible to evaluate the relationship between the timing of transitions between postural configurations and conceptual clusters to view the bottom-up influence of lower-level processes on higher-level interactions. This provides a framework for the analysis of multilevel team dynamics, an area of study with important theoretical implications (Kozlowski, 2015).

Another possible extension is to assess the temporal structure of the thematic data. The current study aggregated all utterances and was therefore blind to differences in the evolution of topic emergence over time. Outstanding questions include whether and if the meso-scale communities reach a stationary distribution or if they continually evolve in time. Additionally, prior work using CRA has shown that in a different consensus building task the content of discussion shifts reliably over time depending on the manner in which information is presented to the teams (Tolston et al., 2017). Expanding the current work with a temporal network approach (Holme & Saramäki, 2012) is a promising direction.

Another future direction is to assess individual team cognitive models against the aggregated model. Moving towards a multivariate framework that can simultaneously decompose structure in data across teams to get an average conceptual model while also placing teams in the conceptual space spanned by the average model to determine interrelations between teams may be an important step in understanding key differences in teams (Thioulouse, Simier, & Chessel, 2004). This would also help determine if there is a heterogeneous distribution of team mental models and cognition. Further, the hierarchical nature of teams means that the aggregation of data from the CRA framework can be conducted in multiple ways to measure different aspects of team interactions, from the individual level, intermediate level of subgroups within teams, within team level, and between team levels (Angus & Wiles, 2018). The current work was based on analyses of data aggregated at the between-team level, but the richness of the data obtained from natural language interaction means there are many more layers to unravel in future work.

A possible limitation of the current approach is the use of modularity quality to determine the optimal network. Though widely used, this metric can result in a number of graphs that have similar quality metrics, but are qualitatively distinct (Good, de Montjoye, & Clauset, 2010). An extension of the current approach would be to evaluate the similarity of the set optimal networks to identify different community structures that have similar partition qualities and to evaluate representative members (e.g., graphs) from multiple communities in the graph relating the different networks. Also of interest is the similarity between the eigenvalues obtained from MDS (Figure 10.2) and the number of utterances in each community (Figure 10.6). This is likely no coincidence and could be explored as an alternative way to choose the number of thematic communities.

Importantly, modifying parameters that led to the graph shown in Figure 10.7 can result in very different graph topologies, with no obvious cutoff parameters with respect to the probability threshold. The graph presented was selected by visually inspecting candidate graphs for interpretability of prominent edges. Future work can extend the current efforts by utilizing surrogate techniques to generate null distributions against which significant similarity can be inferred (Lancaster, Iatsenko, Pidde, Ticcinelli, & Stefanovska, 2018).

The current similarity assessment technique relies on an assumption of linearity (i.e., the meaning of a sentence or utterance is the additive combination of the meaning of all the words contained therein). In other words, the key concepts form a basis of a Euclidean space and can be operated upon with vector addition and scalar multiplication to span the semantic space. We note that this is a rather strong assumption that is certainly incorrect; language is inherently nonlinear and contextual with multiplicative interactions where small changes in the context or composition of the utterance can have large effects on its meaning. Recent trends in NLP have started incorporating contextual information in discovering the meaning of sentences and utterances that can better address this important limitation (Devlin, Chang, Lee, & Toutanova, 2018).

Of final note, the methods introduced in this chapter represent new directions in the assessment of team interactions and cognition. While the underlying method of CRA has been shown to be consistently related to experimental manipulation in ways expected, there are outstanding questions regarding the measurement of semantic data, including validity, accuracy, and reliability. Future work might directly compare CRA using Discursis to other methods of assessing conceptual alignment of utterances in order to better understand the degree to which there is convergence among purportedly similar measurement techniques. Further, while Discursis has been shown to be effective for analysis of small data sets (Tolston et al., 2019), a rigorous analysis of the effect of corpus size on its output would be welcome. Additionally, we emphasize the need to explore how identifying and measuring higher-order observables that exist and operate directly at the level of interest to the researcher (e.g., team-level measurements) can improve our understanding of teams. In particular, comparing the predictive performance of models of team cognition outlined in this chapter, which measure team-level cognition directly, to traditionally obtained estimates of shared mental models, which typically infer team-level cognition from individual-level models, would be quite interesting.

CONCLUSION

Evaluating information flow in distributed team contexts can be difficult. Language analysis is an important key to understanding the relationships between the concepts underlying team decision making in complex tasks. Done carefully, it can yield detailed information about co-occurrence and mixtures of concepts that shows how teams process and group information to make a decision. Here, we have shown that combining CRA with network-based approaches yields consistent structures that allow researchers to understand how teams are dealing with conceptual-linguistic information.

Importantly, aside from team dynamics and team-specific questions, studying behavior in distributed teams provides a critical resource to cognitive science, in that distributed teams often have to talk to one another while they solve problems, and the language they use provides an excellent avenue to map out the cognitive processing of conceptual information that is key to decision making. We think the tools and methods presented in this chapter offer exciting possibilities for this direction.

REFERENCES

Angus, D. (2019). Recurrence methods for communication data, reflecting on 20 years of progress. *Frontiers in Applied Mathematics and Statistics*, *5*, 54.

Angus, D., Smith, A. E., & Wiles, J. (2012a). Conceptual recurrence plots: Revealing patterns in human discourse. *IEEE Transactions on Visualization and Computer Graphics*, *18*(6), 988–997.

Angus, D., Smith, A. E., & Wiles, J. (2012b). Human communication as coupled time series: Quantifying multi-participant recurrence. *IEEE Transactions on Audio, Speech, and Language Processing*, *20*(6), 1795–1807.

Angus, D., Watson, B., Smith, A., Gallois, C., & Wiles, J. (2012). Visualising conversation structure across time: Insights into effective doctor-patient consultations. *PLoS One*, *7*(6), e38014.

Angus, D., & Wiles, J. (2018). Social semantic networks: Measuring topic management in discourse using a pyramid of conceptual recurrence metrics. *Chaos: An Interdisciplinary Journal of Nonlinear Science*, *28*(8), 085723.

Atay, C., Conway, E. R., Angus, D., Wiles, J., Baker, R., & Chenery, H. J. (2015). An automated approach to examining conversational dynamics between people with dementia and their carers. *PLoS One*, *10*(12), e0144327.

Blondel, V. D., Guillaume, J.-L., Lambiotte, R., & Lefebvre, E. (2008). Fast unfolding of communities in large networks. *Journal of Statistical Mechanics: Theory and Experiment*, *2008*(10), P10008.

Brown, V., Tumeo, M., Larey, T. S., & Paulus, P. B. (1998). Modeling cognitive interactions during group brainstorming. *Small Group Research*, *29*(4), 495–526.

Cambria, E., Song, Y., Wang, H., & Howard, N. (2014). Semantic multidimensional scaling for open-domain sentiment analysis. *IEEE Intelligent Systems*, *29*(2), 44–51.

Cannon-Bowers, J. A., & Salas, E. (2001). Reflections on shared cognition. *Journal of Organizational Behavior, 22*(2), 195–202.

Cannon-Bowers, J. A., Salas, E., & Converse, S. A. (1993). Shared mental models in expert team decision making. In N. J. Castellan Jr. (Ed.), *Individual and group decision making: Current issues* (pp. 221–246). Hillsdale, NJ: Erlbaum.

Citera, M. (1998). Distributed teamwork: The impact of communication media on influence and decision quality. *Journal of the American Society for Information Science, 49*(9), 792–800.

Cooke, N. J., Gorman, J. C., & Kiekel, P. A. (2008). Communication as team-level cognitive processing. In M. P. Letsky, N. W. Warner, S. M. Fiore, & C. A. P. Smith (Eds.), *Macrocognition in teams: Theories and methodologies* (pp. 51–64). Hants: Ashgate.

Cooke, N. J., Gorman, J. C., Myers, C. W., & Duran, J. L. (2013). Interactive team cognition. *Cognitive Science, 37*(2), 255–285.

Cooke, N. J., Gorman, J. C., & Rowe, L. J. (2009). An ecological perspective on team cognition. In E. Salas, J. Goodwin, & C. S. Burke (Eds.), *Team effectiveness in complex organizations: Cross-disciplinary perspectives and approaches* (SIOP Organizational Frontiers Series, pp. 157–182). New York: Taylor & Francis.

Cooke, N. J., Neville, K. J., & Rowe, A. L. (1996). Procedural network representations of sequential data. *Human-Computer Interaction, 11*(1), 29–68.

Dale, R., Fusaroli, R., Duran, N. D., & Richardson, D. C. (2013). The self-organization of human interaction. In B. H. Ross (Ed.), *Psychology of learning and motivation* (Vol. 59, pp. 43–95). Waltham, MA: Academic Press.

Dale, R., & Kello, C. T. (2018). "How do humans make sense?" multiscale dynamics and emergent meaning. *New Ideas in Psychology, 50*, 61–72.

Dale, R., & Spivey, M. J. (2006). Unraveling the dyad: Using recurrence analysis to explore patterns of syntactic coordination between children and caregivers in conversation. *Language Learning, 56*(3), 391–430.

DeChurch, L. A., & Mesmer-Magnus, J. R. (2010). Measuring shared team mental models: A meta-analysis. *Group Dynamics: Theory, Research, and Practice, 14*(1), 1–14.

Devlin, J., Chang, M.-W., Lee, K., & Toutanova, K. (2018). Bert: Pre-training of deep bidirectional transformers for language understanding. *arXiv Preprint arXiv:1810.04805.*

Ding, C., & He, X. (2004). K-means clustering via principal component analysis. In *Proceedings of the twenty-first international conference on Machine learning* (pp. 29–36). Banff, Alberta, Canada: Association for Computing Machinery.

Donner, R. V., Donges, J. F., Zou, Y., & Feldhoff, J. H. (2015). Complex network analysis of recurrences. In *Recurrence quantification analysis* (pp. 101–163). London: Springer.

Donner, R. V., Small, M., Donges, J. F., Marwan, N., Zou, Y., Xiang, R., & Kurths, J. (2011). Recurrence-based time series analysis by means of complex network methods. *International Journal of Bifurcation and Chaos, 21*(4), 1019–1046.

Donner, R. V., Zou, Y., Donges, J. F., Marwan, N., & Kurths, J. (2010). Recurrence networks—a novel paradigm for nonlinear time series analysis. *New Journal of Physics, 12*(3), 033025.

Ericsson, K. A., & Simon, H. A. (1998). How to study thinking in everyday life: Contrasting think-aloud protocols with descriptions and explanations of thinking. *Mind, Culture, and Activity, 5*(3), 178–186.

Ferrin, D. L., & Dirks, K. T. (2003). The use of rewards to increase and decrease trust: Mediating processes and differential effects. *Organization Science, 14*(1), 18–31.

Foltz, P. W. (1996). Latent semantic analysis for text-based research. *Behavior Research Methods, 28*(2), 197–202.

Fruchterman, T. M., & Reingold, E. M. (1991). Graph drawing by force-directed placement. *Software: Practice and Experience*, *21*(11), 1129–1164.

Fu, D., Shi, Y. Q., Zou, D., & Xuan, G. (2006). JPEG steganalysis using empirical transition matrix in block DCT domain. In *IEEE 8th workshop on multimedia signal processing, 2006* (pp. 310–313). Piscataway, NJ: IEEE.

Good, B. H., de Montjoye, Y.-A., & Clauset, A. (2010). Performance of modularity maximization in practical contexts. *Physical Review E*, *81*(4), 046106.

Gorman, J. C., Cooke, N. J., Amazeen, P. G., & Fouse, S. (2012). Measuring patterns in team interaction sequences using a discrete recurrence approach. *Human Factors: The Journal of the Human Factors and Ergonomics Society*, *54*(4), 503–517.

Gorman, J. C., Dunbar, T. A., Grimm, D., & Gipson, C. L. (2017). Understanding and modeling teams as dynamical systems. *Frontiers in Psychology*, *8*, 1053.

Gorman, J. C., Foltz, P. W., Kiekel, P. A., Martin, M. J., & Cooke, N. J. (2003). Evaluation of Latent Semantic Analysis-based measures of team communications content. In *Proceedings of the human factors and ergonomics society annual meeting* (Vol. 47, No. 3, pp. 424–428). Los Angeles, CA: SAGE Publications.

Gorman, J. C., Grimm, D. A., Stevens, R. H., Galloway, T., Willemsen-Dunlap, A. M., & Halpin, D. J. (2019). Measuring real-time team cognition during team training. *Human Factors*. Advance online publication. https://doi.org/0018720819852791.

Gorman, J. C., Martin, M. J., Dunbar, T. A., Stevens, R. H., Galloway, T. L., Amazeen, P. G., & Likens, A. D. (2016). Cross-level effects between neurophysiology and communication during team training. *Human Factors*, *58*(1), 181–199.

Holme, P., & Saramäki, J. (2012). Temporal networks. *Physics Reports*, *519*(3), 97–125.

Huang, Z., Xu, L., Wang, L., Zhang, G., & Liu, Y. (2018). Construction of complex network with multiple time series relevance. *Information*, *9*(8), 202.

Iliev, R., Dehghani, M., & Sagi, E. (2014). Automated text analysis in psychology: Methods, applications, and future developments. *Language and Cognition*, *7*, 1–26.

Kivelä, M., Arenas, A., Barthelemy, M., Gleeson, J. P., Moreno, Y., & Porter, M. A. (2014). Multilayer networks. *Journal of Complex Networks*, *2*(3), 203–271.

Kozlowski, S. W. (2015). Advancing research on team process dynamics: Theoretical, methodological, and measurement considerations. *Organizational Psychology Review*, *5*(4), 270–299.

Kozlowski, S. W., & Ilgen, D. R. (2006). Enhancing the effectiveness of work groups and teams. *Psychological Science in the Public Interest*, *7*(3), 77–124.

Lafferty, J. C., Eady, P. M., & Elmers, J. (1974). *The desert survival problem*. Plymouth, MI: Experimental Learning Methods.

Lancaster, G., Iatsenko, D., Pidde, A., Ticcinelli, V., & Stefanovska, A. (2018). Surrogate data for hypothesis testing of physical systems. *Physics Reports*, *748*, 1–60.

Liu, H., & Li, G. (2017). To gain or not to lose? The effect of monetary reward on motivation and knowledge contribution. *Journal of Knowledge Management*, *21*(2), 397–415.

Mancuso, V. F., Finomore, V. S., Rahill, K. M., Blair, E. A., & Funke, G. J. (2014). Effects of cognitive biases on distributed team decision making. *In Proceedings of the Human Factors and Ergonomics Society Annual Meeting*, *58*, 405–409.

McNeese, N. J., & Reddy, M. C. (2015). Concept mapping as a methodology to develop insights on cognition during collaborative information seeking. In *Proceedings of the Human Factors and Ergonomics Society Annual Meeting*, *59*, 245–249.

Mikolov, T., Chen, K., Corrado, G., & Dean, J. (2013). Efficient estimation of word representations in vector space. *arXiv preprint arXiv:1301.3781*.

Mohammed, S., Ferzandi, L., & Hamilton, K. (2010). Metaphor no more: A 15-year review of the team mental model construct. *Journal of Management*, *36*(4), 876–910.

Mohammed, S., Hamilton, K., Sanchez-Manzanares, M., & Rico, R. (2017). Team cognition: Team mental models and situation awareness. In E. Salas, R. Rico, & J. Passmore (Eds.), *The Wiley Blackwell handbook of the psychology of teamwork and collaborative processes*. West Sussex, UK: John Wiley & Sons, Ltd.

Mohammed, S., Klimoski, R., & Rentsch, J. R. (2000). The measurement of team mental models: We have no shared schema. *Organizational Research Methods*, *3*(2), 123–165.

Russell, S. M., Funke, G. J., Knott, B. A., & Strang, A. J. (2012). Recurrence quantification analysis used to assess team communication in simulated air battle management. *Proceedings of the Human Factors and Ergonomics Society Annual Meeting*, *56*, 468–472.

Saggar, M., Sporns, O., Gonzalez-Castillo, J., Bandettini, P. A., Carlsson, G., Glover, G., & Reiss, A. L. (2018). Towards a new approach to reveal dynamical organization of the brain using topological data analysis. *Nature Communications*, *9*(1), 1399.

Salas, E., Sims, D. E., & Burke, C. S. (2005). Is there a "big five" in teamwork? *Small Group Research*, *36*(5), 555–599.

Seber, G. A. F. (2004). *Multivariate observations*. Hoboken, NJ: John Wiley & Sons.

Shockley, K., Santana, M. V., & Fowler, C. A. (2003). Mutual interpersonal postural constraints are involved in cooperative conversation. *Journal of Experimental Psychology: Human Perception and Performance*, *29*(2), 326.

Thioulouse, J., Simier, M., & Chessel, D. (2004). Simultaneous analysis of a sequence of paired ecological tables. *Ecology*, *85*(1), 272–283.

Tolston, M. T., Finomore, V., Funke, G. J., Mancusco, V., Brown, R., Menke, L., & Riley, M. A. (2017). Effects of biasing information on the conceptual structure of team communications. *Advances in Intelligent Systems and Computing*, *488*, 433–455.

Tolston, M. T., Riley, M. A., Mancuso, V., Finomore, V., & Funke, G. J. (2019). Beyond frequency counts: Novel conceptual recurrence analysis metrics to index semantic coordination in team communications. *Behavior Research Methods*, *51*(1), 342–360.

Watson, B. M., Angus, D., Gore, L., & Farmer, J. (2015). Communication in open disclosure conversations about adverse events in hospitals. *Language & Communication*, *41*, 57–70.

Webber, C., & Zbilut, J. P. (1994). Dynamical assessment of physiological systems and states using recurrence plot strategies. *Journal of Applied Physiology*, *76*(2), 965–973.

Wiltshire, T. J., Butner, J. E., & Fiore, S. M. (2018). Problem-solving phase transitions during team collaboration. *Cognitive Science*, *42*(1), 129–167.

Zaff, B. S., McNeese, M. D., & Snyder, D. E. (1993). Capturing multiple perspectives: A user-centered approach to knowledge and design acquisition. *Knowledge Acquisition*, *5*(1), 79–116.

Zhang, J., & Small, M. (2006). Complex network from pseudoperiodic time series: Topology versus dynamics. *Physical Review Letters*, *96*(23), 238701.

Index

D

E

F

G

H

I

J

K

L

M

N

O

P

R